Peter Gille, Yuri Grin (Eds.)
Crystal Growth of Intermetallics

Also of Interest

Structures on Different Time Scales
Theo Woike, Dominik Schaniel (Eds.), 2018
ISBN 978-3-11-044209-0, e-ISBN (PDF) 978-3-11-043392-0,
e-ISBN (EPUB) 978-3-11-043390-6

Multi-Component Crystals. Synthesis, Concepts, Function
Edward R. T. Tiekink, Julio Zukerman-Schpector (Eds.), 2017
ISBN 978-3-11-046365-1, e-ISBN (PDF) 978-3-11-046495-5,
e-ISBN (EPUB) 978-3-11-046379-8

Highlights in Applied Mineralogy
Soraya Heuss-Aßbichler, Georg Amthauer, Melanie John, 2017
ISBN 978-3-11-049122-7, e-ISBN (PDF) 978-3-11-049734-2,
e-ISBN (EPUB) 978-3-11-049508-9

Cementitious Materials. Composition, Properties, Application
Herbert Pöllmann (Ed.), 2017
ISBN 978-3-11-047373-5, e-ISBN (PDF) 978-3-11-047372-8,
e-ISBN (EPUB) 978-3-11-047391-9

Symmetry. Through the Eyes of Old Masters
Emil Makovicky, 2016
ISBN 978-3-11-041705-0, e-ISBN (PDF) 978-3-11-041714-2,
e-ISBN (EPUB) 978-3-11-041719-7

Crystal Growth of Intermetallics

Edited by
Peter Gille, Yuri Grin

DE GRUYTER

Physics and Astronomy Classification Scheme 2010
01.30.Ee, 71.20.Be, 71.20.Eh, 71.20.Lp, 81.10.-h

Editors
Prof. Dr. Peter Gille
Ludwig-Maximilians-Universität München
Department of Earth and Environmental Sciences
Theresienstrasse 41
80333 München
Germany
gille@lmu.de

Prof. Yuri Grin
Max-Planck-Institut für Chemische Physik fester Stoffe
Nöthnitzer Strasse 40
01187 Dresden
Germany
grin@cpfs.mpg.de

ISBN 978-3-11-049584-3
e-ISBN (PDF) 978-3-11-049678-9
e-ISBN (EPUB) 978-3-11-049349-8

Library of Congress Control Number: 2018954381

Bibliographic information published by the Deutsche Nationalbibliothek
The Deutsche Nationalbibliothek lists this publication in the Deutsche Nationalbibliografie; detailed bibliographic data are available on the Internet at http://dnb.dnb.de.

Cover image: By Dr. Rupert Hochleitner; Nickelskutterudit, Fundort: Schneeberg, Sachsen
Typesetting: VTeX UAB, Lithuania
Printing and binding: CPI books GmbH, Leck

www.degruyter.com

Preface

The famous paper in which the later Nobel Prize laureate Percy W. Bridgman, among other things, suggested his well-known crystal-growth method starts with the sentences: *The importance of a knowledge of the physical properties of single crystals of the metals requires no argument. Very little is known of the subject, however.*[1] Nowadays, with respect to intermetallic phases, this can be repeated if we look at the discrepancy between the huge field of various intermetallic systems and the relatively small number of binary and ternary phases investigated so far. In their recently published book on the statistics of intermetallic structures, Steurer and Dshemuchadse[2] noted that less than eight percent of ternary phase diagrams have been studied of a total of approx. 80,000 that can be formed from about 80 metallic elements in the periodic table.

Similar to Bridgman's starting point with the pure metals, single crystals of intermetallic compounds are the necessary prerequisite for any study of their intrinsic properties. Crystal structure information or some thermodynamic knowledge may be obtained from relatively small (tens of μm in size) single crystals or even polycrystalline materials. In contrast, typical measurements of anisotropic magnetic, electrical and thermal properties, as well as structural studies of crystalline surfaces, nowadays desperately need bulky single crystals of appropriate size, although there are very recent developments of techniques for the measurement of μm-sized specimens. The need for single crystals even holds in those cases where possible future applications will be based on polycrystalline materials or nanoparticles, e. g., for catalytic purposes or thermoelectric devices. This is why single-crystal growth should be regarded as the basis of any experimental research in the exciting field of intermetallic compounds and, consequently, as a scientific issue of its own. If this is neglected, even sophisticated physics and materials science using highly developed methods will fail due to badly prepared samples.

As distinguished from semiconductor physics, among intermetallic phases there is no such cornerstone like silicon, but rather, instead, it is the huge variety of combinations of metallic elements that make this group of materials so interesting and different. The crystal structures of intermetallic compounds are often quite complex and not as simple as in the typical, tetrahedrally-bonded semiconductor compounds. Due to the fact that intermetallic compounds are formed by elements located to the left of the Zintl line in the periodic table of elements, the average number of valence electrons per atom is normally less than four. This distinguishes these materials from the normal-valence inorganic compounds. As a consequence, their compositions cannot

1 Bridgman PW. Proc Amer Acad Arts Sci. 1925, 60, 303–83.

2 Steurer W, Dshemuchadse J. Intermetallics. Structures, Properties, and Statistics, Oxford University Press, 2016.

https://doi.org/10.1515/9783110496789-201

be described by applying the usual valence rules. The common feature of the bonding in this family of inorganic substances is the formation of multi-center and multi-atomic bonds. This leads to the tiny influences on the total energy of such structures and opens up a possibility of the local deviations in the crystallographic ordering. Study of all such phenomena often requires materials preparation under closer-to-equilibrium conditions, e. g., by single-crystal growth.

This book consists of three chapters (1, 2, and 11) that form a frame by treating the general aspects of all intermetallic materials, and four chapters that focus on the most important crystal growth methods used nowadays for intermetallics. After these method-driven parts, the subsequent chapters address some groups of intermetallic materials. Here, the material creates the focus, and the reader can learn how an intermetallic compound constrains the preparation method to be used and its special conditions. Looking for the most interesting case studies, we have chosen some especially exiting examples and present, e. g., the just arising field of high-entropy alloys, but not the intermetallic quasicrystals that have already been the subject of other reviews.

The book has been written primarily for senior undergraduate and postgraduate students already doing research in the field of solid-state physics or chemistry and interested in intermetallic alloys or compounds. It should lead them to have a closer look at the preparation conditions of their samples to be studied and, hopefully, to encourage them to start their own crystal growth activities. Anyway, after having read this book, the reader will be aware how much crystal growth conditions do influence the properties of the samples under investigation.

We are indebted to the editorial and production staff of de Gruyter. Special thanks go to Kristin Berber-Nerlinger who had the idea for this book project and convinced the editors. Particularly we are grateful to our colleagues who instantly agreed to contribute to this book when we put them on our wish list. All of them are *the* experts in the specific fields that are presented in the individual chapters. Finally, we would like to express many thanks to Nadja Schedensack of de Gruyter for her considerable amount of patience.

München and Dresden, October 2018 — Peter Gille, Yuri Grin

Contents

List of Contributing Authors

Michael Binnewies
Max-Planck-Institut für Chemische Physik fester Stoffe
Nöthnitzer Strasse 40
01187 Dresden
Germany
michael.binnewies@aca.uni-hannover.de

Janez Dolinšek
University of Ljubljana
Faculty of Mathematics and Physics
and
Jožef Stefan Institute
Jamova 39
1000 Ljubljana
Slovenia
jani.dolinsek@ijs.si

Fainan Failamani
National Institute for Materials Science (NIMS)
International Center for Materials Nanoarchitectonics
and
Center for Functional Sensor & Actuator
Namiki 1-1
Tsukuba 305-0044
Japan
failamani.fainan@nims.go.jp

Zachary Fisk
University of California at Irvine
Department of Physics & Astronomy
290 Rowland Hall
Irvine, CA
USA 92697
zfisk@uci.edu

Michael Feuerbacher
Forschungszentrum Jülich GmbH
Ernst Ruska Centrum für Mikroskopie und Spektroskopie mit Elektronen
and
Peter Grünberg Institut
52425 Jülich
Germany
m.feuerbacher@fz-juelich.de

Peter Gille
Ludwig-Maximilians-Universität München
Department of Earth and Environmental Sciences
Theresienstrasse 41
80333 München
Germany
gille@lmu.de

Yuri Grin
Max-Planck-Institut für Chemische Physik fester Stoffe
Nöthnitzer Strasse 40
01187 Dresden
Germany
grin@cpfs.mpg.de

Gabin Guélou
National Institute for Materials Science (NIMS)
International Center for Materials Nanoarchitectonics
and
Center for Functional Sensor & Actuator
Namiki 1-1
Tsukuba 305-0044
Japan
guelou.gabin@nims.go.jp

Quansheng Guo
National Institute for Materials Science (NIMS)
International Center for Materials Nanoarchitectonics
and
Center for Functional Sensor & Actuator
Namiki 1-1
Tsukuba 305-0044
Japan
guo.quansheng@nims.go.jp

Masato Hedo
University of the Ryukyus
Faculty of Science
1 Senbaru, Nishihara
Okinawa 903-0213
Japan
hedo@sci.u-ryukyu.ac.jp

Zygmunt Henkie
Institute of Low Temperature and Structure Research
Polish Academy of Sciences
ul. Okólna 2
50-950 Wroclaw 2
Poland
z.henkie@int.pan.wroc.pl

Herbert Ipser
University of Vienna
Department of Inorganic Chemistry – Functional Materials
Währingerstrasse 42
1090 Wien
Austria
herbert.ipser@univie.ac.at

Masashi Kakihana
University of the Ryukyus
Faculty of Science
1 Senbaru, Nishihara
Okinawa 903-0213
Japan
kakihana@mail.ryudai.jp

Takao Mori
National Institute for Materials Science (NIMS)
International Center for Materials Nanoarchitectonics
and
Center for Functional Sensor & Actuator
Namiki 1-1
Tsukuba 305-0044
Japan
mori.takao@nims.go.jp

Takao Nakama
University of the Ryukyus
Faculty of Science
1 Senbaru, Nishihara
Okinawa 903-0213
Japan
nakama@sci.u-ryukyu.ac.jp

Yoshichika Ōnuki
University of the Ryukyus
Faculty of Science
1 Senbaru, Nishihara
Okinawa 903-0213
Japan
onuki@phys.u-ryukyu.ac.jp

Andrey Prokofiev
Vienna University of Technology
Institute of Solid State Physics
Wiedner Hauptstrasse 8-10
1040 Wien
Austria
andrey.prokofiev@ifp.tuwien.ac.at

Priscila F. S. Rosa
Los Alamos National Laboratory
Condensed Matter & Magnet Science Group
Bikini Atoll Road
Los Alamos, NM
USA 87545
pfsrosa@lanl.gov

Marcus Schmidt
Max-Planck-Institut für Chemische Physik fester Stoffe
Nöthnitzer Strasse 40
01187 Dresden
Germany
marcus.schmidt@cpfs.mpg.de

Jean-Baptiste Vaney
National Institute for Materials Science (NIMS)
International Center for Materials Nanoarchitectonics
and
Center for Functional Sensor & Actuator
Namiki 1-1
Tsukuba 305-0044
Japan
jean.baptiste.vaney@gmail.com

Yuri Grin

Intermetallic compounds and their single crystals

1 Introduction

The (single) crystals have attracted attention of humans since hundreds—not to say thousands—of years despite any knowledge about their origin and structure. The main reasons for that are the extreme stability of some natural crystalline minerals and their aesthetically appealing shapes (Figure 1). One should mention here at a minimum the use of single crystals as jewels. Recently, there have even been attempts to consider mineral crystals in several countries as a gold-replacing equivalent of value, and so they are discussed as a subject of investment. This clearly overstated approach exemplifies nonetheless another value of single crystals, namely their informational content, i. e., they carry information about the formation, crystal structure and—most important—intrinsic properties of substances. The property issue is especially a key one in respect of the study and the development of materials based on intermetallic compounds.

Figure 1: Crystals of cuprostibite Cu_2Sb (courtesy Dr. R. Cardoso-Gil) [1].

https://doi.org/10.1515/9783110496789-001

2 Chemical bonding in intermetallic compounds

2.1 Definition and the Zintl concept

Intermetallic compounds form the family of inorganic substances which—despite decades of research—still fascinate and challenge working scientists, both experimentally and theoretically. Composed of elements that are located in the periodic table around the Zintl border and on its left side, intermetallic compounds form and exist under the conditions of valence-electron demand. The available amount of electrons (e. g., expressed as an average valence electron concentration) is not sufficient to fulfil the octet rule [2] in the simple way. In consequence, these materials do not follow the usual valence rules and require special concepts for the understanding of their chemical composition and crystal structure [3]. So, if the electronic demand is not very large and the difference between the electronegativities of the components is large enough, the materials can be understood by the so-called Zintl–Klemm concept [4]. The modern variant of the latter assumes charge transfer from the electropositive (cationic) part of the crystal structure to the anionic one and requires that the number of homonuclear (two-centre-two-electron) bonds per atom within the anionic part is defined by the general Pearson rule [5]. This special group of materials within the field of intermetallic compounds is called the 'Zintl phases' (this term was introduced by F. Laves [6]) because the first representatives of this family—phases with a crystal structure of the NaTl type—were discovered by E. Zintl [7]. Usually, they contain alkali, alkaline-earth, or, sometimes, rare-earth metals such as cations, and elements of the groups 13–15 form the anionic part of the crystal structure. For example, the composition and crystal structure of calcium monosilicide CaSi (structure type α-ITl) can be understood assuming the charge transfer from Ca to silicon. In order to obey the octet rule, each silicon atom forms two covalent homonuclear Si–Si bonds within a zig-zag-chain-shaped one-dimensional polyanion Si^{2-}. The latter interact electrostatically (ionic bonding) with the embedded in-between calcium cations:

$$\mathrm{CaSi} = [\mathrm{Ca}^{2+}][(2b)\mathrm{Si}^{2-}].$$

The Zintl–Klemm concept enables chemical understanding of a large family of intermetallic phases as a combination of ionic and covalent interactions in the same structure. But already for the compounds of boron, the condition of 2c–2e bonds in the polyanion cannot be fulfilled more, and only inter-cluster bonds in the polyanion are required to be 2c–2e interactions, within the homonuclear cluster, the 3c–2e bonds are allowed. As an example, the crystal structure of CaB_6 contains calcium cations embedded within cavities of the anionic boron network, which is formed by six-connected B_6 octahedrons. In the structure one recognizes one Ca and six boron atoms, forming one B_6 octahedron, as well as three inter-cluster bonds per crystallographic unit cell. According to the Wade–Mingos rules [8–11], stabilization of such a closo-cluster requires 14 electrons. Together with the six electrons for the three inter-cluster bonds, this yields 20 electrons necessary for stabilization, which is exactly the amount of valence electrons contributed by one Ca and six B atoms:

$$CaB_6 = [Ca^{2+}][(6b)B_6^{2-}].$$

Following this line, further reducing the number of valence electrons per atoms consequently should lead to the formation of multi-centre bonds with three and more atoms participating in the bonding interaction, e. g. in gamma-brass phases. In terms of the electronic band structure, the formation and filling of the 'bands' is introduced. The stabilization of the structural motif is attributed to the Hume–Rothery mechanism via interaction of the Fermi surface with the Brillouin zone [12, 13]. Within this mechanism, a decreasing valence electron concentration, e. g., from $CuZn_3$ to CuZn, allows for narrowing the Brillouin zone to accommodate the valence electrons and is therefore accompanied by a structural change at a critical value. More generally, such a concept results in electron numbers that are sufficient for the stabilization of structural units. However, on the basis of the electronic band structure, it is difficult to interpret the composition of the compound and the coordination behaviour of all atoms within the crystal structure.

2.2 Bonding model in real space

In order to obtain a more general picture of the atomic interactions in intermetallic compounds, a concept is necessary that would allow interpretation of multi-centre interactions maintaining access to their covalency, ionicity and polarity. Recent investigations show that this kind of information can be (also quantitatively) obtained that apply to bonding analysis of quantum chemical techniques in real space. A use of the Quantum Theory of Atoms in Molecules (QTAIM [14]) on the crystal structures of intermetallic compounds yields the effective charges of the atomic species inside. Application of the electron-localizability approach [15], e. g., combining the electron-localizability indicator [16] and QTAIM, enables establishing the number of atoms participating in the bonding interaction, distinguishing between ionic and covalent interactions and quantizing the polarity of the bonds.

The study of the chemical bonding in the family of intermetallic compounds called Laves phases (chemical composition MN_2) by the electron-localizability approach revealed that the diversity of the phases may be understood in terms of the electronegativity difference between M and N atoms (Figure 2). When this difference is small and the charge transfer from M to N is not strong, the corresponding Laves phases are characterized by multi-center (multi-atomic) bonding. In contrast, when N is much more electronegative than M, the charge transfer from M to N is larger, the analysis of chemical bonding indicates the formation of three-dimensional polyanions $(N_2)^{\delta-}$ and ionic interaction between M and the polyanion [17].

The Madelung energies based on the QTAIM charges and the nearest-neighbour-sharing index, based on the delocalization indices, were introduced as measures for

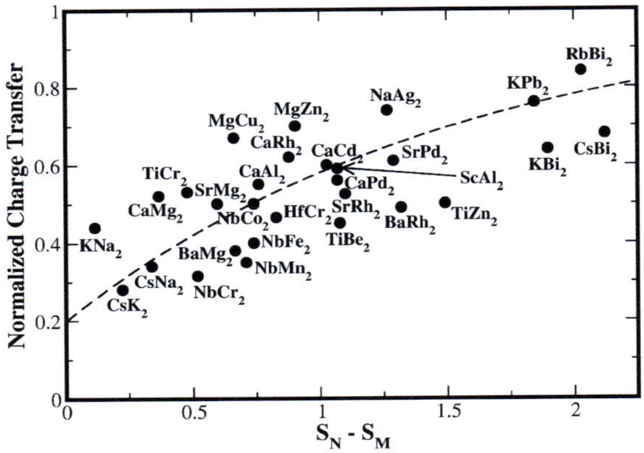

Figure 2: Normalized charge transfer vs. the difference in the electronegativity between the components of binary Laves phases MN_2. Electronegativities after Sanderson are used [17]. Reproduced from Angew. Chem., Int. Ed. Engl. Ormeci A, Simon A, Grin Yu. Structural Topology and Chemical Bonding in Laves Phases, 2010, 49, 8997–9001. Copyright Wiley-VCH Verlag GmbH & Co. KGaA. Reproduced with permission.

ionic and covalent interactions, respectively, in the MgAgAs-type intermetallic compounds (the so-called half-Heusler phases [18]). Due to the low-valence electron concentration, the synapticity of the bonds in these compounds (the number of the contributing atoms) is two and/or five, i. e., the bonding is mostly a multi-center one [19] (Figure 3). The analysis of the bonding characteristics reveals the formation of two chemically different atomic arrangements, either with the late transition metal or with the late main-group element occupying the position with the heterocubic coordination (the other two sites have tetrahedral environment). The increased covalency of interactions are the main "driving force" for the formation of the first type of atomic arrangement. The sequence of the total energies usually matches the sequence of the nearest-neighbour-sharing indices. Ionic interactions in this group of compounds are

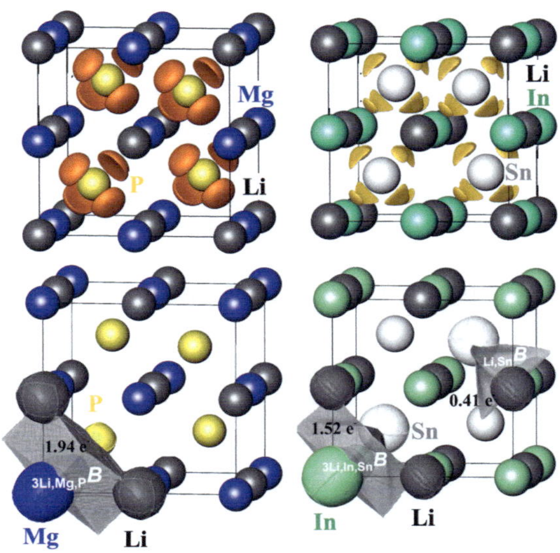

Figure 3: Atomic interactions in MgAgAs-type compounds LiMgP (left) and LiInSn (right): (**top**) Isosurfaces of the distribution of the Electron-Localizability-Indicator (yellow and orange) the covalent bonds. (**bottom**) Basins (B) of the two- and five-center interactions with their populations [19]. Reprinted from Bende D, Grin Yu, Wagner FR. Chemical Bonding in MgAgAs-Type Compounds. Felser C, Hirohata A, Eds. Heusler Alloys. 2016, 133ff, Copyright (2016), with permission from Springer International Publishing.

comparably weak. The MgAgAs-type phases with the main-group element in heterocubic coordination show large QTAIM Madelung energies and reveal—as a rule—small covalency. The energetic sequence of such compounds is caused by the decrease of both ionic and covalent interactions. The compounds containing the late transition metals and the main-group elements realize two different atomic arrangements because two opposing "driving forces" compete with each other. The late transition metal pushes into the heterocubic site in order to optimize its covalent interactions. The ionic interaction, however, becomes maximized, when the late main-group element, which is always anionic, has the heterocubic coordination [20]. The tools developed during these studies enabled quantification of the polarity of the bonds in intermetallic compounds of the half-Heusler family [21] and, furthermore, even prediction of the new MgAgAs-type phases LT-HfPtGe and VIrGe [22].

Application of electron-localizability approach to the intermetallic clathrates confirmed first the usual interpretation in terms of the Zintl–Klemm model. The crystal structure consists of the three-dimensional framework formed by four-bonded atoms of main-group elements (group 13–15 of the periodic table) with two types of cavities filled by alkali, earth-alkali or rare-earth metals. The main-group elements may be partially substituted either by late transition metals or by low-valent main-group elements. QTAIM analysis of the $Ba_8T_8Ge_{40}$ compounds (T – late transition metal, Li, Mg, Al, Ga or Ge) revealed the charge transfer from Ba to the framework (Figure 4, left). The effective charges of barium species are independent on the choice of T and vary between +1.2 and +1.3. The germanium species with homoatomic coordination carry only small negative charge (−0.1 to −0.2). The charges of the germanium species having T ligands vary strongly from −0.05 to −0.60 depending on the polarity of the Ge–T bonds. The T species reveal large variety of the charges between −1.1 (Pt) and +1.5 (Al) for the same reasons. All Ge–Ge and Ge–T interactions have two-centre nature; the contribution of barium here is negligible [23]. On top of these findings which are in agreement with the Zintl–Klemm model, new covalent interactions between Ba and Pt, Ag, Au and Cd were detected by analysis of the distribution of the electron-localizability indicator (Figure 4, right [23, 24]). Presence or absence of such bonds play an important role in the formation of real structure of clathrates, as well as in the physical and chemical behaviour (cf. below).

Recent studies reveal that even for the intermetallic compounds with practically constant composition ('line compounds') the crystal structures often show crystallographic disorder of various kinds caused by special bonding features. As an example, one may mention the local ordering of Au and Ir in Al_3AuIr, the ternary 'substitution' variant of the Al_3Ni_2 type [25]. In the real structure, this ordering is reflected by different positions of the aluminium atoms adjusted to Au or to Ir by multi-centre polar bonds within a close-packed crystal structure. Investigation of such structural details is sometimes possible even using the standard data, but—as a rule—it requires single crystals of high crystallographic quality. In particular, the diffraction experiments up to high-diffraction angles ($\sin\theta/\lambda > 1$) are necessary to obtain sufficient resolution in

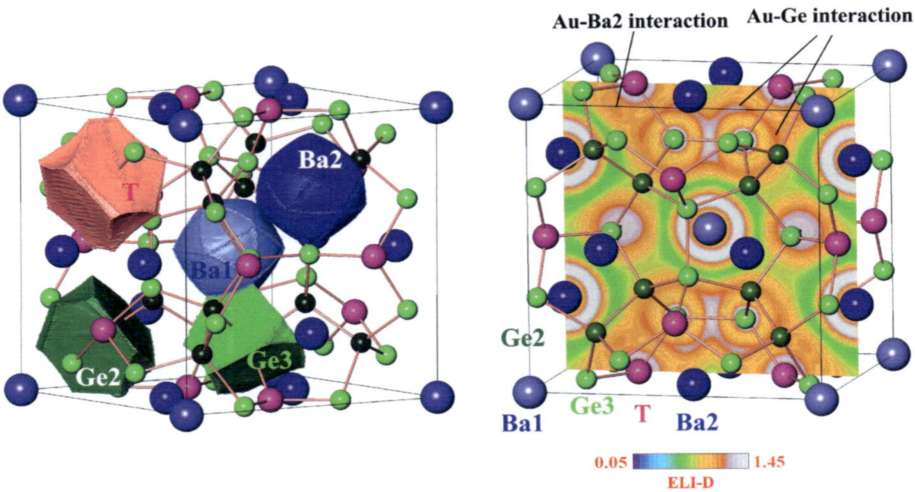

Figure 4: Chemical bonding in intermetallic clathrates $Ba_8T_6Ge_{40}$ from real-space analysis: **(left)** QTAIM atomic basins, for charges, cf. text; **(right)** Distribution of the Electron-Localizability-Indicator (ELI-D) in the (200) plane with maxima visualizing Ge–Ge, Ge–T and T–Ba2 interactions [23]. Adapted from [23].

the difference density maps, enabling extraction of information about the real structure of intermetallic compounds.

3 Single crystals of intermetallic compounds

3.1 Why the single crystals are on demand?

While the natural crystals are mostly obtained for inorganic substances like oxides, sulphides, tellurides, silicates etc., the intermetallic compounds and—in particular—their single crystalline specimens are rather rare in the nature. The huge crystals of cuprostibite (Figure 1) are rather an exception, and the natural origin of this specimen is under discussion. The reason for such a situation is the higher thermodynamic stability, e. g., of oxides or chalcogenides, in comparison with metal-only compounds. Strictly intermetallic minerals, e. g., khatyrkite Al_2Cu [26], cupalite AlCu [26] or icosahedrite $Al_{63}Cu_{24}Fe_{13}$ [27], are known only as polycrystalline matrix or small inclusions therein. One of the reasons for that is the rather complex phase diagrams of intermetallic compounds, which make growth of single crystals a challenge. The next possible reason is the lower thermodynamic stability of the intermetallic compounds in respect of, e. g., oxides or chalcogenides. Thus, the single crystal growth of this materials family requires special precautions and techniques to ensure the size and quality sufficient for the selected application.

In this way, an unusual and sometimes contradictory situation emerges. On the one hand, the single crystals of intermetallic compounds and similar metallic-like substances are required to obtain information about the intrinsic properties of materials. On the other hand, analysis of the literature data in the experimental solid-state physics and chemistry often reveals the specimen's dependence on the properties, and the community accordingly needs time to evaluate the quality experimentally and manufacture the best materials. In other words, the quantification of quality is a crucial issue.

3.2 Criteria for single crystals' quality: residual resistivity

As an important criterion for this purpose, the residual resistivity or residual resistivity ratio RRR is used.

Residual resistivity value is usually obtained by extrapolation of the temperature dependence of electrical resistivity $\rho(T)$ to $T = 0$, e. g., as $\rho(T) = \rho_{res} + AT^2$ [28, 29]. Using this technique, the superconducting transition of the oxide superconductor Sr_2RuO_4 was studied. The single crystals were grown by many runs in a floating-zone image furnace using the feed rods, containing 15 % excess Ru serving as flux, in the mixture of 10 % O_2 and 90 % Ar with a total pressure of 3 bar [30, 31]. The superconducting transition temperature reveals a quite clear dependence on the residual resistivity of the materials (Figure 5). The main explanation of this effect was given by the impurities level (mainly aluminium found by EPMA analysis) [30]. The authors of [31] suggest that the deviation from this trend (point A in Figure 5) are actually coming from the inhomogeneity of defects in crystals. These defects were likely to be tiny amounts of inclusions of $SrRuO_3$ and possibly $Sr_3Ru_2O_7$, but they should not appear to be connected to the defects and impurities that may control residual resistivity and T_C [30].

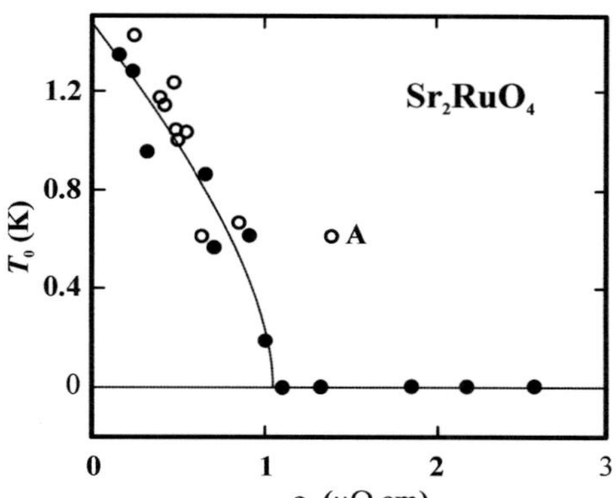

Figure 5: Superconducting transition temperature of Sr_2RuO_4 vs the residual resistivity: filled symbols [30], open symbols [31]. Reprinted with permission from Mao ZQ, Mori Y, Maeno Y. Phys Rev B, 60, 609, 1999. Copyright (1999) by the American Physical Society.

3.3 Criteria for single crystals' quality: residual resistivity ratio (RRR)

The RRR value is usually calculated as $\rho(300\,\mathrm{K})/\rho(2\,\mathrm{K})$. Developed first for the description of the influence of the point defects on the electronic transport of elemental metals, RRR is currently intensively used as a control parameter for the description of the electronic and phononic transport behaviour of single crystals of intermetallic compounds. The following are some examples.

In case of the layered iron germanide YFe_2Ge_2 revealing a non-conventional fragile, non-bulk superconductivity, the application of the RRR does not yield a clear picture on the quality of single crystals. For the polycrystalline non-single-phase samples (obtained from elements by radio-frequency induction melting on a water-cooled copper boat in an argon atmosphere), the superconducting temperature seems to correlate with RRR and achieves the maximum value at RRR ≈ 75 (Figure 6, left) [32]. For the single-crystalline specimens (grown from the tin melt), the RRR values vary between 35 and 65 with, on average, no systematic change in T_C value, indicating that the systematic changes in RRR do not lead to comparable changes in T_C (Figure 6, right [33]). In case of the polycrystalline samples, the RRR is also influenced by the presence of impurity phases, in case of the single-crystalline specimens, the silicon-by-iron substitution is most probably the main control factor [32].

Single crystals of the intermetallic compound UCoGe—a ferromagnetic superconductor—were grown by a modified Czochralski technique [34, 35] (Figure 7, top panel). The as-grown crystals reveal only an onset of the superconducting transition at 0.33 K

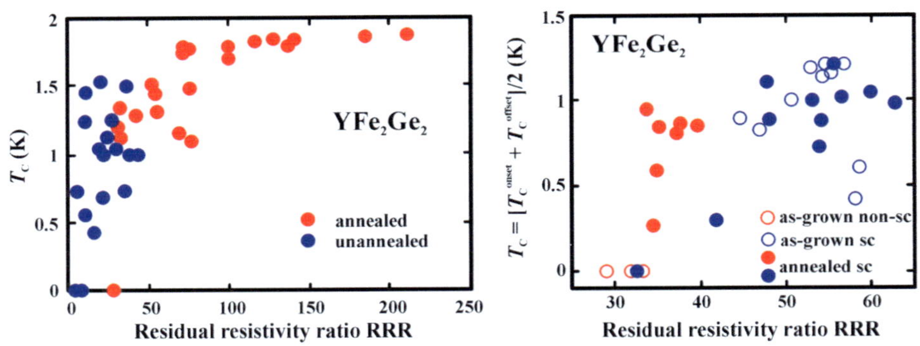

Figure 6: Superconducting transition temperature of YFe_2Ge_2 vs the residual resistivity ratio RRR $\rho(300\,\mathrm{K})/\rho(2\,\mathrm{K})$ (**left**) and $\rho(300\,\mathrm{K})/\rho(1.8\,\mathrm{K})$ (**right**) for the polycrystalline samples (**left** [32]) and single crystals (**right** [33]). Figure 6, left is reprinted with permission from Chen J, Semeniuk K, Feng Zh, Reiss P, Brown Ph, Zou Y, Logg PW, Lampronti GI, Grosche FM. Phys Rev Lett, 116, 127001, 2016. Copyright (2016) by the American Physical Society. Figure 6, right is reprinted from Kim H, Rana S, Mun ED, Hodovanets H, Tanatara MA, Prozorova R, Bud'ko SL, Canfield PC, Crystal growth and annealing study of fragile, non-bulk superconductivity in YFe_2Ge_2. Philos Mag 2015, 95, 804 by permission of the Taylor & Francis Ltd.

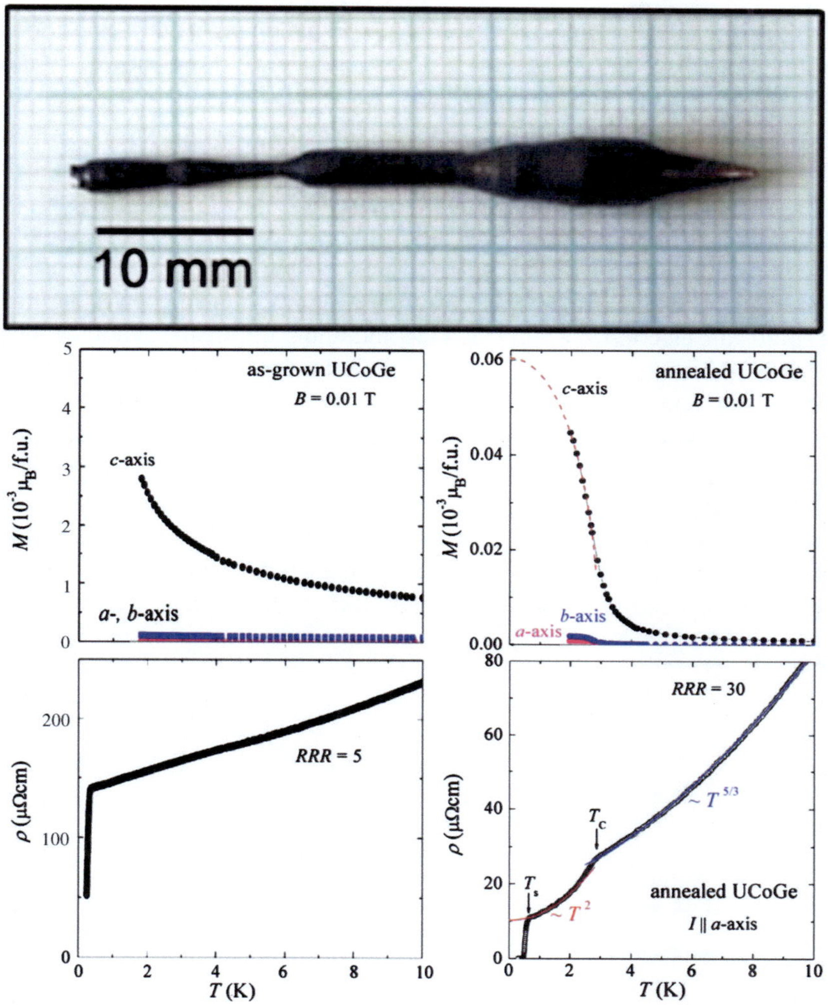

Figure 7: Single crystal (**top panel**) and temperature dependence of resistivity (**middle panels**) and magnetization (**bottom panels**) of UCoGe in correlation to RRR [34]. Reprinted from J Magn Magn Mater 321 2009 Huy NT, Huang YK, de Visser A. Effect of annealing on the magnetic and superconducting properties of single-crystalline UCoGe, 2691–2693, Copyright (2009), with permission from Elsevier.

and indicate a possible magnetic transition below 3 K (Figure 7, bottom left). The RRR value of 5 for the measured crystal was low. The additional annealing at 880 °C (21 d, in analogy with the ferromagnetic superconductor URhGe) improved the RRR value of the single-crystalline material up to 30. These singles crystals revealed a clear magnetic transition at 2.8 K and are superconducting already below 0.65 K (Figure 7, bottom right, [34]). Later, on the single crystals with RRR = 30, even the first observation of quantum oscillations was made [35]. The change in the RRR and other physical behaviours was assumed to be caused by disordered occupation of Co and Ge positions in the crystal structure of the TiNiSi type. On the contrary, all three known crystal-structure determinations of UCoGe using single-crystal X-ray diffraction data did not report any disorder at Co and Ge positions [36–38] despite the crystal growth in Ref. [38] was made by Czochralski technique similar to [34, 35].

Resistivity and specific-heat measurements of the low-carrier unconventional superconductor URu_2Si_2 were performed on various single-crystalline specimens with very different qualities [39]. Single crystals were grown using the Czochralski method in a tetra-arc furnace under an argon-gas atmosphere. The uranium ingot was purified using the solid-state electro-transport (SSE) method under ultrahigh vacuum, which was found to be extremely effective in removing in particular transition metals as impurities in the uranium ingot. The SSE techniques removes the impurities in three ways: sweeping out Fe and Ni by DC electrical current, Mg and Al by diffusion and Mn by evaporation [40]. The crystals with the highest RRR were not obtained in the central core of the crystals but near the surface (Figure 8, bottom left). Both transitions—superconducting and 'hidden'—show clear correlation with the RRR value if evaluated by the specific-heat measurements (Figure 8, top). On the contrary, the superconducting temperature practically does not change with RRR if evaluated by the resistivity

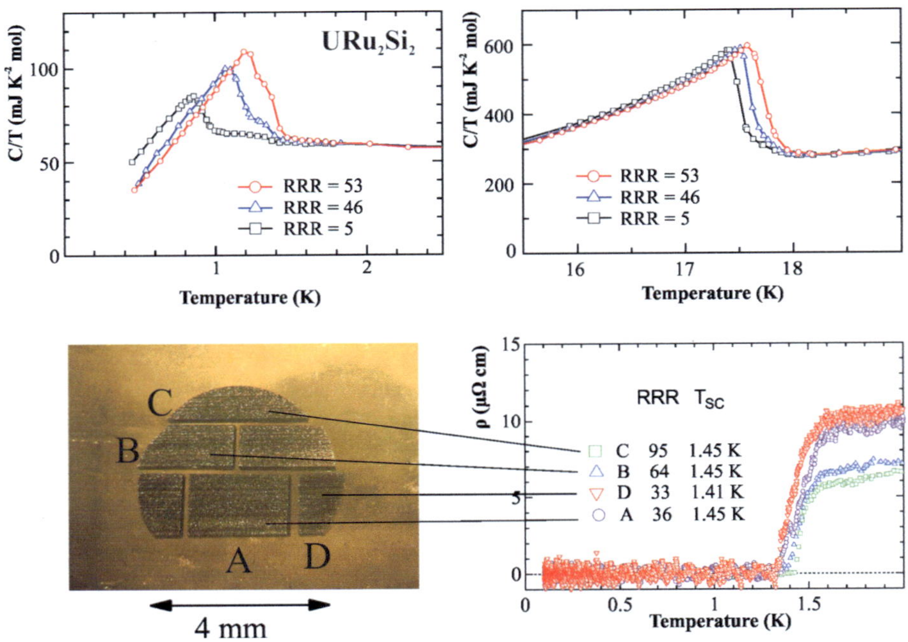

Figure 8: Specific heat vs temperature for single crystals of URu_2Si_2 with various RRR around superconducting (**top left**) and hidden-order transition (**top right**) as well as resistivity around the superconducting transition (**bottom right**) depending on the position of the specimen within the single crystal (**bottom left**) [39]. Adapted from [39] with permission of the Physical Society of Japan.

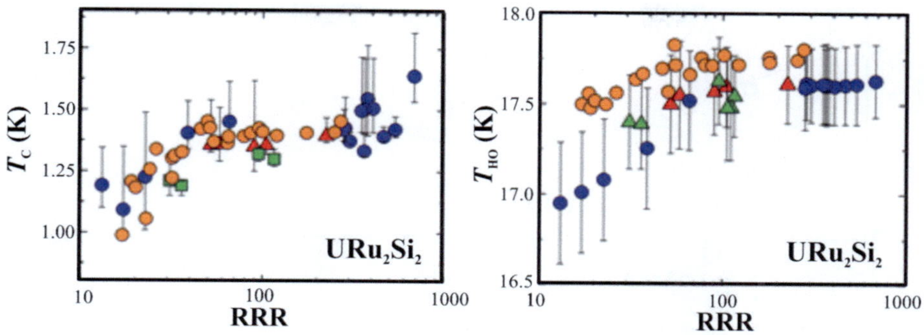

Figure 9: Temperatures of superconducting (**left**) and hidden-order (**right**) transitions vs RRR for single crystals of URu_2Si_2 obtained by different techniques [42].

measurements (Figure 8, bottom). The fact that the highest RRR value is obtained on *y* obtained from the outer side of the grown crystal indicates that, in the case of URu_2Si_2, the RRR value may be induced by the—most probably—substitution by impurities on the transition-metal or p-element sites. In the case of $YbRh_2Si_2$, such a substitution results even in a formation of a homogeneity range $YbRh_{2+x}Si_{2-x}$ ($0.00 \leq x \leq 0.01$) as was shown in Ref [41]. The authors of Ref. [42] obtained mm-sized single crystals of URu_2Si_2 by the many-pass Bridgman technique from the stoichiometric starting mixture of elements and—despite the presence of point 'surface' inclusions of binary phase RuSi (up to 15 µm)—the measured RRR values were above 100. The temperatures for both transitions—semiconducting and 'hidden'—were found to be very close to that of Ref. [39]. In general, the transition temperatures show quite a low RRR dependence which diminishes for RRR > 100 (Figure 9).

The heavy-fermion superconductor UPt_3 forms congruently around 1,700 °C and crystallizes in the hexagonal structure type Mg_3Cd. The single crystals of this substance were grown in a vertical float-zone refining system that operates with electron-beam heating in an ultra-high vacuum (10^{-10} torr) employing high-purity elemental metals as starting materials (total amount of impurities 23 ppm [43]). This was independent proof for the absence of chemical impurities. The superconducting temperature was found to clearly correlate to residual resistivity (Figure 10, left). Using transmission electron microscopy, planar defects were identified as principally responsible for the scattering. Somewhere unexpectedly, RRR was reduced by increasing the annealing temperature (Figure 10, right), in particular because the latter was far from the melting point (800–1,250 °C being less than 70 % of the melting temperature).

Summarizing the previous examples, the residual resistivity and residual resistivity ratio RRR are important parameters used for the characterisation of the electronic transport behaviour in single crystals of the intermetallic and metal-like materials. Beside the point defects as basic reasons, also substitution by minority components on selected sites in the crystal structure (URu_2Si_2, single-crystalline YFe_2Si_2), planar defects, most likely stacking defects (UPt_3), inclusions of $SrRuO_3$ and possibly $Sr_3Ru_2O_7$

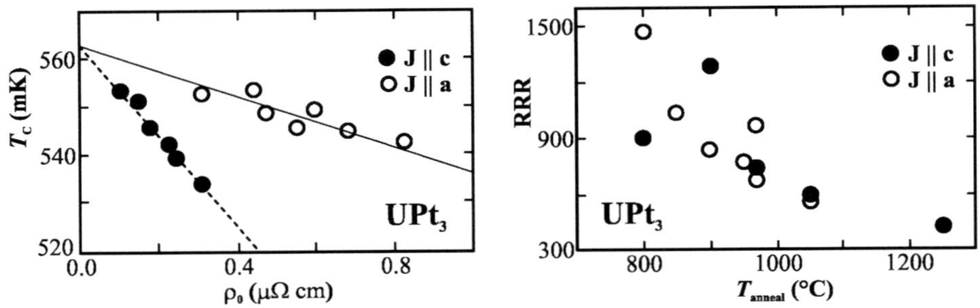

Figure 10: Single crystals of UPt_3: suppressing of the temperature of superconducting transition vs residual resistance for various current directions (**left**) and decreasing of RRR vs annealing temperature (**right**) [43]. Reprinted with permission from Kycia JB, Hong JI, Graf MJ, Sauls JA, Seidman DN, Halperin WP. Phys Rev B 58, R603 1998. Copyright (1998) by the American Physical Society.

in Sr_2RuO_4, partial substitutional disorder (UCoGe), and even impurity phases (polycrystalline YFe_2Si_2) were introduced as reasons regarding atomic resolution for the reduction of RRR.

3.4 Quality of single crystals, structure-property relation and chemical bonding

The proof of such suggestions is a complex task, which can be realized only by combination of several experimental techniques. One of the consequently realized studies of this kind is the investigation of the magnetism of α–$TmAlB_4$. The material was simultaneously synthesized and the mm-sized single crystals of were grown from the aluminium flux according to the reaction

$$Tm_2O_3 + 4Al + 8B \rightarrow 2TmAlB_4 + Al_2O_3.$$

In the crystal structure of the $YCrB_4$ type, there is only one crystallographic position for the magnetic component—thulium. Despite that, the magnetic specific heat did not show the expected clear lambda-like peak, but rather suggested several different magnetic interactions in the single crystal (Figure 11, top, green crosses). Precise crystal structure determination revealed residual electron density in the structure that was successfully interpreted by the presence of two different structural motifs in the same single crystal—the majority one from the $YCrB_4$ type and minority one from the $ThMoB_4$ type (Figure 11, middle). This was also confirmed by investigation by the transmission electron microscopy. In the course of the study, a successful growth of a single crystal without such structural variation (Figure 11, bottom) yields the expected clear picture of the magnetic specific heat (Figure 11, top, filled circles). From the point of view of the crystal's symmetry, its translational part is violated. In this case, the violation is relatively week (1.5 % of the minority pattern) and happens only along one direction.

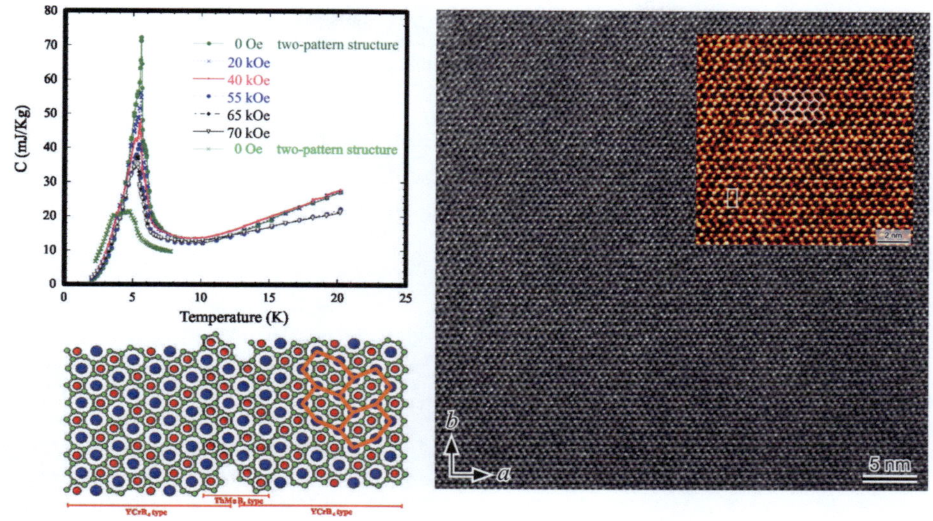

Figure 11: Magnetism and real structure of the single crystal of α–$TmAlB_4$: (**top**) magnetic specific heat vs temperature (green crosses – single crystal with two structural motifs, points – single crystal without variation in structural pattern); (**middle**) intergrowth of the structural patterns of $YCrB_4$ and $ThMoB_4$ types in one crystal; (**bottom**) HRTEM image in the [001] direction of the single crystal without structural variation [44, 45]. Top and right panels reproduced from Mori T, Kuzmych-Ianchuk I, Yubuta K, Shishido T, Okada S, Kudou K, Grin Yu. J Appl Phys 2012, 111, 07E127, with the permission of AIP Publishing.

The ytterbium analogue of the later compound $YbAlB_4$ also exists in two modifications with the crystal structures of the $YCrB_4$ (α) and $ThMoB_4$ (β) types. The second modification of this material β-$YbAlB_4$ is known as the first ytterbium-containing ambient pressure heavy-fermion superconductor [47]. The single crystals of both modifications were obtained by the two-step reaction of Yb_2O_3 with elemental boron to the precursor 'YbB_x' ($1 \leq x \leq 3$) with the subsequent treatment with the molten aluminium and grown from the excess-aluminium flux [47]. Also in this case, the precise crystal structure determination of the α-$YbAlB_4$ revealed residual electron density in the (001) plane (Figure 12, top left). Contrary to the thulium compound, the overlap of three structural patterns (Figure 12, top right) is necessary to describe this density distribution. The presence of these patterns in the single crystal investigated is confirmed by the HRTEM image ([001] zone, Figure 12, bottom left), revealing the violation of the translational symmetry in the [100] direction. In the case of β-$YbAlB_4$, the violation is more pronounced—it appears already in two directions (Figure 12, bottom right). The reasons for the local deviations from the translational symmetry can be found in the chemical bonding in the crystal structures of $YCrB_4$ and $ThMoB_4$ types. The covalent bonds between the boron, as well as between boron and aluminium atoms (red and green in Figure 11, middle), lead to the formation of the three-dimensional framework bearing the ytterbium cations in the channel-like cavities (blue in Figure 11, mid-

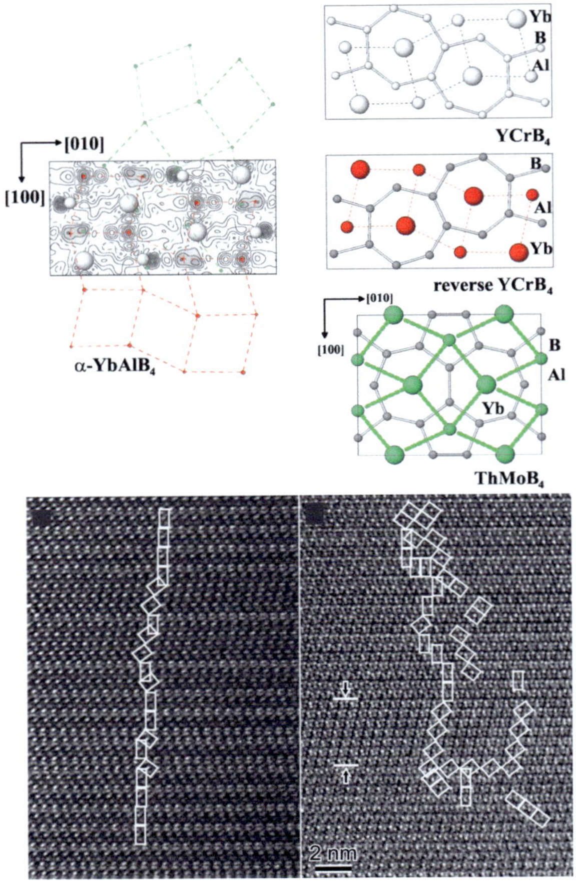

Figure 12: Real structure of single crystal of α-YbAlB$_4$: (**top left**) distribution of the residual electron density in the (001) plane; (**top right**) three structural motifs required for the interpretation of the residual electron density; (**bottom**) HRTEM images in the [001] direction of the α- (**left**) and β-YbAlB$_4$ (**right**) single crystals with one- and two-dimensional violation of the translational symmetry, respectively [46]. Reprinted from Yubuta K, Mori T, Okada S, Prots Yu, Borrmann H, Grin Yu, Shishido T. High-resolution electron microscopy and X-ray diffraction study of intergrowth structures in α- and β-type YbAlB$_4$ single crystals. Philos. Mag 2013, 93, 1054 by permission of the Taylor & Francis Ltd.

dle [44]). The relative positions of these channels that cause the difference between the $YCrB_4$ and $ThMoB_4$ patterns change only the ionic interactions in the outer coordination shells and thus do not influence strongly the total energy of the structure. Thus may be the reason that crystals with either type of structure can be obtained by the same growth experiment from the aluminium flux and that both atomic arrangements can co-exist in the same crystal, although in different ratios.

The ternary compounds with the crystal structure of $HoCoGa_5$ type [48] attract the attention of the solid-state physicists since the discovery of heavy-fermion superconductivity in $PuCoGa_5$ [49]. In particular, the $CeTIn_5$ compounds (T – transition metal) offer a wide playground for investigation of the interplay of heavy-fermion superconductivity and magnetism [50, 51]. The underlying crystal structure is rather simple and contains seven atoms in the unit cell (Figure 13, top right). In an attempt to directly visualise the crystal structure of $CeIrIn_5$, the topography measurements by STM were conducted [52]. Rather unexpectedly, the image in the (113) plane revealed two

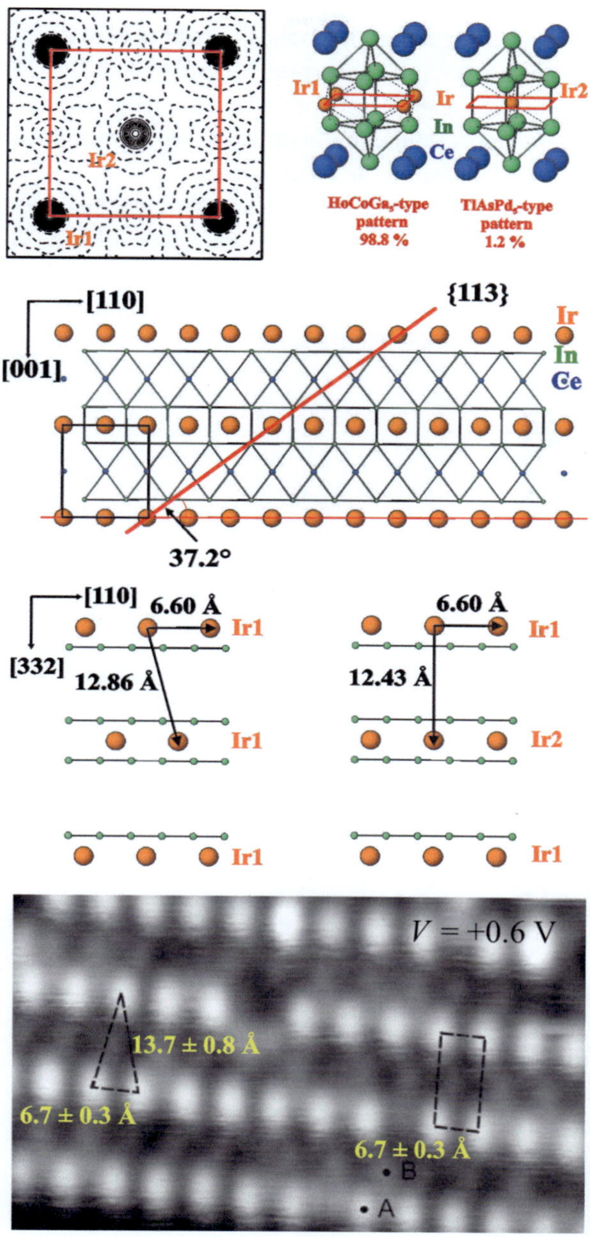

Figure 13: Real structure of single crystal of $CeIn_5Ir$: **(top left)** distribution of the residual electron density in the (002) plane; **(top right)** two structural motifs required for the interpretation of the residual electron density; **(middle)** projection of the crystal structure along [1–10] and with two atomic arrangements in the plane (113) forming the surface found in the STEM experiments **(bottom)** [52]. Reprinted from [52] with permission of the Physical Society of Japan.

different patterns instead of the one possible if the pristine crystal structure of the $HoCoGa_5$ type (Figure 13, bottom) is assumed. Precise crystal-structure determination of the same single crystal of $CeIrIn_5$ revealed Ir atoms in two different positions (in the ratio 98.5:1.5 and the total occupancy of 1), which reflect the structure types $HoCoGa_5$

and $TlAsPd_5$ (Figure 13, top left). This finding reveals the coexistence of two similar but different structural motifs in one-and-the same single crystal (Figure 13, middle). The reason for that feature is the similar total energies of both structural patterns which differ only slightly in bonding interactions. While, in the $HoCoGa_5$-type motif, the direct Ce–Ir bonding was found, it is not present in the $TlAsPd_5$ one.

The most impressive current example of the violation of translational symmetry in a single crystal is represented by the intermetallic clathrate $Ba_{7.81}Ge_{40.67}Au_{5.33}$ [53]. Based on the detailed study of the phase diagram of the ternary system Ba–Ge–Au, the cm-sized single crystal of the clathrate phase with remarkable crystallinity (mosaicity of 0.01°) was grown by the Bridgman technique (Figure 14, top).

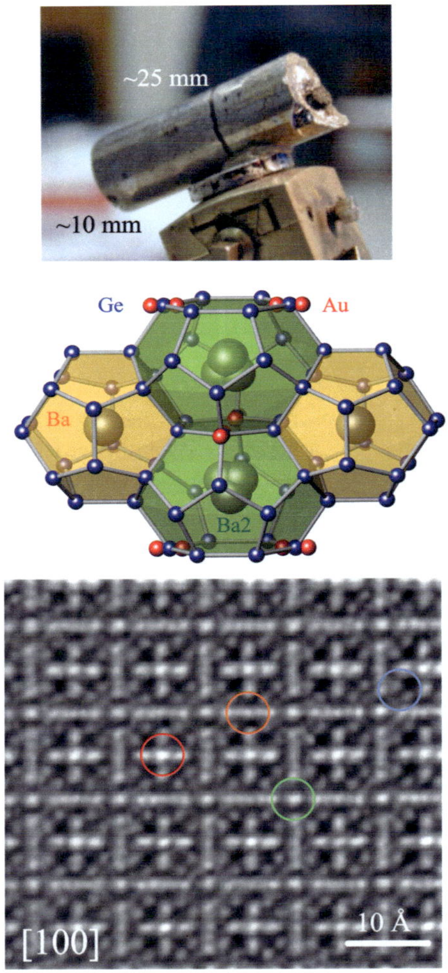

$Ba_{7.81}Ge_{40.67}Au_{5.33}$

Figure 14: Violation of the translational symmetry in of the intermetallic clathrate $Ba_{7.81}Ge_{40.67}Au_{5.33}$: **(top)** cm-sized single crystal grown by Bridgman method; **(middle)** off-center positions of Ba and Au atoms in the crystal structure from the single-crystal diffraction experiment; **(bottom)** atomic-resolution TEM image showing the real atomic structure [53]. Middle and bottom panels are reproduced from [53]. Top photo courtesy Prof. Dr. C. Krellner.

Detailed investigation of the crystal structure made on a small piece of the large specimen revealed marked local displacements of the Ba and Au atoms from the ideal crystallographic positions caused by the special bonding feature—a direct dative Au–Ba interaction [24] (Figure 14, middle). Atomic-resolution TEM study confirmed very clearly the violations of the translational symmetry in the crystal structure (Figure 14, bottom). Inelastic neutron-scattering experiments reveal the very unusual behaviour of such pseudo-translational system: surprisingly, the thermal transport in this substance in dominated by phonons with long lifetimes. These results challenge the current picture of thermal transport in clathrates, underline the inability of the modern theoretical techniques to reproduce the experimental data and represent crucial experimental contribution to theoretical developments [53].

4 How much a single crystal may be a single crystal?

The findings described in the previous paragraphs exemplify the various expectations of single crystals of intermetallic compounds from the crystallographic point of view and from the point of view of the study of their physical and chemical behaviours. For high-quality investigation of the crystal structure usually only very small specimens (µm-size) are required. The experience shows that the small size should preferably have a natural origin, i. e., the really suitable single crystals for structure analysis seldom can be obtained by reducing the size of the pristine specimen by cutting, crushing etc. For the studies of properties, the single crystals should be larger and normally have a mm-size. Only recently, the measurement technologies have been developed to perform property studies on mm-sized single crystals [54]. In order to establish the often target structure-property relation for the large crystals, the crystallographic characterisation should be made on the same crystal that was used for the property investigation. The precise crystal structure determinations revealed already for very small single crystals of intermetallic compounds local violations of point and—crucial—translational symmetry caused obviously by the presence of covalent bonding in these materials, which allows variation of local atomic arrangements.

Bibliography

[1] Grin Yu, Cardoso-Gil R, Caroca-Canales N, Bravo G, Bobnar M, Veremchuk I. Int Conf Appl Mineral Adv Mater. Castellaneta Marina, Italy, 2017 Sci Res Abstr 2017, 6, 15.

[2] Pearson WB. Acta Crystallogr 1964, 17, 1.

[3] Grin Yu. In: Reedijk J, Poeppelmeier K, eds. Comprehensive Inorganic Chemistry II, vol. 2. Oxford, Elsevier, 2013. 359ff.

[4] Klemm W, Bussman E. Z Anorg Allg Chem 1963, 319, 297.

[5] Kniep R. In: Kauzlarich S, ed. Chemistry, Structure and Bonding of Zintl Phases and Ions. New York, VCH, 1996. xvii.
[6] Laves F. Naturwissenschaften 1941, 29, 241.
[7] Zintl E, Dullenkopf W. Phys Chem B 1932, 16, 195.
[8] Wade K. Adv Inorg Radiochem 1976, 18, 1.
[9] von Schnering HG. Angew Chem, Int Ed Engl 1981, 20, 33.
[10] Mingos DMP. Acc Chem Res 1984, 17, 311.
[11] Simon A. Angew Chem, Int Ed Engl 1988, 27, 159.
[12] Hume-Rothery W. Materials Science and Engineering. NY, McGraw Hill, 1967. 3ff.
[13] Mizutani U. Hume-Rothery Rules for Complex Metallic Alloys. London, Taylor & Francis, 2010.
[14] Bader RFW. Atoms in Molecules – A Quantum Theory. Oxford, Clarendon Press, 1995.
[15] Wagner FR, Bezugly V, Kohout M, Grin Yu. Chem Eur J 2007, 13, 5724.
[16] Kohout M. Faraday Discuss 2007, 135, 43.
[17] Ormeci A, Simon A, Grin Yu. Angew Chem, Int Ed Engl 2010, 49, 8997.
[18] Bende D, Grin Yu, Wagner FR. Chem Eur J 2014, 20, 9702.
[19] Bende D, Grin Yu, Wagner FR. In: Felser C, Hirohata A, eds. Heusler Alloys. Springer, 2016, 133ff.
[20] Bende D, Wagner FR, Grin Yu. Inorg Chem 2015, 54, 3970.
[21] Wagner FR, Bende D, Grin Yu. Dalton Trans 2016, 45, 3236.
[22] Bende D, Wagner FR, Sichevych O. Angew Chem, Int Ed Engl 2017, 56, 1313.
[23] Ormeci A, Grin Yu. J Thermoelectr 2015, 6, 16.
[24] Zhang H, Borrmann H, Oeschler N, Candolfi C, Schnelle W, Schmidt M, Burkhardt U, Baitinger M, Zhao J-T, Grin Yu. Inorg Chem 2011, 50, 1250.
[25] Kadok J, de Weerd MC, Boulet P, Gaudry E, Grin Yu, Fournée V, Ledieu J. Inorg Chem 2015, 54, 7898.
[26] Razin LV, Rudashevskij NS, Vyalsov LN. 1985, 114, 90.
[27] Bindi L, Steinhardt PJ, Yao N, Lu PJ. Science 2009, 324, 1306.
[28] Maeno Y, Hashimoto H, Yoshida K, Nishizaki S, Fujita T, Bednorz JG, Lichtenberg F. Nature 1994, 372, 532.
[29] Maeno Y, Yoshida K, Hashimoto H, Nishizaki Sh, Ikeda Sh, Nohara M, Fujita T, Mackenzie AP, Hussey NE, Bednorz JG, Lichtenberg F. J Phys Soc Jpn 1997, 66, 1405.
[30] Mackenzie AP, Haselwimmer RKW, Tyler AW, Lonzarich GG, Mori Y, Nishizaki S, Maeno Y. Phys Rev Lett 1998, 80, 161.
[31] Mao ZQ, Mori Y, Maeno Y. Phys Rev B 1999, 60, 610.
[32] Chen J, Semeniuk K, Feng Zh, Reiss P, Brown Ph, Zou Y, Logg PW, Lampronti GI, Grosche FM. Phys Rev Lett 2016, 116, 127001.
[33] Kim H, Rana S, Mun ED, Hodovanets H, Tanatara MA, Prozorova R, Bud'ko SL, Canfield PC. Philos Mag 2015, 95, 804.
[34] Huy NT, Huang YK, de Visser A. J Magn Magn Mater 2009, 321, 2691.
[35] Aoki D, Sheikin I, Matsuda TD, Taufour V, Knebel G, Floquet J. J Phys Soc Jpn 2011, 80, 013705.
[36] Canepa F, Manfrinetti P, Pani M, Palenzona A. J Alloys Compd 1996, 234, 225.
[37] Pospisil J, Prokes K, Reehuis M, Tovar M, Poltierova-Vejpravova J, Prokleska JJ, Sechovsky V. Phys Soc Jpn 2011, 80, 084709.
[38] Samsel-Czekala M, Elgazzar S, Oppeneer PM, Talik E, Walerczyk W, Troc R. J Phys Condens Matter 2010, 22, 015503.
[39] Matsuda TD, Hassinger E, Aoki D, Taufour V, Knebel G, Tateiwa N, Yamamoto E, Haga Y, Onuki Y, Fisk Z, Floquet J. J Phys Soc Jpn 2011, 80, 114710.
[40] Haga Y, Honma T, Yamamoto E, Ohkuni H, Onuki Y, Ito M, Kimura N. Jpn J Appl Phys 1998, 37, 3604.

[41] Wirth S, Ernst S, Cardoso-Gil R, Borrmann H, Seiro S, Krellner C, Geibel C, Kirchner S, Burkhardt U, Grin Yu, Steglich F. J Phys Condens Matter 2012, 24, 294203.

[42] Gallagher A, Nelson WL, Chen KW, Besara T, Siegrist Th, Baumbach RE. Crystals 2016, 6, 128.

[43] Kycia JB, Hong JI, Graf MJ, Sauls JA, Seidman DN, Halperin WP. Phys Rev B 1998, 58, R603.

[44] Mori T, Borrmann H, Okada S, Kudou K, Leithe-Jasper A, Burkhardt U, Grin Yu. Phys Rev B 2007, 76, 064404.

[45] Mori T, Kuzmych-Ianchuk I, Yubuta K, Shishido T, Okada S, Kudou K, Grin Yu. J Appl Phys 2012, 111, 07E127.

[46] Yubuta K, Mori T, Okada S, Prots Yu, Borrmann H, Grin Yu, Shishido T. Philos Mag 2013, 93, 1054.

[47] Nakatsuji S, Kugai K, Machida Y, Tayama T, Sakakibara T, Karaki Y, Ishimoto H, Yonezawa S, Maeno Y, Peason E, Lonzarich GG, Balicas L, Lee H, Fisk Z. Nat Phys 2008, 4, 603.

[48] Grin YuN, Yarmolyuk YaP, Gladyshevski EI. Sov Phys Crystallogr 1979, 24, 137.

[49] Curro NJ, Caldwell T, Baue ED, Morale LA, Graf MJ, Bang Y, Balatsky AV, Thompson JD, Sarrao JL. Nature 2004, 434, 622.

[50] Sarrao JL, Thompson JD. J Phys Soc Jpn 2007, 76, 051013.

[51] Thompson JD, Fisk Z. J Phys Soc Jpn 2012, 81, 011002.

[52] Wirth S, Prots Yu, Wedel M, Ernst S, Kirchner S, Fisk Z, Thompson JD, Steglich F, Grin Yu. J Phys Soc Jpn 2014, 83, 061009.

[53] Lory P-F, Pailhès S, Giordano VM, Euchner H, Nguyen HD, Ramlau R, Borrmann H, Schmidt M, Baitinger M, Ikeda M, Tomeš P, Mihalkovi M, Allio C, Johnson MR, Schober H, Sidis Y, Bourdarot F, Regnault LP, Ollivier J, Paschen S, Grin Yu, de Boissieu M. Nat Commun 2017, 8, 1.

[54] Ronning F, Helm T, Shirer KR, Bachmann MD, Balicas L, Chan MK, Ramshaw BJ, McDonald RD, Balakirev FF, Jaime M, Bauer ED, Moll P. Nature 2017, 548, 313.

Herbert Ipser

Phase diagrams for crystal growth of intermetallics

1 Introduction

1.1 What is a phase diagram?

When a single crystal of an intermetallic compound, or better, of an intermetallic phase,[1] is grown out of a melt by the so-called Czochralski method [1, 2], it is crucial to know which of its particular compositions is in thermodynamic equilibrium with the corresponding liquid phase at the selected temperature and pressure. Of course, the same is true for crystals that are grown by the self-flux technique [3, 4], and, even in the Bridgman method, one has to consider the equilibrium between a liquid phase and the growing crystal [5]. In order to find out, one needs to know the relevant phase diagram of the materials system under consideration.

By definition, a phase diagram describes the state of a materials system in thermodynamic equilibrium as a function of temperature, pressure and composition. In other words, phase diagrams are maps of the equilibrium phases associated with various combinations of temperature, pressure, and composition. Here it is important to note that a phase diagram a priori always refers to a state of thermodynamic equilibrium. Of course, one can include metastable phases in the diagram, but this has to be made explicitly clear in the representation. On the other hand, equilibrium phase diagrams can still be used to understand non-equilibrium situations that may develop during various metallurgical reactions, as long as local equilibria are maintained.

A two-dimensional representation of the phase equilibria as a function of temperature and pressure is only possible for a one-component, or unary, system. Probably all of the readers will have seen such a one-component phase diagram, e. g., the one of pure water. For a full representation of the phase equilibria in a two-component, or binary, system, one already requires three dimensions: pressure, temperature and composition. Therefore, it is quite common to keep the pressure constant and depict a two-dimensional temperature-composition diagram. The situation becomes even more complicated for three-component, i. e., ternary, and higher-order systems. A full representation of a ternary system as a function of composition, temperature and pressure would require a four-dimensional diagram, and, even if the pressure is kept con-

1 Since the term "compound" frequently suggests a material with a strictly stoichiometric composition, the term "intermetallic phase", or for short "intermetallic", is preferred here.

https://doi.org/10.1515/9783110496789-002

stant, we still need three dimensions. Thus, it is quite common to show ternary phase equilibria in two dimensions, i. e., as isothermal sections, isopleths,[2] or liquidus projections, as will be discussed in detail in subchapter 4. A so-called Scheil diagram [6, 7] may additionally help to understand the sequence of invariant reactions in a ternary or higher-order system.

Many books or book chapters on phase diagrams and their interpretation have been published over recent decades. The references [8–13] serve as a few examples of which the book by Rhines [9] is the author's personal favorite.

1.2 Why do we need phase diagrams?

A striking example of the importance of knowing the phase equilibria is the case of the ternary phase "$YbInCu_4$", which shows a valence transition for Yb. In a study by Kindler et al. [14], a transition temperature of 66.9 K was reported, whereas earlier investigations had found a temperature of 45 K [15] or temperatures between about 40 and 80 K depending on the value of x in $Yb_{1\pm x}In_{1\pm x}Cu_4$ [16]. This latter paper shows already the importance of knowing the phase diagram since obviously this ternary phase is highly nonstoichiometric with a possible exchange of Yb and In ions on the corresponding lattice sites. Consequently, the valence transition temperature of Yb varies with its concentration in the phase, a fact that had apparently not been clear to all investigators.

The Cr–Te binary system may serve as another example of the importance of knowing the phase diagram when synthesizing intermetallic phases and studying their properties. Up to the 1970s, there were many reports of the magnetic properties of the NiAs-type compound "CrTe" (see, for example [17, 18]), and single crystals of "CrTe" were even grown by the Bridgman method [19]. A few years later, it became clear that exactly stoichiometric CrTe does not exist in reality. Instead, a two-phase mixture of a non-stoichiometric phase $Cr_{1-x}Te$ (showing various superstructures depending on composition) with Cr is stable at 50 % Te at all temperatures up to 1,180 °C [20]. Therefore, magnetic properties have been measured either of heterogeneous samples or of a homogeneous single-phase sample, though with a different composition.

1.3 The phase rule

The so-called phase rule, or Gibbs' phase rule, is a useful tool to check for the correctness and consistency of a phase diagram. A derivation can be found in many text-

2 An isopleth is a vertical section through a three-dimensional temperature-composition space diagram of a ternary system, frequently keeping the composition of one component or the ratio of the compositions of two components constant.

books on physical chemistry (see, e. g., Ref. [21]). It gives the “degrees of freedom” F, i. e., the number of variables that can be varied independently in a system composed of C components and containing P phases, without changing its equilibrium. On the other hand, it tells us also the maximum number of phases P that can co-exist in a system composed of C components.

$$F = C - P + 2 \tag{1}$$

An application of the phase rule will be subsequently demonstrated for unary and binary systems.

2 One-component (unary) systems

The art of crystal growth by the Czochralski method has been perfected for elemental Si in the electronics industry where huge single crystals of extremely high purity, two meters long and 30 cm in diameter, are the standard today. Nevertheless, as this book deals with crystal growth of intermetallics, phase diagrams of unary systems will be treated only briefly, mostly in order to demonstrate a few principles.

2.1 Pressure-composition phase diagrams of unary systems

Figure 1 shows a typical pressure-composition phase diagram of a one-component system. It shows us which phase, or which combination of phases, is stable at a certain pressure-temperature combination. In Figure 1 there are one solid, one liquid and one gas phase, however, many substances exist in more than one solid modification, which means that for each of them a corresponding phase field must appear in the diagram. The boundary line between the solid and liquid phases is usually called the melting curve, the one between the solid and gas phases is the sublimation curve and

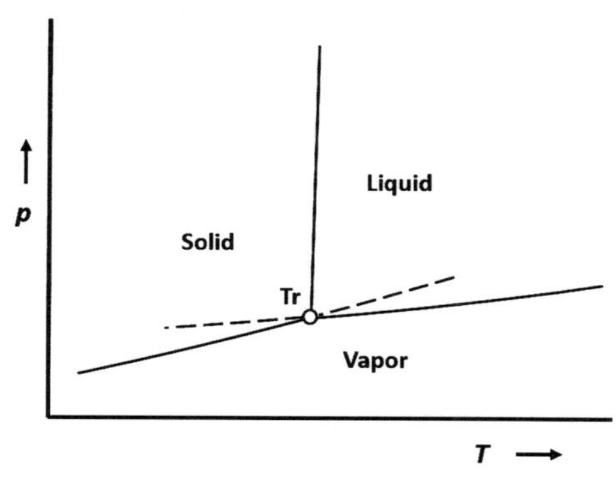

Figure 1: Unary *p-T* phase diagram; Tr is the triple point.

the one between the liquid and gas phases is the boiling curve. The latter one ends in a so-called critical point (not shown in Figure 1) where the liquid phase and gas phase become undistinguishable since their densities become equal.

Applying the phase rule to one of the single-phase fields in Figure 1, one has $C = 1$ (one component) and $P = 1$ (one phase). According to eq. (1), this results in $F = 2$, i. e., one has two degrees of freedom, one can vary both the pressure and the temperature independently without changing the equilibrium (one phase). For any point on one of the phase boundaries along which two phases are in equilibrium, we have now $P = 2$ which results in $F = 1$. One can vary only one of the two variables independently without changing the equilibrium: if one changes, e. g., the temperature, one has to change the pressure correspondingly, or vice versa; otherwise one of the two phases disappears. The triple point Tr (see Figure 1) is a very special case where the three phases are in equilibrium. Applying the phase rule, one obtains $F = 0$, and one has a so-called invariant equilibrium. With zero degrees of freedom, both variables are fixed, i. e., only at a well-defined temperature and pressure can three phases be in equilibrium. In other words, three phases are the maximum number that can be in equilibrium in a one-component system.

2.2 Thermodynamic background

As just pointed out, a phase diagram represents the state of a materials system in thermodynamic equilibrium or, in other words, it is itself a representation of the thermodynamic situation in the system. It is well-known that the thermodynamic equilibrium is characterised by a minimum of the Gibbs energy of the system. Thus, it will always be the case that the most stable phase, or the most stable combination of phases, at a given temperature and pressure will be the one with the lowest Gibbs energy.

In a one-component system, the Gibbs energy is a function of temperature and pressure only:

$$G = f(p, T) \tag{2}$$

Thus, this produces a surface in a corresponding diagram where G is plotted as a function of p and T. In a one-component system, there is a Gibbs-energy surface for each phase, the vapor phase, the liquid phase and one or more solid phases. These surfaces intersect each other and that portion of each surface that is the lowest of all surfaces in any region gives the p-T range in which the corresponding phase is stable. If the intersection lines of the Gibbs-energy surfaces are projected on the p-T plane, one obtains the unary phase diagram as it is usually shown (Figure 1).

If in eq. (2) the pressure is kept constant, the three-dimensional G-p-T representation reduces to two dimensions, and, from the intersection of the Gibbs energy curves, one can clearly see the temperature where the phase transition takes place at the given

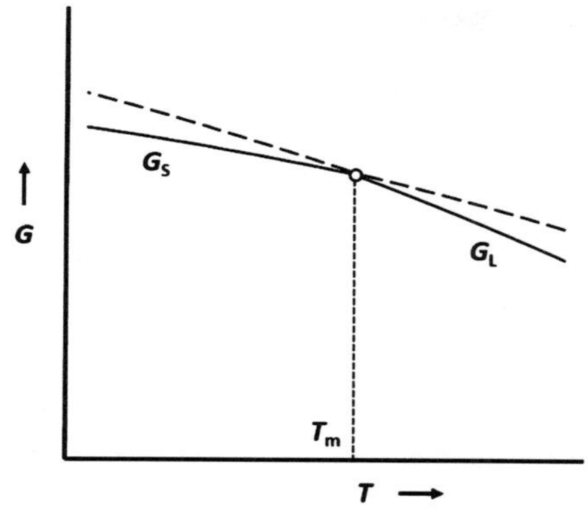

Figure 2: Gibbs-energy curves of the solid and the liquid phase as a function of temperature at constant pressure. The intersection of the two curves defines the melting point (at this particular pressure).

pressure. This is shown for the melting process in Figure 2. Of course, there is the possibility that the substance exists in two or more solid modifications depending on temperature (just think of pure iron which exists in three modifications depending of temperature, α-, γ-, and δ-Fe). Then it is the relative position of the corresponding Gibbs-energy curves that will tell if all of the modifications are stable or if some are metastable. Consider the situation that there exist two solid forms of a pure substance, α and β. Figure 3(a) shows the case where both of these modifications are stable, depending on temperature. In Figure 3(b), one can see that the Gibbs energy of modification β is at no temperature lower than that of α or the liquid. Consequently, β is metastable over the entire temperature range at the selected pressure. (However, it may still crystallize as metastable β out of the melt if the crystallization of α is suppressed for any reason.)

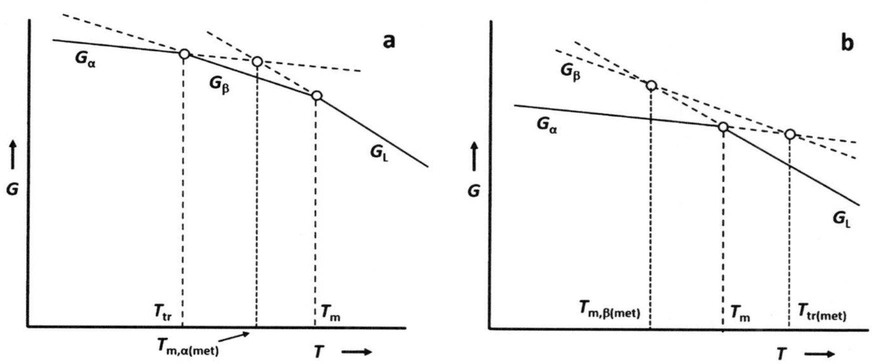

Figure 3: Gibbs-energy curves of two solid phases and the liquid phase as a function of temperature at constant pressure; **(a)** β is stable over a certain temperature interval; **(b)** β is metastable over the entire temperature interval (curvature of the Gibbs-energy curves neglected). T_m = melting temperature; T_{tr} = transition temperature; (met) = metastable.

3 Two-component (binary) systems

In a two-component, or binary, system, three variables are necessary for a full description, i. e., the temperature, the pressure and the composition, which necessitates a three-dimensional diagram. As this is not very practical, the usual representation is a two-dimensional diagram temperature vs. composition where the pressure is kept constant.

Very frequently, the pressure is not even accurately defined. This is a minor problem as long as only equilibria between condensed phases are considered since their pressure dependence is usually rather small if not negligible. The situation changes, of course, drastically if equilibria with the gas phase come into play, and this may be the reason why the gas phase is excluded in many of the traditional binary phase diagrams, although one of the components has a high vapor pressure (see, as just one example, the system Ni–Se [22, 23]). That the pressure can play a significant role may be demonstrated by the Ni–P system where the phase equilibria change drastically when the pressure is varied [24].

Nevertheless, the following discussion will be restricted to the usual representation of binary phase diagrams in the form of temperature-composition diagrams for which the pressure is assumed to be constant.

3.1 The lever rule

A useful tool for the interpretation of binary phase diagrams is the so-called lever rule. It enables calculating an estimate of the relative amounts of the two phases in equilibrium in a two-phase field. A derivation of it can be found in some of the books on phase diagrams [8–13]. Figure 4 shows schematically a two-phase field between a liquid phase L and a solid phase β. At the temperature T_1, there is a tie line connecting phase L of composition x_B^L with phase β of composition x_B^β. For an alloy with the gross

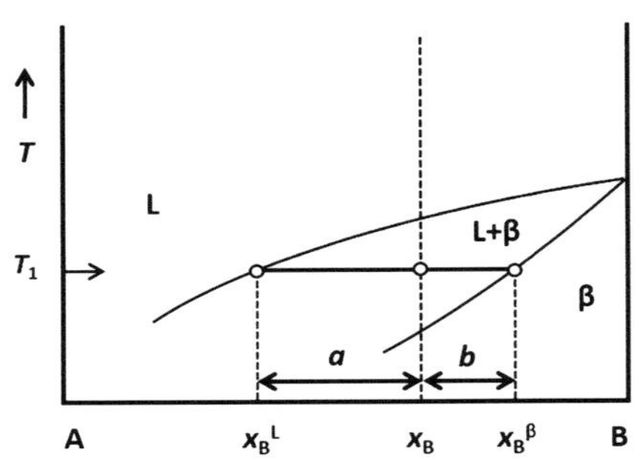

Figure 4: Demonstration of the application of the lever rule (see text).

composition x_B, the molar ratio n_L/n_β of the two phases at T_1 is given by

$$\frac{n_L}{n} = \frac{b}{a} = \frac{x_B^\beta - x_B}{x_B - x_B^L} \tag{3}$$

If the gross composition is not given in mole fractions (atomic percent) but in mass percent, then one would obtain the mass ratio of the two phases instead of the molar ratio.

3.2 Binary isomorphous systems

Figure 5 shows the temperature-composition diagram of an isomorphous binary system consisting of the components A and B, meaning that the two components show complete mutual solubility in the liquid and the solid state. If we want to apply the phase rule in this case, we have to modify eq. (1) because we have already fixed the pressure and thus used one of the degrees of freedom:

$$F' = C - P + 1 \tag{4}$$

where F' is now the degrees of freedom after fixing the pressure. For any temperature-composition combination in one of the single-phase fields L or α, we have $C = 2$ (two components) and $P = 1$ (one phase) which results in $F' = 2$. That proves that both the temperature and the composition can be varied independently without changing the equilibrium (one single phase). Within the two-phase field (L + α), we have $P = 2$ (two phases) which results in $F' = 1$. This indicates that we have only one degree of freedom left: either one can choose the temperature independently, then the compositions of the two phases L and α are fixed by the corresponding tie line; or one can choose the composition of one of the two phases in equilibrium, but then the temperature as well as the composition of the other phase are automatically fixed.

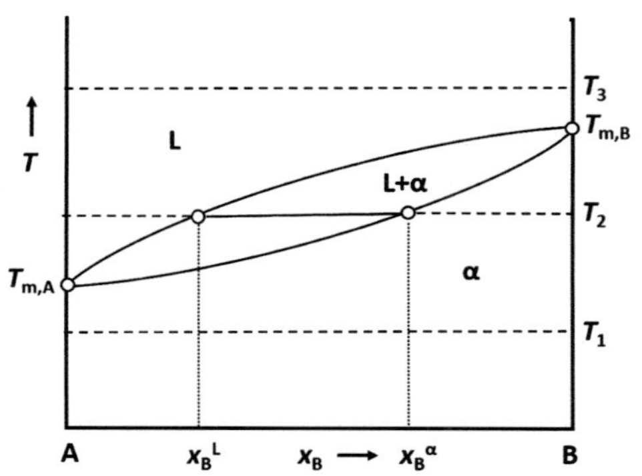

Figure 5: Isomorphous binary temperature-composition phase diagram.

In a binary system, the Gibbs-energy function of each phase is a function of pressure, temperature and composition, resulting in a four-dimensional diagram, which is clearly impossible to visualise. To reduce this to two dimensions, it is customary to show the Gibbs energy as a function of composition at constant pressure and temperature. Therefore, it is necessary to investigate the corresponding Gibbs-energy curves at various temperatures to arrive at the usual binary temperature-composition phase diagram.

Let us, as an example, investigate the competition between the solid and the liquid phases in the isomorphous binary system shown in Figure 5. Instead of the absolute Gibbs energy (which is difficult to calculate), we will use the Gibbs energy of mixing ΔG^{Φ} of each phase Φ:

$$\Delta G^{\Phi} = G^{\Phi} - x_{\mathrm{A}} G_{\mathrm{A}}^{*} - x_{\mathrm{B}} G_{\mathrm{B}}^{*} \tag{5}$$

where G^{Φ} is the (absolute) Gibbs energy of the phase Φ, and G_{A}^{*} and G_{B}^{*} are the (absolute) Gibbs energies of the two pure components in their reference states. As reference state, we select the pure component in its stable state at the given pressure and temperature. Thus, if for example the component A is solid at the temperature under consideration, then solid A is the corresponding reference state; if it is liquid, then liquid A is the reference state. With this definition, the Gibbs energies of mixing of the liquid and the solid phases α (where α defines a certain crystal structure) are given by:

$$\Delta G^{\mathrm{L}} = G^{\mathrm{L}} - x_{\mathrm{A}} G_{\mathrm{A}}^{*} - x_{\mathrm{B}} G_{\mathrm{B}}^{*} \tag{6a}$$
$$\Delta G^{\alpha} = G^{\alpha} - x_{\mathrm{A}} G_{\mathrm{A}}^{*} - x_{\mathrm{B}} G_{\mathrm{B}}^{*} \tag{6b}$$

Considering the situation at the temperature T_1, both components are solid at this temperature, thus, according to the definition above, the reference states for both the liquid and the solid phase are the pure solid components in the crystal structure α. With this, eqs. (6) become:

$$\Delta G^{\mathrm{L}} = G^{\mathrm{L}} - x_{\mathrm{A}} G_{\mathrm{A}}^{\alpha} - x_{\mathrm{B}} G_{\mathrm{B}}^{\alpha} \tag{7a}$$
$$\Delta G^{\alpha} = G^{\alpha} - x_{\mathrm{A}} G_{\mathrm{A}}^{\alpha} - x_{\mathrm{B}} G_{\mathrm{B}}^{\alpha} \tag{7b}$$

For pure A where $x_{\mathrm{A}} = 1$ and $x_{\mathrm{B}} = 0$, ΔG^{L} becomes equal to the Gibbs energy of melting of pure A, $\Delta_{\mathrm{m}} G_{\mathrm{A}}$, at the temperature T_1:

$$\Delta G^{\mathrm{L}} = G^{\mathrm{L}} - 1 \cdot G_{\mathrm{A}}^{\alpha} - 0 \cdot G_{\mathrm{B}}^{\alpha} = G^{\mathrm{L}} - G_{\mathrm{A}}^{\alpha} = \Delta_{\mathrm{m}} G_{\mathrm{A}} \tag{8a}$$

and ΔG^{α} becomes zero:

$$\Delta G^{\alpha} = G^{\alpha} - 1 \cdot G_{\mathrm{A}}^{\alpha} - 0 \cdot G_{\mathrm{B}}^{\alpha} = G^{\alpha} - G_{\mathrm{A}}^{\alpha} = 0 \tag{8b}$$

since G^{α} and G_{A}^{α} are identical at this composition. Identical arguments are valid for pure B where $x_{\mathrm{A}} = 0$ and $x_{\mathrm{B}} = 1$, i. e., $\Delta G^{\mathrm{L}} = \Delta_{\mathrm{m}} G_{\mathrm{B}}$ and $\Delta G^{\alpha} = 0$. This situation

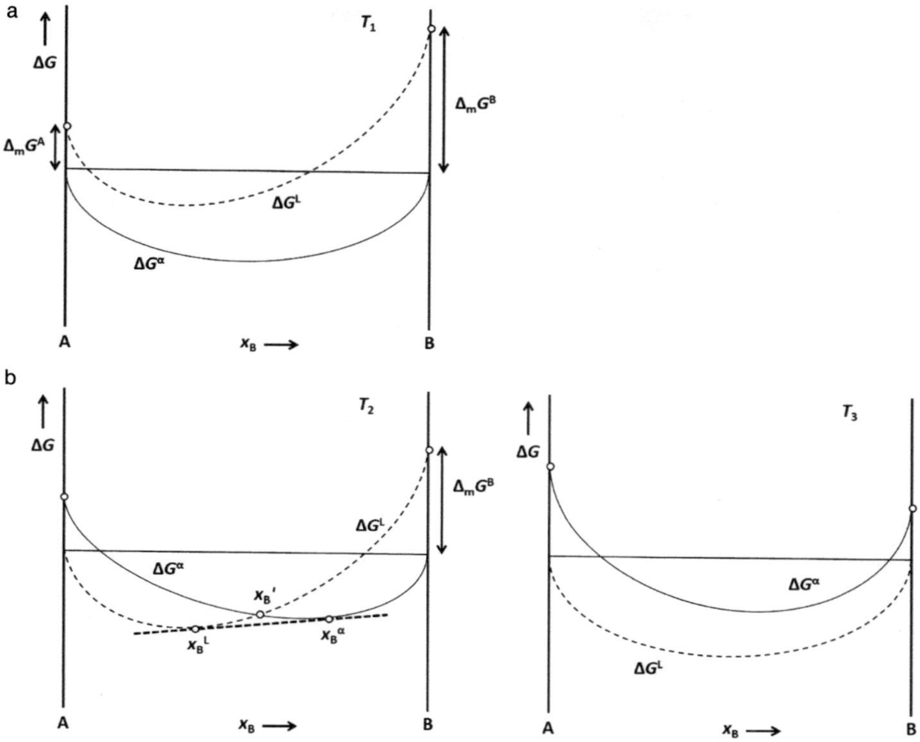

Figure 6: Gibbs energy of mixing of the liquid phase and the solid phase α in the binary system A–B at the three temperatures T_1, T_2 and T_3 (see Figure 5).

is shown schematically in Figure 6(a). One can see that over the entire composition range ΔG^α has the most negative value, indicating that at T_1 the solid phase α is stable for all compositions (see Figure 5).

At the temperature T_2, component A is liquid and component B is solid; therefore, these are also the corresponding reference states. The curve for the Gibbs energy of mixing of the liquid phase, ΔG^L, starts at a value of zero on the side of pure A ($x_B = 0$) and ends at a positive value on the side of pure B ($x_B = 1$) which corresponds again to the Gibbs energy of melting of pure B at T_2. On the other hand, the curve for the Gibbs energy of mixing of the solid phase α, ΔG^α, starts at a positive value at $x_B = 0$ and ends at zero for $x_B = 1$. The two curves intersect at a certain composition x_B' (see Figure 6(b)), thus for $0 \leq x_B \leq x_B'$ the Gibbs energy of mixing of the liquid phase has the most negative value whereas for $x_B' \leq x_B \leq 1$ it is the Gibbs energy of mixing of solid α which exhibits the most negative value.

However, one has to keep in mind that one can draw a common tangent to the two curves touching at the compositions x_B^L and x_B^α. This tangent signifies the Gibbs energy of a mechanical mixture of the two phases L and α with concentrations x_B^L and x_B^α, respectively. It is obvious that in this composition range the mixture of L and α has a lower Gibbs energy than either one of the two phases alone, with the consequence that a two-phase field shows up in the phase diagram at the temperature T_2, where its phase boundaries x_B^L and x_B^α are defined by the contact points of the tangent.

At the temperature T_3, both components are liquid which means that the curve for the Gibbs energy of mixing of the liquid, ΔG^L, starts at zero at $x_B = 0$ and ends at zero at $x_B = 1$. The curve for ΔG^α, on the other hand, starts at a positive value on both sides. Furthermore, ΔG^L is more negative than ΔG^α over the entire composition range, indicating that the system A–B is a homogeneous liquid phase for all compositions at T_3.

It should be pointed out that the curve of the Gibbs energy of mixing for any phase starts with a slope of minus infinity on either side. This is caused by the entropy term in the equation

$$\Delta G^\Phi = \Delta H^\Phi - T \cdot \Delta S^\Phi \tag{9}$$

where the Gibbs entropy of mixing consists of the ideal term $\Delta S^{\Phi,id}$ and the excess term $\Delta S^{\Phi,xs}$,

$$\Delta S^\Phi = \Delta S^{\Phi,id} + \Delta S^{\Phi,xs} \tag{10}$$

with

$$\Delta S^{\Phi,id} = -R \cdot (x_A \ln x_A + x_B \ln x_B) \tag{11}$$

It is this latter function, which ends with a slope of minus infinity on either side of the binary system. Thus, the curve of ΔG^Φ starts always with a negative slope on either side, even in the case of a miscibility gap, which means that there is always some solubility, be it extremely small. Thus, thermodynamics dictates that absolute insolubility, or immiscibility, is not possible.

3.3 Eutectic, peritectic and monotectic systems

If in a binary system the two components exhibit complete mutual solubility in the liquid phase but not in the solid, then this can result either in a eutectic- or in a peritectic-type phase diagram. A eutectic system is characterised by a minimum in the melting temperature at some intermediate composition (Figure 7), whereas the temperature of the peritectic reaction is usually between the melting points of the two pure components (Figure 8). If the liquid phase shows some extent of immiscibility too, then one obtains a monotectic system (Figure 9).

Very similar arguments as those outlined in Figure 6, as well as eqs. (5)–(8), can be used to describe these more complicated binary systems by their Gibbs-energy curves. If we take the example of the eutectic system in Figure 7, there are now three different phases competing in stability, i. e., we have to consider three different Gibbs-energy

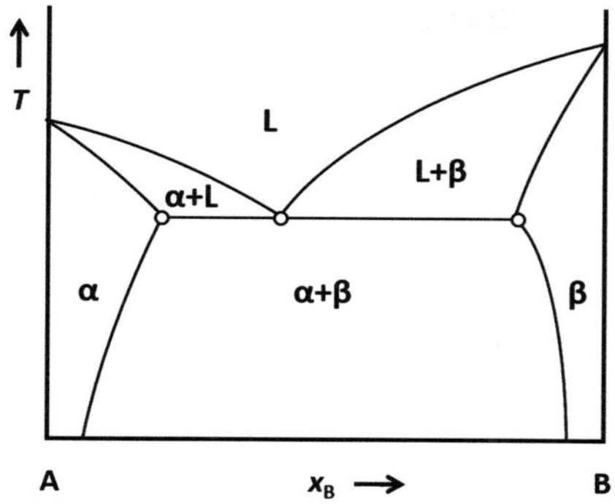

Figure 7: Example of a eutectic phase diagram.

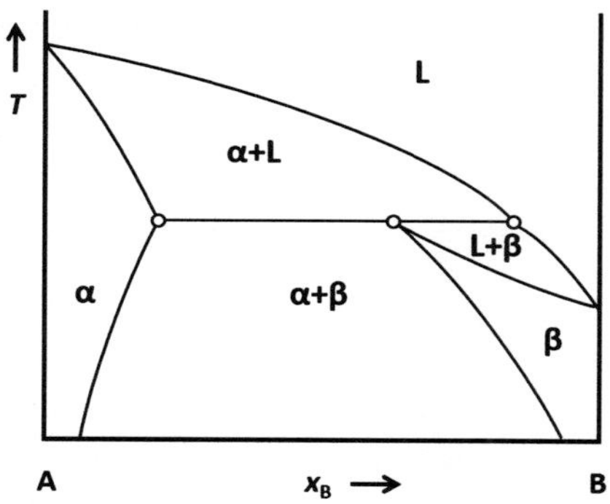

Figure 8: Example of a peritectic phase diagram.

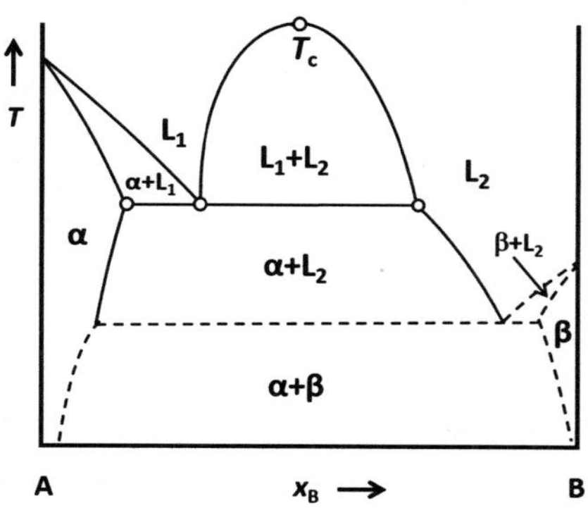

Figure 9: Example of a monotectic phase diagram, completed by a eutectic reaction at lower temperatures; T_c is the critical point where the two liquid phases become undistinguishable on heating.

curves, one for each of the solid phases α and β (assuming that they have different crystal structures), and one for the liquid phase. Figure 10 shows the special situation of the exact eutectic temperature where we have now a common tangent to all three curves, indicating that three different phases with exactly defined compositions are in equilibrium with each other.

Application of the Gibbs phase rule (eq. (3)) at the eutectic temperature T_E, with $C = 2$ and $P = 3$, results in $F' = 0$, i. e., we have zero degrees of freedom. For three

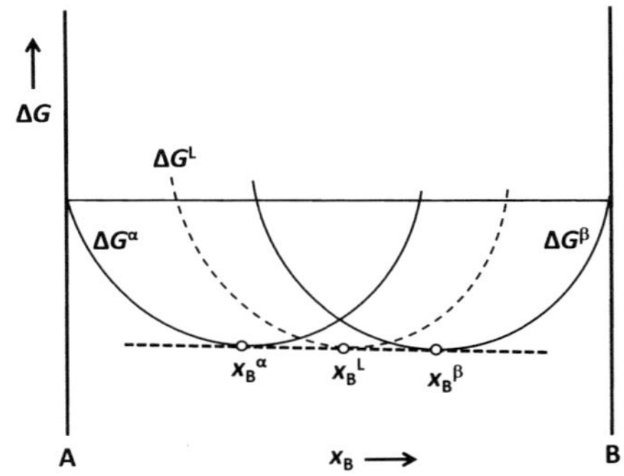

Figure 10: Gibbs energy of mixing of the liquid phase and of the solid phases α and β in the binary eutectic system A–B (see Figure 7) at the exact eutectic temperature (schematic).

phases to be in equilibrium in a binary system at defined pressure, this is only possible at one particular temperature with exactly defined compositions of the three phases. No variables are left that could be changed independently. Therefore, such an equilibrium is called invariant, or the eutectic reaction (which is usually written for the cooling process), $L \leftrightarrows \alpha + \beta$, is called an invariant reaction, respectively.

3.4 Binary systems with intermetallic phases

If intermetallic phases exist in binary systems, then there are two possibilities how a phase can be in equilibrium with the liquid phase. Either a phase shows a congruent melting point, i. e., there is a temperature (usually a maximum) where solid and liquid phase have the same composition, or the phase melts incongruently, i. e., it decomposes on heating into a liquid and another solid phase, both having different compositions. There are yet other possibilities how a phase can disappear on heating. The phase can decompose into two other solid phases with different compositions (without ever being in equilibrium with a liquid), a reaction that is called a peritectoid; or the phase can transform congruently into another solid phase. It is clear that in both cases it will be impossible to grow a single crystal out of the melt.

Let us start with the simple case of one congruently melting intermetallic phase in a binary system A–B, as shown in Figure 11. When using the Gibbs energies of mixing to understand such a diagram, one has to keep in mind that for each phase appearing in the phase diagram one Gibbs-energy curve of mixing has to exist. Assuming that the pure components A and B have different crystal structures, there are now four different curves for the Gibbs energies of mixing of the four phases involved: liquid, α, the intermetallic phase γ and β. Very frequently, the Gibbs energy of mixing of a solid intermetallic phase is called the Gibbs energy of formation. However, when comparing with tabulated values, one has to be careful to consider the corresponding reference states used.

Similarly, as in the case of the systems discussed above, it is the competition between the stabilities of the four phases that is reflected in the $T-x$ phase diagram. This

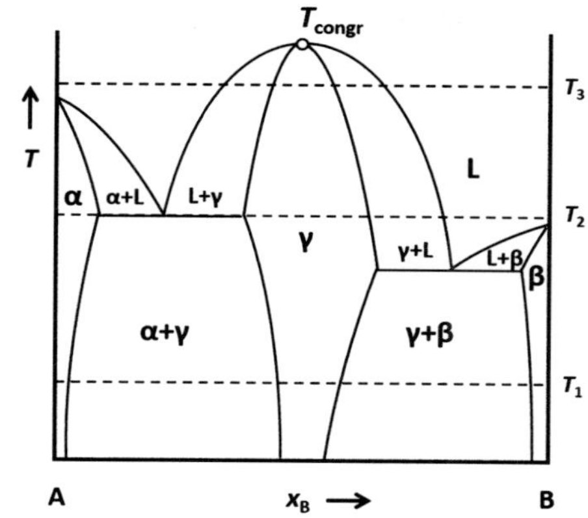

Figure 11: Binary system A–B with a congruently melting intermetallic phase γ.

is discussed briefly for the three temperatures marked in Figure 11. At temperature T_1, both A and B are solid, i. e., the Gibbs energy of mixing curve for the phase α starts at zero for $x_B = 0$ and ends at some positive value for $x_B = 1$ where it is not stable; and it is exactly the opposite for the phase β. In addition, there is a corresponding curve for the intermetallic phase γ that starts at some positive values on either side since neither pure A nor pure B are stable in the crystal structure of γ at T_1. The contact points of the common tangents for the curves of ΔG^{α} and ΔG^{γ}, as well as for ΔG^{γ} and ΔG^{β}, will define the phase boundaries of the two-phase fields $(\alpha + \gamma)$ and $(\gamma + \beta)$. The Gibbs-energy curve of the liquid does not interfere at all at this temperature.

At temperature T_2, there are again three phases that compete in the phase diagram, but now it is solid phase α and intermetallic phase γ, as well as the liquid. The situation is somewhat special, as T_2 is exactly the temperature of the eutectic between α and γ, i. e., we have a three-phase equilibrium which is indicated by a common tangent to the three Gibbs-energy curves ΔG^{α}, ΔG^{L} and ΔG^{γ} where the contact points define the compositions of the three phases in equilibrium. For B-rich compositions, the two-phase field $(\gamma + L)$ is defined by the contact points of the common tangent to the Gibbs-energy curves ΔG^{γ} and ΔG^{L}, as shown in Figure 12.

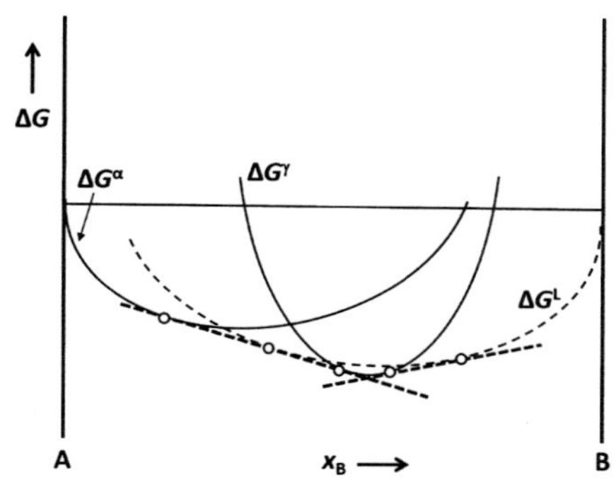

Figure 12: Gibbs-energy curves (schematic) of the solid phase α, the intermetallic phase γ and the liquid competing at temperature T_2 (see Figure 11).

The situation is rather simple for the temperature T_3 where only two phases compete, the liquid and the intermetallic phase γ; the two-phase fields ($L + \gamma$) on either side are defined by the common tangents to the curves of ΔG^{γ} and ΔG^{L} where the curve for ΔG^{γ} will be lower over a certain composition range around $x_B = 0.5$.

When growing a single crystal of the intermetallic phase γ out of the liquid phase, the ideal temperature and composition would be the congruent melting point where the solid and the liquid phase have exactly the same composition. This should result in a single crystal with a well-defined composition and no composition gradients. If this is not possible for experimental reasons (e. g., the temperature is too high), then single crystals can be grown at lower temperatures on both sides of the maximum. However, it is clear that the composition of the solid crystal will be defined by the tie line between the intermetallic phase γ and the liquid at this particular temperature. Since in the process of crystal growing from the liquid the temperature is frequently lowered very slowly during the growth process [25], this causes also a concentration gradient in the single crystal because the phase boundary shifts noticeably with the temperature in most cases.

Figure 13 shows the situation of an intermetallic phase β that melts incongruently, i. e., it decomposes on heating in an invariant peritectic reaction. The principles are very much the same as shown in Figures 11 and 12: there is the same number of Gibbs-energy curves as in Figure 12; it is just the different relative positions of these curves that leads to a different phase diagram. It is clear that, in the case of incongruent melting as in Figure 13, a single crystal can only be grown out of a melt that is richer in component B. The temperature window, within which crystal growth from the melt is possible, is limited by the points a and b, i. e., it is only possible at temperatures below the peritectic reaction and above the temperature of the eutectic between the intermetallic phase β and the solid solution γ. Depending on the particular phase diagram, this range can become quite narrow. The problem of a composition gradient in the single crystal will be the same as in the case of a congruent melting phase where a temperature lower than the congruent melting point has been selected.

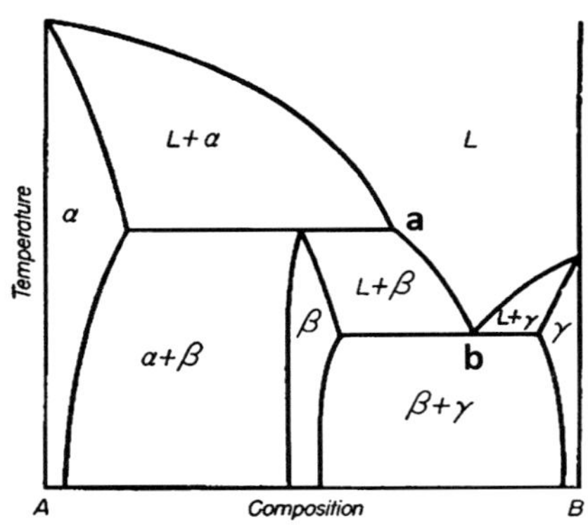

Figure 13: Binary system with an intermetallic phase β that melts incongruently, i. e., it decomposes on heating in a peritectic reaction (from Ref. [9]; copyright McGraw-Hill).

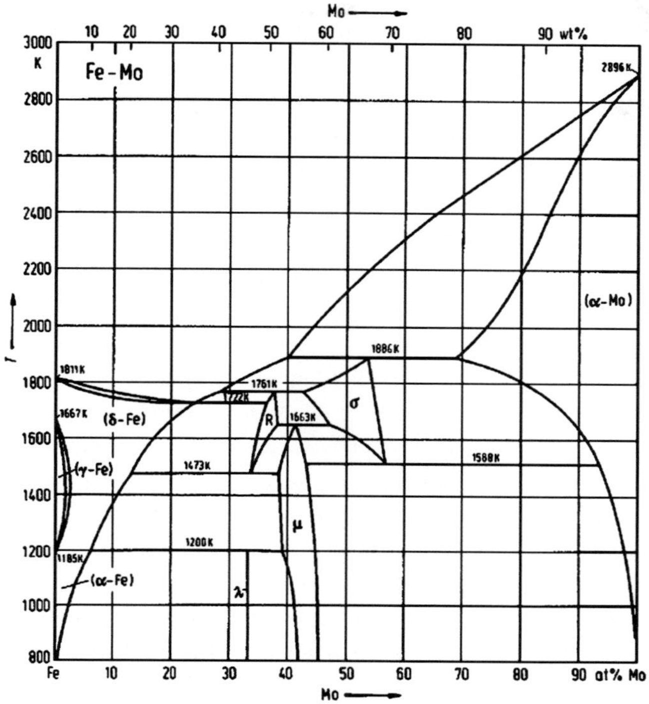

Figure 14: Binary Fe–Mo phase diagram showing two intermetallic phases (λ and μ) which decompose in peritectoid reactions into two other solid phases; none of them is ever in equilibrium with the liquid (from Ref. [26]; copyright Springer).

Of course, in many binary systems more than just one intermetallic phases exist. In this case, one has to consider the competition between all the corresponding Gibbs-energy curves in order to define the phase boundaries at various temperatures.

As mentioned above, a special case is the peritectoid reaction where a phase decomposes on heating into two other solid phases. Figure 14 shows the Fe–Mo phase diagram where both the phase λ and the phase μ are never in equilibrium with the liquid phase; on heating, λ decomposes at about 1,200 K into the solid solution (α-Fe) and the phase μ which itself decomposes at 1,643 K into the two solid phases R and σ. Thus, it is impossible to grow bulk single crystals of λ or μ from a binary liquid. One would have to resort to other methods, which, however, would result in much smaller single crystals.

3.5 Non-stoichiometry

It is a fact that the majority of intermetallic phases exhibits some noticeable non-stoichiometry, i. e., the phases are stable over a certain composition range, which is usually dependent on temperature. Although some intermetallics may appear as so-called line compounds in the corresponding phase diagrams, non-stoichiometry is rather the rule than the exception. On closer look, even these so-called line compounds are usually stable over a narrow, but still appreciable, composition range.

Non-stoichiometry can be caused by various mechanisms, i. e., by various types of defects in the crystal structure. In most cases, it is point defects that are responsible for the deviation from the exact stoichiometric composition. Point defects can be substitutional defects and interstitial defects, as well as vacancies. (There are also some cases where two-dimensional defects, e. g., shearing of blocks of crystal planes, can be responsible for non-stoichiometry.)

Take the example of the β'–AgMg phase in the Ag–Mg binary system (Figure 15): β'–AgMg crystallizes in the simple CsCl structure, with Ag atoms on one type of lattice sites and Mg atoms on the other one. Non-stoichiometry is caused by mutual substitution of the two types of atoms. (Defects that are responsible for non-stoichiometry are usually called constitutional defects. In addition to these constitutional defects, there are also so-called thermal defects present in the lattice at finite temperatures, i. e., even at the exactly stoichiometric composition, some of the Ag and Mg atoms exchange places, and their number increases with temperature.) Growing a single crystal of β'–AgMg would theoretically be possible within a wide composition range between

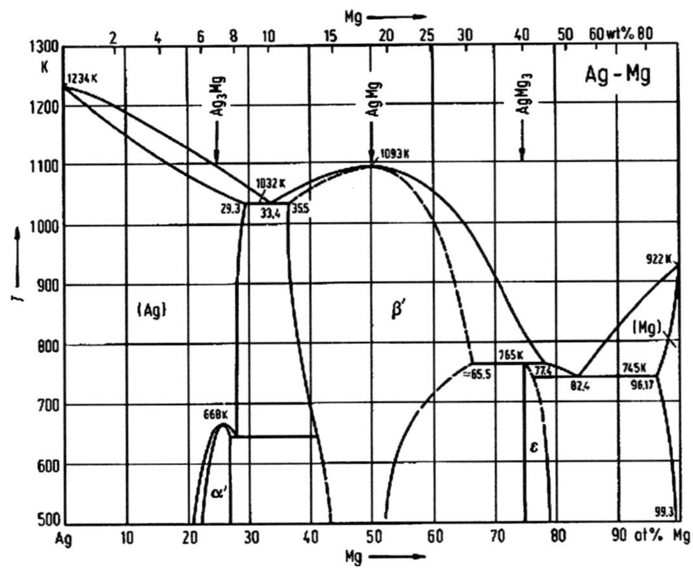

Figure 15: Binary Ag–Mg phase diagram showing an intermetallic phase, β'–AgMg with a wide homogeneity range; in addition, there appears α'–Ag_3Mg which transforms congruently into the solid solution of fcc (Ag) as well as ε–$AgMg_3$ which decomposes in a peritectic reaction (from Ref. [27]; copyright Springer).

35.5 and about 65.5 at% Mg, out of melts containing between 33.4 and 77.4 at% Mg. However, the composition of the single crystal will be strongly dependent on the temperature since it is fixed by the corresponding tie line between the liquid and solid β'–AgMg. Any change in temperature during the experiment will result in a considerable concentration inhomogeneity of the crystal.

A more complicated mechanism of non-stoichiometry is operative in the intermetallic phase β'–GaNi (Figure 16) although it crystallizes basically in the same structure (CsCl-type) as β'–AgMg. Whereas a deviation from the exactly stoichiometric 1:1 composition to the Ni-rich side is caused by substitution, i. e., by Ni atoms on Ga lattice sites, a deviation to the Ga-rich side is caused by Ni vacancies. As can be seen from the phase diagram, a wide composition (between about 22 and 68 at% Ni according to Figure 16) and temperature range could be utilized to grow single crystals from the melt. However, the composition of the single crystal will again be defined by the corresponding tie line between liquid and solid GaNi phase at the experimental temperature [28].

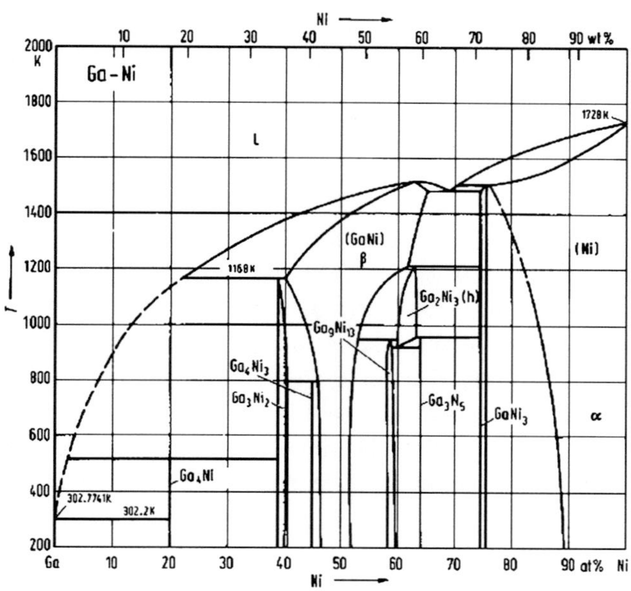

Figure 16: Binary Ga–Ni phase diagram (from Ref. [29]; copyright Springer).

Yet another mechanism of non-stoichiometry can be observed for the intermetallic phase NiSb (Figure 17) which crystallizes in the NiAs structure type: the deviation to the Sb-rich side is caused by vacancies on the Ni sublattice whereas interstitial Ni atoms (in the trigonal-bipyramidal or double-tetrahedral sites) are responsible for a Ni surplus. Again, there is a wide temperature and composition range available for

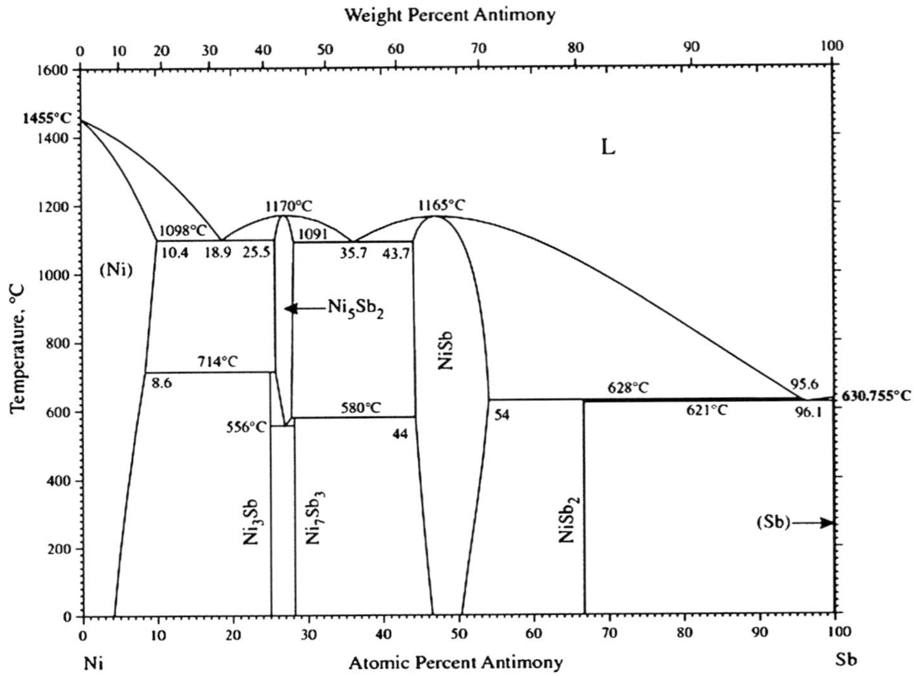

Figure 17: Binary Ni–Sb phase diagram (from Ref. [30]; copyright Springer).

growing single crystals of NiSb but the composition of the crystal will always be defined by the tie line between melt and solid NiSb at the corresponding temperature.

These three examples should suffice to demonstrate that non-stoichiometry is a nearly ubiquitous phenomenon in intermetallic phases. Growing a single crystal with a well-defined composition from the liquid will not be simple since the composition will always be given by the phase boundary of the phase in equilibrium with the liquid. As the phase boundary usually shifts with temperature, any variation of the temperature during the growth process will consequently lead to concentration inhomogeneities in the crystal.

4 Three-component (ternary) systems

For the complete representation of the phase diagram of a ternary system A–B–C, one would need four dimensions: two composition variables, the temperature and the pressure. Thus, it is customary to keep at least the pressure constant, which allows showing a three-dimensional composition-temperature space diagram based on a Gibbs triangle for the representation of the composition. This is shown for a simple

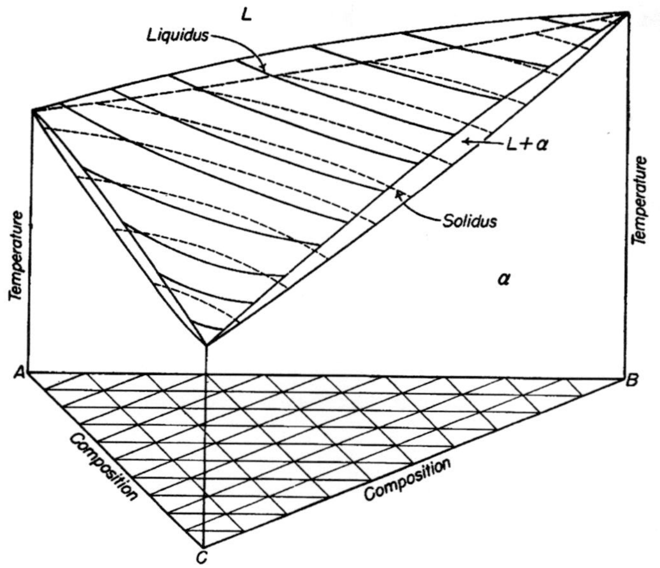

Figure 18: Ternary isomorphous system as a temperature-composition space diagram (from Ref. [9], copyright McGraw-Hill).

isomorphous ternary system A–B–C in Figure 18. However, even this is rather difficult to visualise, especially if there are many phases and reactions in the system, therefore, one tends to further reduce the representation to two dimensions: isothermal sections (where the temperature is kept constant) or so-called isopleths (which are vertical sections through the ternary space diagram). These two possibilities are shown in Figure 19, representing the corresponding sections through the space diagram in Figure 18. Frequently, isopleths are shown for the special cases where the concentration of one component is kept constant or where the concentration ratio x_i/x_j of two of the three components is kept constant.

Another possible representation in two dimensions are the so-called liquidus projections which could be described as maps of the liquidus landscape, showing hills and valleys (especially if isothermal lines are added for regular temperature intervals), and showing also all invariant reactions in which the liquid phase is involved. A very simple liquidus projection for the ternary isomorphous system in Figure 18 is shown in

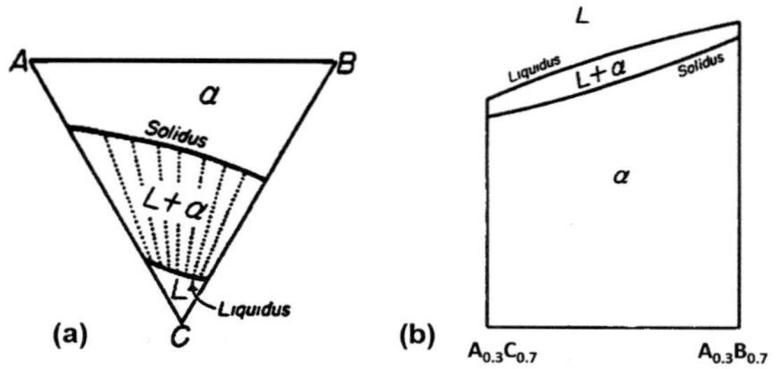

Figure 19: Isothermal section through the ternary isomorphous system A–B–C in Figure 18 at a temperature above the melting point of component C and below the melting points of components A and B (**a**); isopleth through the ternary system at a constant concentration of $x_A = 0.3$ (**b**) (from Ref. [9], copyright McGraw-Hill).

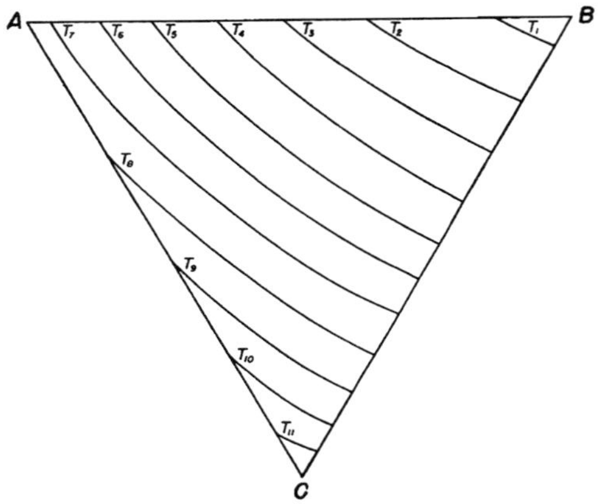

Figure 20: Liquidus projection of the ternary isomorphous system in Figure 18; T_1 is the highest and T_{11} the lowest temperature (from Ref. [9], copyright McGraw-Hill).

Figure 20 indicating isothermal lines at regular temperature intervals, which enables seeing the slope of the surface.

Any isothermal section shows exactly the phase equilibria at the given temperature, especially if the tie lines for the two-phase fields are given, as can be seen in Figure 19(a). On the other hand, the isopleths show, in the normal case, only phase fields but not the phase equilibria. Since the tie lines are usually not in the plane of representation (except in the special case of a quasi-binary system), one knows that in a certain temperature-composition range two or three phases are in equilibrium with each other; however, one knows nothing about the compositions of the phases that are in equilibrium (see Figure 19(b)).

For single crystal growth, one has to consider ternary systems with a ternary intermetallic. Figure 21 shows, as an example, two isothermal sections of such a system with a phase δ that melts congruently; it has the nominal composition ABC, however with a noticeable homogeneity range. The three binary systems A–B, A–C, and B–C

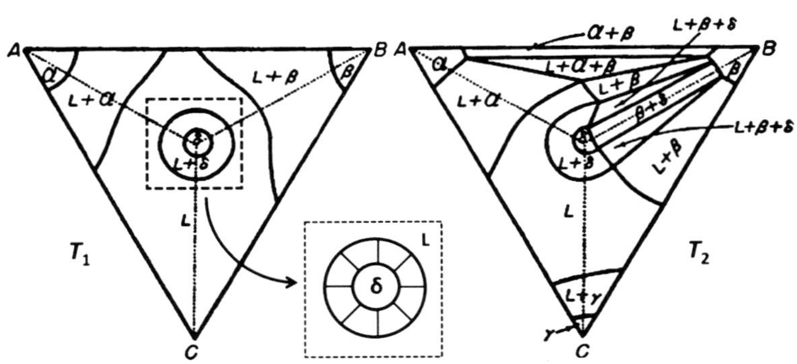

Figure 21: Isothermal sections through a ternary system with a ternary intermetallic phase δ (nominal composition ABC) at two temperatures T_1 and T_2; quasi-binary sections A–δ, B–δ and C–δ are indicated by dotted lines (adapted from Ref. [9], copyright McGraw-Hill).

form simple eutectics. T_1 is a temperature above the melting point of component C and above all eutectic temperatures but below the melting points of A and B; T_2 is below the melting points of all three components and below the eutectic in the system A–B but still above the eutectics in A–C and B–C. (In addition, it is indicated by dotted lines that there are also three quasi-binary sections A–δ, B–δ and C–δ where all tie lines are within these particular section so that they can be treated like binary phase diagrams.)

One can see that at temperature T_1, the phase δ is in equilibrium with liquid only, and the inset shows a magnification of the equilibrium between the phase δ and the liquid including a few of the tie lines. Thus, for growing a single crystal of δ by the Czochralski method, one can start with liquid of any composition along the liquidus line. However, the composition of the crystal that starts growing is fixed by the conjugate phase boundary of the solid phase, and there will be slight differences in the composition depending on the concentration of the starting liquid. At temperature T_2, the intermetallic phase is in equilibrium with the liquid phase as well as with the solid solution β, which restricts the choice of compositions of the liquid from which a single crystal could be grown.

As in the case of binary intermetallic phases, ternary intermetallics may also be formed by a ternary peritectic reaction. In this case, the liquid reacts with two different solid phases (α and β) to form a third solid phase γ: $L + \alpha + \beta = \gamma$. The conditions for growing single crystals out of the melt will depend on the shape of the liquidus surface in equilibrium with the solid phase γ.

5 Experimental methods

The experimental methods for the determination of phase diagrams can be roughly divided into two groups: dynamic methods (thermal analysis) and static or non-dynamic methods. According to IUPAC,[3] thermal analysis is defined as "a group of techniques in which a physical property of a substance and/or its reaction product is measured as a function of temperature while the substance is subjected to a controlled temperature program" [31]. Thus, the well-known experimental methods like the observation of cooling (or heating) curves, DTA (differential thermal analysis) and DSC (differential scanning calorimetry) would fall into this category (where DSC is usually a DTA method with the possibility to derive quantitative information). Although these are the standard methods for determining phase transformations in metallic multicomponent systems, there exist many more experimental methods where a property is measured as a function of temperature in order to derive information about phase

3 IUPAC: International Union of Pure and Applied Chemistry.

transitions: this can be the volume (thermodilatometry), the lattice parameter (dynamic high-temperature X-ray diffraction) and the electrical conductivity, as well as several others.

The group of non-dynamic—or static—methods comprises all those experiments where samples with a well-known composition are brought into thermodynamic equilibrium at a defined temperature and are then investigated for their phase composition. This can happen at the temperature of equilibration, which is usually experimentally more difficult, or the samples are quenched and studied at room temperature, assuming that the high-temperature equilibrium can be frozen in. Unfortunately, this is not always the case as some transformations are too fast to be suppressed by quenching. The standard methods for the investigation of equilibrated samples are powder X-ray diffraction (XRD), optical microscopy and scanning electron microscopy (SEM) coupled with EDX (Energy-Dispersive X-ray spectroscopy) or WDS (Wavelength-Dispersive X-ray Spectroscopy). Whereas powder XRD in combination with modern evaluation software enables identification of the phases present in the sample, as well as estimating their relative amounts, SEM gives also quite accurate information on the elemental composition of the various phases.

The best results in phase-diagram investigation are obtained by a combination of dynamic and static methods.

5.1 Dynamic methods

Several books have been published on the topic of thermal analysis within the past decades [32–36]. Probably one of the most popular methods in the investigation of phase diagrams is DTA or its more quantitative version, DSC. A sample and a reference substance are heated or cooled with a defined heating rate (mostly in the range between 1 and 20 K/min), and both the temperature (of the sample or of the reference), as well as the temperature difference between them, are registered, usually by thermocouples. A substance that does not show any transformation within the investigated temperature range usually serves as a reference. For accurate temperature measurements, it is important to standardise the thermocouples, which is done by measuring the melting temperatures of various standard substances.

The evaluation and interpretation of the corresponding DTA curves for binary systems is relatively straightforward, although one has to follow a number of well-defined rules [37, 38]. As a rule of thumb, invariant reactions are evaluated as the peak onset, both on heating and on cooling. For non-invariant reactions, e. g., the transition through a two-phase field, peak onset and maximum indicate beginning and end of the two-phase field. Generally, the heating curve is the more reliable one as supercooling may be significant on cooling, and the temperatures evaluated from the cooling curves can be significantly too low.

On the other hand, the evaluation of the DTA curves for ternary or higher-order systems can become quite tricky and requires a good understanding of the method, as well as of the principles of phase equilibria in these systems. Whereas invariant reactions are again evaluated from the peak onset as in the case of binary systems, the interpretation of the non-invariant effects becomes much more involved since one has to consider the additional existence of three-phase fields in the case of ternary systems, four-phase fields in quaternary systems and so on.

The quality of the DTA curves can be influenced by many external parameters depending on the particular instrument and the characteristics of the sample. One important parameter is the heating rate: A fast heating rate will nearly always induce a temperature gradient in the sample leading to apparent transformation temperatures that can be off by several degrees. For very accurate studies, the same transformation should actually be measured at various heating rates, and the correct value of the transition temperature is obtained by extrapolation to a zero heating rate. Of course, in most cases this will not be possible but, at least, measurement and standardization of the instrument should be made at the same, not too high, heating rate.

In addition, asymmetries in the furnace and the atmosphere in the furnace can play a significant role for the quality of the DTA curves, as can the size and the form of the sample (powder, bulk piece) and the type of crucible containing the sample.

For more details on DTA and DSC, the reader is referred to the corresponding literature.

5.2 Static methods

As pointed out above, a sample with well-defined composition is equilibrated by annealing at a given temperature and is afterwards investigated for its phase composition by XRD and/or SEM. It is clear that the annealing time will depend on the (relative) temperature, i. e., the lower the temperature the more difficult will it be to establish thermodynamic equilibrium. Therefore, many phase diagrams in literature are shown by dashed lines at low temperatures, and many homogeneity ranges of intermetallic phases at room temperature are probably too wide as they are indicated in the literature.

Modern methods of powder XRD and the corresponding software for evaluation enable a good estimate of the relative amounts of phases that are present in a sample, at the same time giving information on their crystal structures. Quite accurate homogeneity ranges for binary intermetallic phases at a given temperature can be obtained by following their lattice parameters as a function of composition: Since these have to remain constant in two-phase regions, kinks in their composition dependence clearly indicate the phase boundaries. As this is mostly done at room temperature, one has to rely on quenching the samples from higher temperatures, assuming that the correct phase equilibrium can be frozen in (which may not always be the case).

SEM methods, in combination with EDX or WDS, also provide the compositions of the phases that are present, giving again valuable information on phase boundaries in the case of samples containing two or more equilibrium phases in binary and higher-order systems.

A very elegant method, especially for binary systems, are the so-called diffusion couples. In principle, the two pure elements are brought into intimate contact with each other and are annealed at a selected temperature where diffusion is fast enough. After a certain time, the diffusion couple is quenched and investigated for the composition of the reaction products, mostly by a so-called line scan using SEM. In the ideal case, all intermetallic phases that are stable in the binary system at the equilibration temperature should have formed, and the phase boundaries at the annealing temperature can be determined from the compositions of the phases in contact with each other. In reality, the amount of any phase formed in a diffusion couple will depend on the (relative) diffusion rates of the two elements within it, and it may happen that one or the other phase appears only as a very thin layer that could easily be overlooked.

Of course, a diffusion couple must not necessarily start with the pure elements. It is also possible to bring two intermediate phases in a binary system into close contact with each other and anneal this couple at a defined temperature. In this case, all phases with compositions in between the two starting phases should be formed.

Diffusion couples in ternary and higher order systems can also be used but the interpretation of the results may be more complicated [39]. As an example, one can start in a ternary system with a binary phase in one of the limiting binary systems and anneal it at a defined temperature in contact with the third element. After quenching the diffusion couple, the change of composition along the couple is analysed by SEM. However, due to the one more degree of freedom compared to a binary system, one will find both single-phase and two-phase fields. In addition, in most cases the diffusion path in a ternary system will not be a straight line connecting the starting compositions, which makes the interpretation of the results even more complicated.

In the literature, diffusion multiples have also been proposed for higher order systems [40], but the interpretation becomes still more complicated, and the information on the phase diagram may be limited.

6 Phase-diagram calculation

As pointed out at the beginning of this chapter, a phase diagram describes the state of a materials system in thermodynamic equilibrium as a function of temperature, pressure and composition. That is, if one knows the Gibbs energy of all phases in a system as a function of temperature, pressure, and composition, one can calculate the phase diagram. The principle is exactly the same as demonstrated in Figures 6, 10, and 12 where the composition dependence of the Gibbs energy of different phases at constant pressure and temperature was used to derive the phase equilibria in a binary system at this particular temperature. On the other hand, a phase diagram will also provide some information on the (relative) stability of the different existing phases and enable some conclusions with respect to the thermodynamic properties.

This was realised nearly 50 years ago when Kaufman and Bernstein published their book "Computer Calculation of Phase Diagrams" [41]. Their approach, which has been further developed by Kaufmann and many other scientists, came to be known as the CALPHAD method (where CALPHAD stands for CALculation of PHAse Diagrams). Whereas in the 1970s it was simple and homemade computer codes that were used for these calculations, a number of commercial software packages are available today for calculating phase diagrams [42–45]. Together with these software packages, one can buy various types of databases that contain the thermodynamic properties of the elements and of many known intermediate phases in parametric form. The corresponding data can serve as immediate input in these software packages. The most recent book on the CALPHAD method and its application is by Lukas et al. [46].

For the preparation of the databases, it is necessary to convert the thermodynamic properties into equations that provide them as functions of temperature, pressure and composition (although the pressure is frequently neglected if calculations are to be performed for ambient pressure). To arrive at such equations, various thermodynamic models are used, i. e., Dilute Solution Models, Ideal Solution Models and various Non-ideal Solution Models. This enables the storage of parameters in the databases rather than listing the numerical values of properties. Of additional interest for scientists is the availability of a so-called optimiser that is included in most of the software packages. In simple words, such a program enables optimising the phase diagram based on input of experimental thermodynamic data, as well as experimental phase-diagram information. The result is the best possible agreement between phase diagram and thermodynamics of a system. (It should be noted that the process of optimisation frequently requires a critical judgement of the reliability of experimental literature data, especially if differing data are available from different sources. This necessitates, of course, a good knowledge of the thermodynamic background as well as of phase equilibria in binary and higher-order systems.)

The process of phase-diagram calculation itself is fully automated today, i. e., the user can withdraw the necessary parameters from the database, use them as input into the particular software, and calculate the phase equilibria in the concentration, temperature and pressure ranges of interest. Moreover, the calculations provide additional information, as for example, a so-called phase fraction diagram [47] that gives the relative amounts of various phases as a function of temperature, both under equilibrium and non-equilibrium conditions. All this is of particular value for users from industry.

Although the phase diagrams of most binary metallic systems are known today (at least, to some extent), knowledge of ternary metallic phase diagrams is much more limited, and the phase equilibria in quaternary and higher-order metallic systems are, with very few exceptions, mostly unknown. For these cases, the CALPHAD method provides the possibility to extrapolate the phase equilibria from binary to ternary and higher-order systems, based on proper thermodynamic models, always considering the (limited) experimental information available in the literature. It is clear that the obtained higher-order phase diagrams may not be fully correct, but it is at least possible to arrive at an estimate, which enables planning and performing well-directed experiments to verify the results of the calculations.

Bibliography

[1] Czochralski J. Ein neues Verfahren zur Messung der Kristallisationsgeschwindigkeit der Metalle. Z Phys Chem 1917, 92, 219–21.

[2] Gille P, Bauer B. Single crystal growth of $Al_{13}Co_4$ and $Al_{13}Fe_4$ from Al-rich solutions by the Czochralski method. Cryst Res Technol 2008, 43, 1161–7.

[3] Nakamura N, Shimotomai M. Growth of $YBa_2Cu_3O_x$ single crystals by a self-flux method with alkali chlorides as additives. Physica C 1991, 185–189, 439–40.

[4] Anand VK, Johnston DC. Antiferromagnetism in $EuCu_2As_2$ and $EuCu_{1.82}Sb_2$ single crystals. Phys Rev B 2015, 91, 184403.

[5] Bridgman PW. Certain physical properties of single crystals of tungsten, antimony, bismuth, tellurium, cadmium, zinc, and tin. Proc Am Acad Arts Sci 1925, 60, 305–83.

[6] Scheil E. Darstellung von Dreistoffsystemen. Arch Eisenhüttenwes 1935, 9, 571–3.

[7] Lukas HL, Henig ETh, Petzow G. 50 years reaction scheme after Erich Scheil. Z Metallkde 1986, 77, 360–7.

[8] Zernike J. Chemical Phase Theory. Deventer, Netherlands, N.V. Uitgevers-Maatschappij – Ae.E. Kluwer, 1955.

[9] Rhines FN. Phase Diagrams in Metallurgy: Their Development and Application. New York, McGraw-Hill, 1956.

[10] Prince A. Alloy Phase Equilibria. Amsterdam, Netherlands, Elsevier, 1966.

[11] Gordon P. Principles of Phase Diagrams in Materials Systems. New York, McGraw-Hill, 1968.

[12] Predel B, Hoch M, Pool M. Phase Diagrams and Heterogeneous Equilibria. Berlin, Germany, Springer, 2004.

[13] Campbell F, ed. Phase Diagrams: Understanding the Basics. Materials Park, ASM International, 2012.

[14] Kindler B, Finsterbusch D, Graf R, Ritter F, Assmus W, Luthi B. Mixed-valence transition in $YbInCu_4$. Phys Rev B 1994, 50, 704–7.

[15] Severing A, Gratz E, Rainford BD, Yoshimura K. Study of the valence transition in ytterbium indium copper ($YbInCu_4$) by inelastic neutron scattering. Physica B, Condens Matter 1990, 163, 409–11.

[16] Felner I, Nowik I. First-order valence phase transition in cubic $Yb_xIn_{1-x}Cu_2$. Phys Rev B 1986, 33, 617–9.

[17] Haraldsen H, Kowalski E. Magnetochemische Untersuchungen. XVII. Das magnetische Verhalten der Chalkogenide des zweiwertigen Chroms. Z Anorg Allg Chem 1935, 224, 329–36.

[18] Guillaud Ch, Barbezat S. Ferromagnetic properties of the definite compound CrTe. Compt Rend 1946, 222, 386–8.

[19] Hirone T, Chiba Sh. On the magnetic anisotropy of single crystal of chromium telluride. J Phys Soc Jpn 1960, 15, 1991–1994.

[20] Ipser H, Komarek KL, Klepp KO. Transition metal-chalcogen systems VIII: the Cr-Te phase diagram. J Less-Common Met 1983, 92, 265–82.
[21] Atkins P, de Paula J. Physical Chemistry. Oxford, UK, Oxford University Press, 2014.
[22] Komarek KL, Wessely K. Die Systeme Nickel-Selen und Kobalt-Nickel-Selen. Übergangsmetall-Chalkogensysteme, 2. Mitt. Monatsh Chem 1972, 103, 923–33.
[23] Predel B. Ni–Se (Nickel-Selenium), In: Madelung O, ed. Landolt-Börnstein – Group IV Physical Chemistry subvolume 5I. Berlin-Heidelberg, Germany, Springer, 1998.
[24] Ritscher A, Schmetterer C, Ipser H. Pressure dependence of the tin–phosphorus phase diagram. Monatsh Chem 2012, 143, 1593–602.
[25] Schwerin J, Müller D, Kiese S, Gille P. Single crystal growth in the Ga-Pd system. J Cryst Growth 2014, 401, 613–6.
[26] Predel B. Fe–Mo (Iron-Molybdenum), In: Madelung O, ed. Landolt-Börnstein – Group IV Physical Chemistry subvolume 5E. Berlin-Heidelberg, Germany, Springer, 1998.
[27] Predel B. Ag–Mg (Silver-Magnesium), In: Madelung O, ed. Landolt-Börnstein – Group IV Physical Chemistry subvolume 5A. Berlin-Heidelberg, Springer, 1998.
[28] Kaisermayr M, Combet J, Ipser H, Schicketanz H, Sepiol B, Vogl G. Nickel diffusion in B2–NiGa studied with quasielastic neutron scattering. Phys Rev B 2000, 61, 12038.
[29] Predel B. Ga–Ni (Gallium-Nickel), In: Madelung O, ed. Landolt-Börnstein – Group IV Physical Chemistry subvolume 5F. Berlin-Heidelberg, Springer, 1998.
[30] Okamoto H. Ni-Sb (Nickel-Antimony). J Phase Equilib Diffus 2009, 30, 301–2.
[31] Stahl JW. Nomenclature of thermometric and enthalpimetric methods in chemical analysis. Pure Appl Chem 1994, 66, 2487–92.
[32] Gabbott P, Ed. Principles and Applications of Thermal Analysis. Oxford, UK, Wiley-Blackwell, 2008.
[33] Brown ME, Ed. Introduction to Thermal Analysis: Techniques and Applications. Dordrecht, Netherlands, Kluwer Academic Publishers, 2001.
[34] Haines PJ. Thermal Methods of Analysis: Principles, Applications and Problems. Dordrecht, Springer, 1995.
[35] Wunderlich B. Thermal Analysis. San Diego, CA, USA, Academic Press, 1990.
[36] Hemminger WF, Cammenga HK. Methoden der Thermischen Analyse. Berlin, Springer, 1989.
[37] Gäumann A. Differential-thermal analysis (DTA) measurements and phase diagrams. Chimia 1966, 20, 82–8.
[38] Boettinger WJ, Kattner UR, Moon K-W, Perepezko J. DTA and Heat-Flux DSC Measurements of Alloy Melting and Freezing. Gaithersburg, MD, USA, NIST, 2006; can be downloaded from the NIST homepage: https://www.nist.gov/publications/nist-recommended-practice-guide-dta-and-heat-flux-dsc-measurements-alloy-melting-and.
[39] Kodentsov AA, Bastin GF, van Loo FJJ. The diffusion couple technique in phase diagram determination. J Alloys Compd 2001, 320, 207–17.
[40] Shuro I, Kobayashi S, Nakamura T, Tsuzaki K. Determination of α/γ phase boundaries in the Fe–Cr–Ni–Mn quaternary system with a diffusion-multiple method. J Alloys Compd 2014, 588, 284–9.
[41] Kaufman L, Bernstein H. Computer Calculation of Phase Diagrams. New York, Academic Press, 1970.
[42] Thermocalc Software. Solna, Sweden. (Accessed May 24, 2017, at http://www.thermocalc.com/)
[43] Pandat Software from Computherm LLC. Madison, WI, USA. (Accessed May 24, 2017, at http://www.computherm.com/)
[44] Factsage Software from GTT Technologies and Thermfact and CRCT. Herzogenrath, Germany, and Montreal, Canada. (Accessed May 24, 2017, at http://www.factsage.com/)

[45] MTDATA Software from the National Physical Laboratory. Tedington, UK. (Accessed May 24, 2017, at http://www.npl.co.uk/science-technology/mathematics-modelling-and-simulation/mtdata/)
[46] Lukas H, Fries SG, Sundman B. Computational Thermodynamics: The Calphad Method. Cambridge, UK, Cambridge University Press, 2007.
[47] Schmid-Fetzer R. Phase diagrams: the beginning of wisdom. J Phase Equilib Diffus 2014, 35, 735–60.

Priscila F. S. Rosa and Zachary Fisk

Flux methods for growth of intermetallic single crystals

1 Introduction

The flux growth of intermetallics from solution is but one aspect of the solution growth of crystals, a subject with a long history and an activity wonderfully present in the natural world. Quartz—one of the thermodynamically stable crystalline forms of silica (SiO_2) and one of the most common minerals in the Earth's crust—grows from solution when molten magma cools slowly. Pure quartz crystals are transparent, but the presence of natural impurities in the crystal structure leads to beautiful colored gems, including amethyst and agate.

Advances in materials characterization have made the study of intermetallics in single-crystal form the accepted norm and, among the techniques for producing single crystals, flux methods are particularly attractive for their simplicity and low implementation cost. Good reviews of the techniques exist in the literature [1]. The purpose here is to present an introduction to the technique useful for those interested in actively pursuing it, along with some tricks of the trade that have arisen from experience. This chapter is not meant to be an exhaustive treatment, but rather a presentation through specific examples along with a view to where one goes in thinking about directions to investigate.

2 Typical growth

The simple principles involved in flux growth are easily appreciated by considering examples of the so-called self-flux growth, in which crystals are grown from melts containing only the elements present in the grown crystal. The simplest case to consider is a binary phase diagram, for example the La–Sn phase diagram (Figure 1). $LaSn_3$ is the most Sn-rich compound in the phase diagram, and this is a compound that can be readily grown from Sn-rich solution.

$LaSn_3$ seems to be congruently melting in the phase diagram, so one could attempt to grow single crystals of it by the Bridgman technique, i. e., slow cooling a stoichiometric melt. Slow cooling of a melt richer in Sn than 1:3, however, allows the growth of separated single crystals as the melt composition slides down the liquidus with cooling and precipitation of $LaSn_3$. More specifically, one obtains cubic single crystals of $LaSn_3$ by heating a mixture of La pieces and Sn shots with ratio La:Sn = 1:10 to 1150 °C, followed by slow cooling to 400 °C.

https://doi.org/10.1515/9783110496789-003

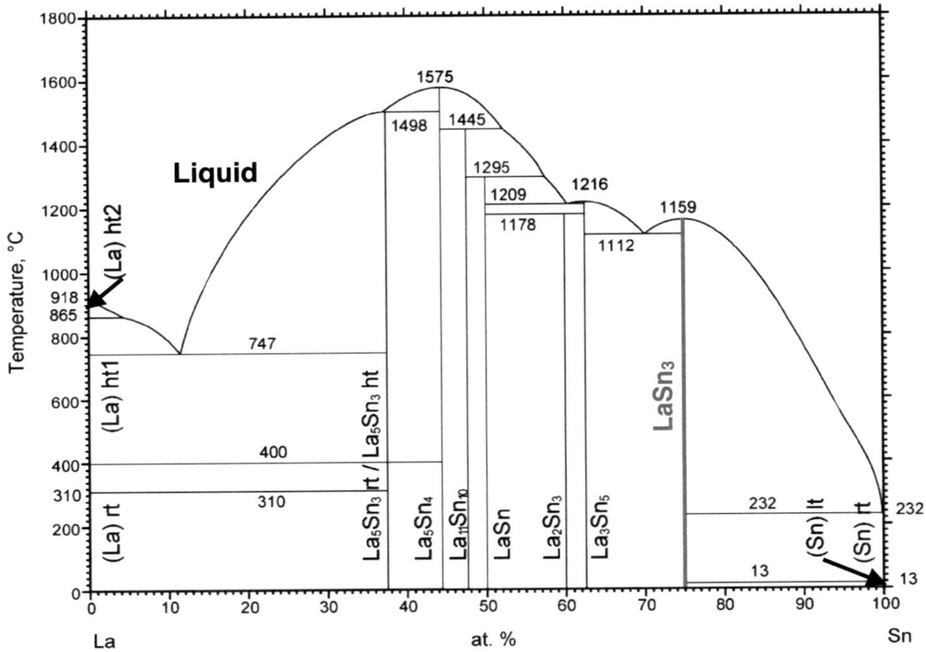

Figure 1: La–Sn binary phase diagram adapted from Ref. [2].

An important point here is that the crystals that precipitate from this melt will have a composition corresponding to the stoichiometry on the Sn-rich boundary of the $LaSn_3$ phase at the temperature of growth. Many materials have considerable range of stoichiometry over which they exist, and this range of stoichiometry generally varies with temperature. This can result in grown crystals from a given melt composition having a range of stoichiometry. A good example of this is found in the material $YbIn_xCu_{5-x}$ which crystallizes in the cubic Laves phase structure with generic formula AB_2. In this structure, the A-atoms form a cubic diamond lattice, which can be regarded as consisting of two fcc lattices shifted by 1/4 of the cube diagonal. It is possible to form chemically ordered compounds in which each of these fcc lattices is occupied by only one atomic species, forming a compound with formula $AA'B_4$. In the case of $YbIn_xCu_{5-x}$, one can look at the compound as $YbIn_xCu_{1-x}Cu_4$, with a mixture of In and Cu on the second fcc lattice. It proved possible to find an In–Cu self-flux composition for this compound which resulted in crystals with stoichiometry $YbInCu_4$ [3] (Figure 2). This compound exhibits a first order valence transition near 42 K which has a width in stoichiometric crystals of approximately 1 K [4], whereas this transition becomes spread over tens of K when the 1:4 In:Cu stoichiometry is not precise. It is worth noting that the compound exists all the way to stoichiometry $YbCu_5$. The isostructural ferromagnet $YbInNi_4$ forms stoichiometrically at the 1:4 In:Ni ratio and can be grown from an In–Ni flux.

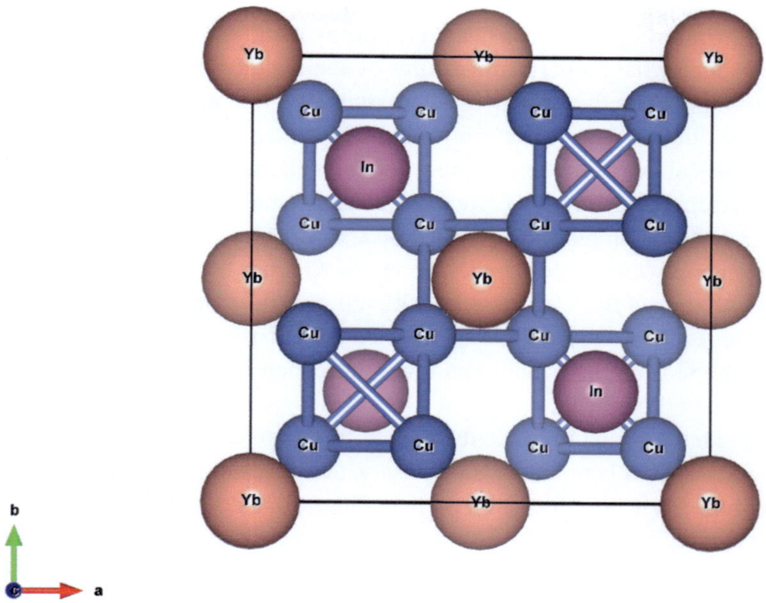

Figure 2: $YbInCu_4$ crystal structure.

Occasionally, it is also possible to tune the self-flux ratio as a means to change the stoichiometry and the structure of the resulting crystal. An example of this is found in the ternary phase diagram Eu–In–Sb using a combination of In and Sb as the self-flux. At a ratio Eu:In:Sb = 1:5:1.2, the In-rich growth provides sizeable crystals of $Eu_{11}InSb_9$, an orthorhombic compound without inversion symmetry [5]. As the In concentration is increased, needle-like orthorhombic crystals of inversion-symmetric $Eu_5In_2Sb_6$ are favored and a ratio of 1:16:1.8 provides millimeter-long needles as the main phase. This compound is an antiferromagnetic insulator recently synthesized in single crystalline form [6]. There are several ternary phase diagrams that have been mapped out, but Eu–In–Sb is not one of them. Due to the increased complexity of ternary diagrams compared to binary ones, researchers often have to make educated guesses and learn from trial-and-error.

3 Containers

In any flux growth, it is necessary to have the melt heated in a container that is not attacked, or at least not strongly attacked, by the melt. Further, an environment that is inert relative to the melt is needed. An inert environment is often easily provided by sealing the elements of the melt at the desired relative concentrations in a quartz glass (also known as fused silica) tube, either in a vacuum or a partial pressure of argon or other inert gas. Pure quartz glass can be heated to approximately 1200 °C without softening and used at somewhat higher temperatures by sealing the quartz glass off with an inert gas partial pressure that produces approximately 1 atm at the working temperature. Given the difficulty in exactly controlling this during sealing, short working times at temperatures above the softening point of quartz glass are prudent. To further protect the tube, a thin coating layer of carbon may be deposited on the quartz glass inner walls. Carbon not only protects the quartz tube against reactive elements but also reacts with oxygen and water impurities [7]. To check whether the inert atmosphere survived the growth, a useful trick is to write with a pencil the sample identification (e. g., batch number) on the crucible holding the elements. If oxygen is present, it will react with the graphite from the pencil, and the sample identification will no longer be visible at the end of the growth.

It is usual to place the melt ingredients in an appropriate crucible which is then sealed in quartz glass. This is not always necessary, and there are some advantages when the quartz tube itself can serve as a crucible. Clearly one does not want the flux to attack the tube, and, in many melts at not too high temperatures, the activities of the melt constituents are sufficiently reduced within the melt that chemical attack of the quartz glass does not occur. For example, $CeCoIn_5$ crystals can be grown from In-rich melts using quartz glass as the container [8]. The idea here is that quartz has a smooth surface which perhaps provides fewer nucleation sites than a sintered ceramic crucible. Another example of crystals readily grown in quartz glass without crucibles are the 3-4-13 Remeika phase stannides, such as $Ce_3Rh_4Sn_{13}$ [9]. Experience shows, however, that rare-earth-rich melts cannot be heated directly in quartz glass.

In a majority of cases, one needs to contain the melt in an appropriate crucible. Alumina crucibles are generally used if possible. These are available commercially and made from 99.8 % purity Al_2O_3 in many sizes, 2 ml and 5 ml being the most popular ones. They have very good thermal shock resistance and can be used up to quite high temperatures. Rare-earth tetra- and hexaborides, for example, have been grown from Al in alumina crucibles from temperatures as high as 1600 °C in a tube furnace in an argon inert atmosphere. Yttria-stabilized zirconia is a more chemically inert crucible than alumina, but its thermal shock resistance is poor and its use requires a more careful heating protocol. Graphite crucibles are relatively easy to machine and have good thermal shock resistance, but they require pre-firing in vacuum to remove absorbed water vapor. Beryllia crucibles also require pre-firing to remove absorbed water vapor, a problem alumina crucibles do not have. Further, beryllia crucibles are more inert than alumina ones, but the health risks associated with beryllia make their use problematic. It is worth noting that some combinations of elements have extremely exothermic interactions that are sufficient to crack even alumina crucibles during heating. Fluxes containing early rare-earth elements and platinum are an example. An important point to consider when using crucibles sealed in quartz glass is the extremely low thermal expansion of quartz glass relative to that of other materials. A crucible that is essentially wedged into the bottom of a quartz tube, however slightly, will crack the quartz tube upon heating. This cracking can be prevented by supporting the crucible off the bottom of the tube with either small pieces of quartz or quartz wool, or flattening the bottom of the quartz tube that has been heated to the softening point against a metal flat.

In some cases unexpected interaction with the crucible occurs. In an attempt to prepare Ce–Co–Pb intermetallics from molten Pb as a flux, a new tetragonal compound was grown with composition Ce_5Pb_3O [10]. No Co was found in the crystals. The source of the oxygen appears to have been the alumina crucible used to contain the melt in the growth cooling down from 1150 °C. This serendipitously found new material was interesting in having one of the highest known magnetic ordering temperatures for a Ce intermetallic, ordering ferrimagnetically at 46 K. Another example of reactions with the crucible was found in an attempt to prepare Ca–Fe–Si intermetallics. After the growth, crystals of $Ca_3Al_{2-x}Fe_x[SiO_4]_3$ were found on the walls of the crucible as a result of the reaction with Al_2O_3. $Ca_3Al_{2-x}Fe_x[SiO_4]_3$, also known as hessonite, is a naturally occurring garnet found in ore deposits in China, Brazil and the Alps.

4 Separation of crystals from the flux

A typical flux growth is from fairly dilute solution, several percent concentration of product relative to the dominant metal making up the flux. This tends to yield separated crystals when the residual flux is removed, as illustrated in Figure 3. While chemical leaching of flux is often possible, it is not always trivial to find a chemical etch that removes the flux without attacking the crystals.

A very effective way to separate grown crystals from a growth melt is via filtration at the end of the growth cycle, while the melt is still liquid. This can in many cases be accomplished by inserting a quartz wool plug into the quartz tube above the crucible holding the melt. After quickly removing the quartz tube from the furnace using a stainless steel tong, while the melt is still liquid, and inverting it into a simple chemical

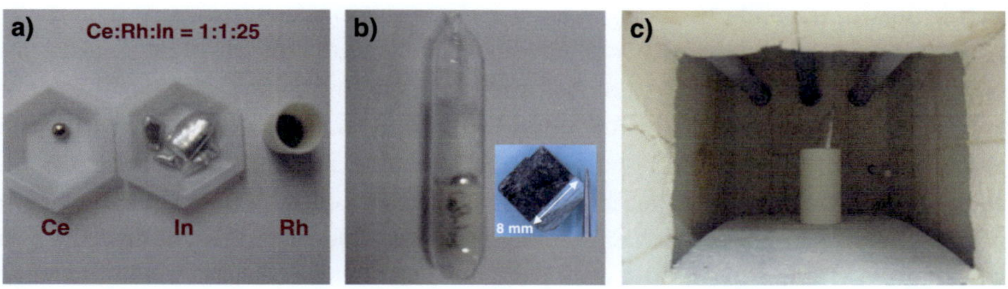

Figure 3: a) Initial elements used in the growth of $CeRhIn_5$: an arc-melted button of Cerium, Indium pieces, and Rhodium powder. **b)** Encapsulated alumina crucible holding the elements. A picture of the obtained crystal is shown on the right. **c)** Internal arrangement of a typical box furnace.

centrifuge, a short spin will push the liquid through the quartz wool plug, leaving grown crystals behind. To prevent the quartz tube from shattering during the spin, an appropriate refractory cushion should be placed in the centrifuge cup and care should be taken when the quartz tube is sealed to provide a reasonably robust end nub. It is also possible to load the quartz tube with the filter plug against a flattened end, then to support the crucible in the tube with a second plug below the crucible to hold it in place while sealing off the quartz tube below it. The flattened end is generally more resistant to breaking than a pointed nub, but experience usually teaches one how to prepare a robust nub on a quartz tube.

5 General considerations

As another example, consider the Al–La phase diagram shown in Figure 4. To grow the most Al-rich phase in this system (La_3Al_{11}) using a quartz encapsulated alumina crucible, the composition of the melt needs to be such that the liquidus temperature is below 1200 °C, a composition more Al-rich than approximately $La_{0.05}Al_{0.95}$. Another consideration comes into play here, namely that hot Al attacks quartz glass. This has two aspects. The first is that during the growth, the Al vapor pressure, while fairly small below 1200 °C, is still enough to slowly attack the quartz. This can be slightly ameliorated by sealing the quartz tube off with a partial pressure of argon inside, but with Al-rich melts one generally carries out the growths with ramp down times of a few days rather than weeks. The second point is that when spinning off the melt, the hot Al will often crack the quartz, exposing the hot crystals to oxidation. One approach to dealing with this is to load the tube with an inverted alumina crucible containing the quartz wool plug above the crucible containing the melt, so that the spun melt is contained in the second crucible. In this case, one also can place a second quartz wool plug above the catch crucible to cushion the force of the catch crucible against the quartz tube during the spin. A more elaborate procedure is to have commercially fabricated a fitted catch crucible which can be manufactured containing a frit that acts as the filter for the melt.

When inserting the quartz wool plug above the growth crucible in the quartz containing tube, it invariably happens that some small amount of quartz wool dust contaminates the growth load. This of course introduces an unwanted impurity into the melt that often has little effect on the product, but it acts as well to provide extra nucleation sites for crystallization. This problem can be avoided by carrying through the crystal growth to full solidification, reloading the solidified growth in its crucible into a fresh quartz tube with a quartz wool plug, and then heating this up to a liquid stage and spinning.

The self-flux growth is a common one, but in many cases it is not applicable for the growth of the crystals one might be interested in. Often one can find an appropri-

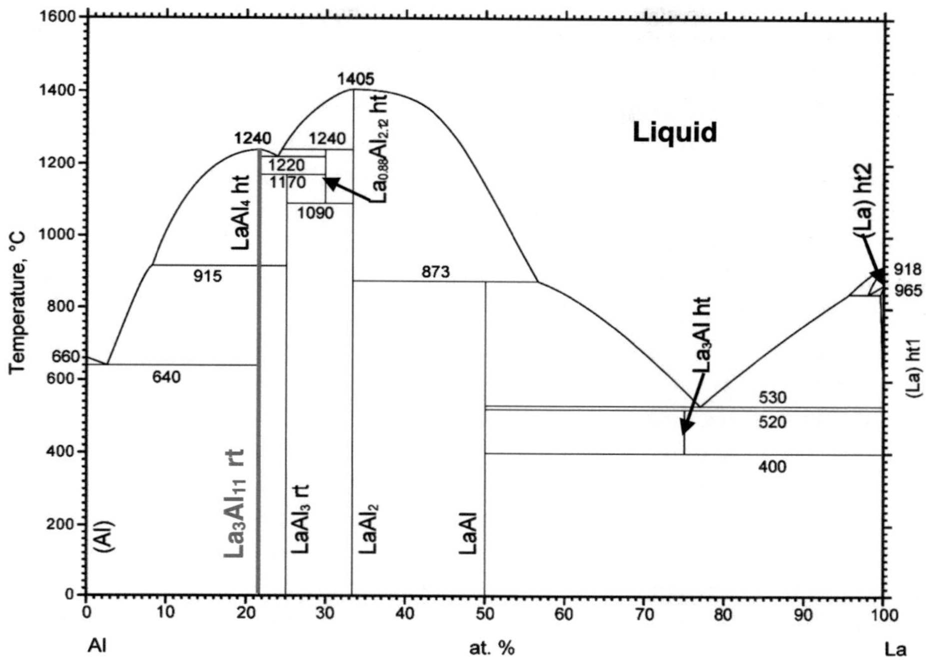

Figure 4: Al–La binary phase diagram adapted from Ref. [11].

ate flux that allows growth of crystals at much lower temperature than possible with a self-flux. This is particularly true for very refractory materials. An example is the growth of borides from molten Al, for instance the light rare earth hexaborides. Here the solubility of the hexaborides in molten Al is quite low, and the melts typically have concentration of the hexaboride of order 10^{-3} molar. The growths from Al have been from temperatures of 1500 °C in tube furnaces with the protective atmosphere provided by a mullite or alumina tube through which ultra-high-purity argon is flowing at a slow rate. Typically 50-ml alumina crucibles hold the melts, and it is inconvenient to try to seal these afterwards into quartz glass for spinning, so chemical etching with NaOH solution is used to remove the Al. This does not attack the hexaboride. Al impurities do not substitute into the hexaboride lattice, but in larger grown crystals one often finds Al lamellae coherently growing with the hexaborides, which can be mechanically removed by polishing. A more benign leaching can be accomplished by growing from an Al–Ga melt, which can be leached using H_2O.

Besides the case of refractory materials, there are examples in which self-flux is not applicable because either a more stable ternary compound forms or self-flux involves more risks and perhaps yields off-stoichiometry crystals. An example of the former is found in the tetragonal compound $ThCo_{2-x}Sn_2$. Attempts to grow this material via Sn-rich melts yields the growth of cubic $Th_3Co_4Sn_{13}$, but Indium-flux provides plate-like crystals of $ThCo_{2-x}Sn_2$ that enable the determination of the magnetic anisotropy below the antiferromagnetic transition at $T_N = 78$ K [12]. An example of the latter is found in the tetragonal compound $BaFe_2As_2$. This material can be grown via self-flux (FeAs), which requires a substantial amount of Arsenic in each growth and may lead to off-stoichiometry crystals. Another approach is to use Sn-flux, which leads to Sn incorporation in the structure and extrinsic physical properties. A successful route to solve these issues is, again, to use Indium as the flux. $BaFe_2As_2$ single crystals grown with Indium display the lowest residual resistivity among all known growth methods and yield the highest superconducting transition temperatures when Fe is substituted with Co, Ni, or Cu [13].

In cases where one is able to grow a compound from a flux that does not incorporate into the grown crystal, it is often the case that the congruently melting compounds in the phase diagram containing the crystal grow rather than the incongruently melting ones. In the case of rare earth borides, this results in the hexaborides coming down in Al melts for the light rare earths and the tetraborides for the rare earths Gd and beyond [14]. With the growth of rare earth tetraborides, a further interesting complication in flux growth from Al is a competition between the growth of the simple rare earth tetraboride and the growth of the phase $RAlB_4$ (R being a heavy rare earth element). The precise growth conditions favoring one or the other phase have not been completely worked out, although very high temperature growth appears to favor the pure tetraboride. There is even a further aspect to the $RAlB_4$ growth, namely, two possible different crystallographic arrangements of the planar boron sheets in the structure exist and intergrowths of the two types are known, as well as the pure end types [15]. These high temperature growths from Al flux can be carried out in open-top alumina crucibles or with lids (in an inert atmosphere), taking advantage of the fact that molten Al is generally covered by a very thin layer of its oxide which restricts evaporation of Al (at 1500 °C, the vapor pressure of Al is 1 mbar). At temperatures over approximately 1600 °C, Al will dissolve its oxide covering, resulting in the growth of sapphire needles and significant Al evaporation.

A variant of the simple Al-flux growth of rare earth hexaborides has been used to grow GdB_6, which does not grow from Al-flux as usually carried out. Instead, an arc melted pill of GdB_6 was placed in an alumina crucible containing a flux of composition $Ga_{0.95}Al_{0.05}$. Ga does not dissolve B significantly at temperatures below 1500 °C, and GdB_6 has a lower density than this flux so that it will float on the surface of the flux in which GdB_6 will be only sparingly soluble at temperatures near 1500 °C. By establishing a temperature gradient such that the bottom of the crucible is at a lower temperature than the top, GdB_6 crystals grew in the lower part of the crucible. This particular growth was carried out using an induction heating coil in an evacuated bell jar.

Other borides such as ZrB_2 can be grown from Al flux. This particular compound is only sparingly soluble in Al presumably due to its extremely high melting temperature, requiring similarly high temperatures (1500 °C) as in the growth of rare earth hexaborides [16]. Some early rare earth tetraborides can be grown from rare-earth-rich eutectics with Co or Ni at temperatures below 1200 °C. Such eutectics are readily attacked by dilute HCl which can be used to isolate the crystals grown from them.

Al flux has also been used to grow UBe_{13} single crystals, the original choice of flux partly motivated by the simple binary Al–Be phase diagram containing no binary phases. While originally carried out in beryllia crucibles, it was subsequently found that the Al melt containing Be did not attack alumina, and so use of beryllia was not necessary. Here it turned out that some Al is incorporated into the UBe_{13} structure, resulting in a higher superconducting transition temperature than material not containing Al [17].

Indium is an attractive flux for many systems, most notably for the growth of very high quality crystals of $CeTIn_5$ (T = Co, Rh and Ir). Isostructural rare earth and actinide phases with Ga similarly grow from Ga [18]. Indium and Ga are easy to deal with because of their low melting points, making centrifuging off of the molten flux possible at low temperatures. This also is the case with the slightly higher melting metals Bi, Sn and Pb. Additionally, very high-quality crystals of UPt_3 grow from Bi flux [19].

Indium-rich growths have been carried out at temperatures up to 1500 °C for the refractory compounds $YbRh_2Si_2$ and URu_2Si_2 [20, 21]. The vapor pressure of In at 1500 °C is near 13 torr, making evaporation of In a problem. In these cases, Ta tubing sealed off in an argon arc furnace was used as a container and the grown crystals subsequently leached out from the solidified growth melt with dilute HCl. Ta is easy to use as a container because the ductility of the metal tubing allows crimping the ends in a vice which can then be melted shut. W and Mo are more chemically inert than Ta, but much less ductile. For these, premade crucibles are necessary with fitting lids that can be e-beamed shut. It is possible to make a jig that can be mounted in an arc furnace to hold such crucibles in a cold clamp making possible fusing the lid to the crucible body with the arc melter stinger.

In addition to the advantage flux growth has in producing crystals grown at temperatures often well below their melting points, flux growth also can provide an effective way to prepare crystals whose component elements have high vapor pressures at the compound's melting temperature. The divalent rare earths and the alkaline earth elements are examples of this. Sulfur and phosphorus also have very high vapor pressures at the melting points of phosphide and sulfide compounds of interest. The filled Skutterudite phosphides have been nicely grown from Sn flux. Canfield and coworkers have found transition metal-S fluxes containing surprising concentration of S with low vapor pressures, as well as Bi–S fluxes that can be safely used with high S content [22].

Cu can also be a useful flux but has the dual problems of high melting temperature, making spinning off the flux somewhat problematic, plus being chemically fairly robust so that etching out crystals from solidified melts where it was used as a flux requires some care to avoid dissolving the crystals. Nice crystals of $TmIr_2$ and $YbIr_2$ have been grown from Cu [23]. These were leached out using HNO_3, the $TmIr_2$ and $YbIr_2$ not being attacked by this acid. The growth was not possible in a simple Ta tube, the Ir in the melt attacking Ta at high temperature. In this case, an yttria-stabilized zirconia crucible was used to hold the melt.

In situations where it is desired to grow doped crystals or mixed crystals such as $CeRh_{1-x}Co_xIn_5$ or $Yb_{1-x}Sm_xB_6$, it is necessary to examine the actual composition of the grown crystals since even in a given batch large variations in x within the grown crystals can be found. In cases where one can pre-melt (via arc melting for example), the elements to be mixed prior to loading the growth, better uniformity of the grown product is generally found. Holding the melt at a high soak temperature before the growth ramp down helps, but good mixing is not always achieved. Occasionally, the actual concentration in a crystal may differ substantially from the nominal concentration, even when the overall concentration in a crystal is homogeneous. This is the case of $CeRhIn_{5-x}Sn_x$ in which the actual Sn concentration is 60 % of the nominal one.

Although the focus of this chapter is on the growth of bulk single crystals, we end it with a remark on recent efforts to synthesize intermetallic nanowires with diameters ranging from 25 to 250 nm [24]. The so-called metallic-flux nanonucleation technique is based on the conventional flux-growth technique discussed here, but is performed in a nanometric template that mediates the preferential nucleation of the single crystals in the desired geometry. This template is an Al_2O_3 membrane fabricated by a hard-anodization process, and fixed on the base of an alumina crucible enclosing the initial metals. It has been shown that simple binary materials, such as the RIn_3 (R = rare-earth) compounds discussed in the very beginning of this chapter, can be synthesized using this technique.

6 Conclusion

The study of materials in the form of single crystals is now the norm in solid state chemistry and condensed matter physics. Flux growth is a very cost effective technique for producing single crystals, as well as being a worthwhile route in the search for new materials. In a certain sense, finding a useful flux for the growth of particular crystals amounts to exploring parts of the corresponding phase diagrams, and this is an on-going activity that is not well explored in multi-component systems. The use of simple mechanical methods for separating crystals from their growth melts has greatly expanded the use of the technique and made available high-quality materials to an expanded community.

Bibliography

[1] General reviews of the flux growth methods can be found in: Fisk Z, Remeika JP. Growth of single crystals from molten metal fluxes. In: Gschneidner, KA Jr, Eyring L, eds. Handbook on the

Physics and Chemistry of Rare Earths. Elsevier, Amsterdam, 1989, 12, 53; Canfield PC, Fisk Z. Growth of single crystals from metallic fluxes. Philos Mag B 1992, 65, 1117; Kanatzidis MG, Pöttgen R, Jeitschko W. The metal flux: a preparative tool for the exploration of intermetallic compounds. Angew Chem, Int Ed Engl 2005, 44, 6996.

[2] Okamoto H. La-Sn (Lanthanum-Tin). J Phase Equilib 2002, 23, 289.

[3] Sarrao JL, Immer CD, Benton CL, Fisk Z, Lawrence JM, Mandrus D, Thompson JD. Evolution from first-order valence transition to heavy-fermion behavior in $YbIn_{1-x}Ag_xCu_4$. Phys Rev B 1996, 54, 12207.

[4] Sarrao JL, Ramirez AP, Darling TW, Freibert F, Migliori A, Immer CD, Fisk Z, Uwatoko Y. Thermodynamics of the first-order valence transition in $YbInCu_4$. Phys Rev B 1998, 58, 409.

[5] Xia SQ, Hullmann J, Bobev S, Ozbay A, Nowak ER, Fritsch V. Synthesis, crystal structures, magnetic and electric transport properties of $Eu_{11}InSb_9$ and $Yb_{11}InSb_9$. J Solid State Chem 2007, 180, 2088–94.

[6] Rosa PFS et al. (unpublished).

[7] Harrison MJ, Graebner AP, McNeil WJ, McGregor DS. Carbon coating of fused silica ampoules. J Cryst Growth 2006, 290, 597–601.

[8] Petrovic C, Pagliuso PG, Hundley MF, Movshovich E, Sarrao JL, Thompson JD, Fisk Z, Monthoux P. Heavy-fermion superconductivity in $CeCoIn_5$ at 2.3 K. J Phys Condens Matter 2001, 13, L337.

[9] Remeika JP, Espinosa GP, Cooper AS, Barz H, Roweil JM, McWhan DB, Vandenberg JM, Moncton DE, Fisk Z, Woolf LD, Hamaker HC, Maple MB, Shirane G, Thomlinson W. A new family of ternary intermetallic superconducting/magnetic stannides. Solid State Commun 1980, 34, 923–6.

[10] Macaluso RT, Moreno NO, Fisk Z, Thompson JD, Chan JY. Structure and magnetism of Ce_5Pb_3O. Chem Mater 2004, 16, 1560–3.

[11] Okamoto H. Al-La (Aluminum-Lanthanum). J Phase Equilib, 2000, 21, 205.

[12] Rosa PFS, Garitezi TM, Fisk Z, Pagliuso PG. 3*d* magnetism in $ThCo_2Sn_2$ single crystals. J Phys Conf Ser 2015, 592, 012053.

[13] Garitezi TM, Adriano C, Rosa PFS, Bittar EM, Bufaical L, de Almeida RL, Granado E, Grant T, Fisk Z, Avila MA, Ribeiro RA, Kuhns PL, Reyes AP, Urbano RR, Pagliuso PG. Synthesis and characterization of $BaFe_2As_2$ single crystals grown by the In-flux technique. Braz J Phys 2013, 43, 223–9.

[14] Fisk Z, Cooper AS, Schmidt PH, Castellano RN. Preparation and lattice parameters of the rare earth tetraborides. Mater Res Bull 1972, 7, 285–8.

[15] Macaluso RT, Nakatsuji S, Kuga K, Thomas EL, Machida, Y, Maeno Y, Fisk Z, Chan JY. Crystal structure and physical properties of polymorphs of $LnAlB_4$ (Ln = Yb, Ly). Chem Mater 2007, 19, 1918–22.

[16] Higashi I, Takahashi Y, Atoda T. Crystal growth of borides and carbides of transition metals from molten aluminum solutions. J Cryst Growth 1976, 33, 207–11.

[17] Ott HR, Rudigier H, Fisk Z, Smith JL. UBe_{13}: an unconventional actinide superconductor. Phys Rev Lett 1983, 50, 1595.

[18] Sarrao JL, Morales LA, Scott BL, Stewart GR, Wastin F, Rebizant J, Boulet P, Colineau E, Lander GH.: Nature 2002, 420, 297–9.

[19] Stewart GR, Fisk Z, Willis JO, Smith JL, Possibility of coexistence of bulk superconductivity and spin fluctuations in UPt_3. Phys Rev Lett 1984, 52, 679.

[20] Krellner C, Taube S, Westerkamp T, Hossain Z, Geibel C. Single-crystal growth of $YbRh_2Si_2$ and $YbIr_2Si_2$. Philos Mag 2012, 92, 2508–23.

[21] Baumbach RE, Fisk Z, Ronning F, Movshovich R, Thompson JD, Bauer ED. High purity specimens of URu_2Si_2 produced by a molten metal flux technique. Philos Mag 2014, 94, 3663–71.

[22] Lin X, Bud'ko S, and Canfield PC. Development of viable solutions for the synthesis of sulfur bearing single crystals. Philos Mag 2012, 92, 2436–47.
[23] Willis JO, Smith JL, Fisk Z, Unusual magnetic behavior of $TmIr_2$ and $YbIr_2$. J Magn Magn Mater 1985, 47, 581–2.
[24] Rosa PFS, de Oliveira LAS, de Jesus CBR, Moura KO, Adriano C, Iwamoto W, Garitezi TM, Granado E, Saleta ME, Pirota KR, Pagliuso PG, Exploring the effects of dimensionality on the magnetic properties of intermetallic nanowires. Solid State Commun 2014, 191, 14–8.

Peter Gille

Single crystal growth of intermetallics by the Czochralski method

1 Introduction

Today, the Czochralski method may be regarded to be the most prominent technique for single-crystal growth due to its use for the production of huge amounts of large-scale silicon crystals for semiconductor devices and photovoltaics. In both cases, the productivity of the large companies is measured in thousands of tons of single crystals per year, each of them in the typical size of eight to twelve inches in diameter and about two metres in length.

Contrary to this, nobody is interested in really large single crystals of intermetallic phases, and to the best of the author's knowledge, so far there has been no initiative for any device to be industrially produced from single-crystalline intermetallics. Intermetallic compounds grown nowadays are the subject of basic research in order to study their intrinsic properties. Thus, the goals in Czochralski growth of intermetallics cannot be compared with those in industrial use for semiconductors or oxides. Instead, a phase-pure material with structural perfection as high as possible is usually the main target and single crystals of a few cm^3 are considered to be "large samples" in the research community of intermetallic compounds. A lot of exciting physics can be done with really tiny samples; and the most challenging needs with respect to the size of crystals come from suggested studies that use neutrons for the investigation of bulk single crystals and from surface physics and chemistry or surface crystallography where typically well-oriented crystallographic faces of about one cm^2 will be sufficient.

Although laboratory apparatuses for Czochralski crystal growth of intermetallics look much different compared to their industrial counterparts, research can very much benefit from the knowledge about Czochralski technology that has been gained over decades of industrial development. For example, with respect to technical issues, hydrodynamics of the melt and segregation effects in the growing crystals, the Czochralski method is the best studied technique for bulk single-crystal growth.

2 Czochralski method

In 1916, Jan Czochralski observed that a part of a metallic melt wetting a thin-diameter glass carrier can be pulled above the average level of the melt surface. By increasing the distance to the level of the hot melt, it was crystallized and could be further pulled upwards without losing contact to the meniscus of the extended melt. Moreover, he

https://doi.org/10.1515/9783110496789-004

found that the pulling rate of the thin single-crystalline wires can be increased only up to some limit. He concluded that this limit is given by the maximum crystallization velocity of the individual metals. These observations were published to suggest "A new technique for measuring the crystallization rate of metals" which is the translated title of the first paper that appeared in the German language in 1918 [1]. Soon, this principle was used to grow the first metallic single crystals for basic research, enabling, e. g., the determination of elastic constants, thermal expansion coefficients and electrical conductivity. The real breakthrough of the Czochralski method was not earlier than at the end of the 1940s when Teal and Little at Bell Labs started to apply the Czochralski method to semiconductor single-crystal growth, at first to grow germanium crystals [2]. A comprehensive description of the interesting history of developing the Czochralski method until the present status has been written recently by Uecker [3].

2.1 Principle of the method

Crystal growth using the Czochralski method means pulling a single-crystalline ingot from an appropriate melt that is close to thermodynamic equilibrium with the growing solid. Instead of Czochralski's glass carrier that was taken to initiate heterogeneous nucleation, nowadays a well-oriented native seed from a former growth run is used to be wetted by the melt and to start crystal growth. The melt is stored in a crucible that should be inert with respect to the melt. During the whole process, the growing crystal has no contact to the crucible or some other material except the melt. This is regarded one of the main advantages because any contact to a container could cause thermomechanical stress to the crystal upon cooling due to differences in the thermal expansion. Usually, slow rotation of the crystal and counter-rotation of the crucible is part of the typical conditions, e. g., in order to ensure rotational symmetry of the temperature field in the crystal as well as in the melt. Often, crystal rotation is also used to assist materials transport in the melt adjacent to the growing interface. A sketch of the typical crystal growth setup of the Czochralski method is shown in Figure 1.

There have been other single-crystal growth techniques that look similar to the Czochralski method, e. g., the Kyropoulos method and top-seeded solution growth (TSSG) [4, 5], that use a seed crystal that is brought into contact to the top of the liquid and crystal growth occurs under the melt surface. The decisive feature of the Czochralski method is that during the whole process crystal growth proceeds at the three-phase coexistence line where the growing crystal is in contact to the melt meniscus and the surrounding gas atmosphere. The liquid phase wetting the growing crystal is held by capillary forces which was already mentioned in Czochralski's original paper [1]. Thus, the shape of the crystal, i. e., the diameter of the actually crystallizing layer, is determined by the curvature of the meniscus. Therefore, it is well-known that diameter control can be done either by slight changes in the absolute temperature of the melt or by slightly varying the pulling rate of the crystal. Both parameters

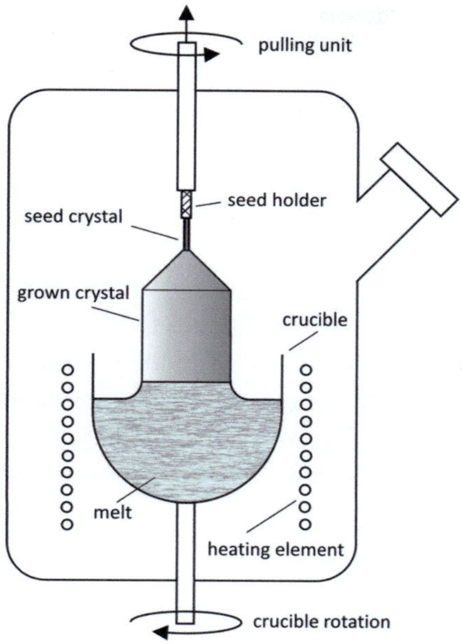

Figure 1: Schematic sketch of a Czochralski growth setup in a vacuum chamber.

cause the same result: Increasing the heating power or the pulling rate, both to a small extent, shifts the solid/liquid interface to a slightly higher position which forces the meniscus to change its shape (see Figure 2). The latter is a consequence of the limited mass of liquid that can be held by capillary forces against gravity. In industrial crystal growth processes with large volumes of the melt, diameter control by varying pulling rates is the preferred measure because of its faster response. This is no longer an argument when dealing with small amounts of the melt in the growth of intermetallics. And, what is more important, when in any case using slow pulling for reasons that are explained later, a change in this rate would hardly influence the diameter since this effect is simply driven by the changed yield of latent heat at the growth interface.

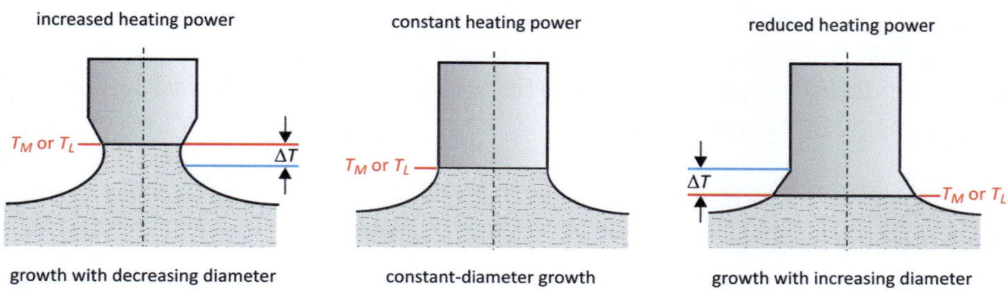

Figure 2: Diameter control by various heights of the meniscus. The actual diameter is determined by the three-phase coexistence line at the solid/liquid interface at equilibrium temperature of T_M (melting point) or T_L (liquidus temperature).

Growing at rates of less than one mm/h (instead of, e. g., approx. 50 mm/h in silicon crystal growth), latent-heat release can be neglected compared to the total heating power. Thus, in Czochralski growth of intermetallics, diameter control can be effectively done by slight changes of the total heat supply.

2.2 Technical issues

Among the crystal growth techniques used for intermetallic compounds, the Czochralski method is considered to be the most complicated with respect to the necessary technical effort which is as high as that for advanced systems in the floating-zone technology. None of the commercially available crystal growth apparatuses is well-adapted to all requirements of the various intermetallic compounds that are in the focus of basic research.

2.2.1 Crucibles

In classic Czochralski growth, the starting melt is stored in a crucible. Compared to crucible-free growth methods like the floating-zone method (see Chap. 5), this is often considered as a drawback. Chemical corrosion of the crucible may pollute the melt. Consequently, the choice of an appropriate crucible material is an important issue. Most of the liquid metals do react with fused silica and with graphite, but, for a large variety of metallic melts, crucibles made of simple oxides like corundum or stabilized zirconia are quite inert. These crucibles are offered in suitable shapes from various suppliers. For example, sintered Al_2O_3 crucibles that are available in 99.7 % purity (with MgO being the additional phase) have been used by the author for all Al-, Ga-, and In-based solutions of a large number of intermetallic phases. Due to the low price of these crucibles, single-use is no problem. Prior to use, Al_2O_3 crucibles should be cleaned in diluted nitric acid, carefully rinsed in distilled water and baked in oxidic atmosphere at temperatures exceeding 1000 °C.

An interesting approach to solving the contact problem between melt and crucible is the so-called cold-crucible technique that can be used with RF-heated systems and will be treated in the next paragraph.

2.2.2 Heating systems

The appropriate choice of the technique for heating the melt is the most pivotal decision and cannot easily be changed later in an established crystal growth setup. The simplest type is a cylindrical resistance heater made from a metallic heating element

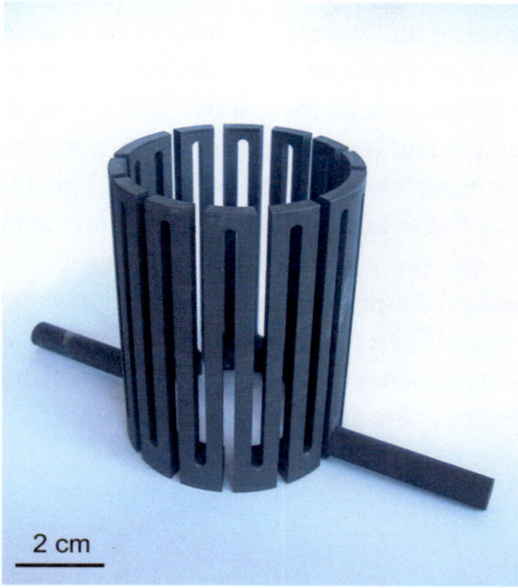

Figure 3: High-current resistance heater for Czochralski growth made from high-purity graphite.

or from a meander-shaped graphite element, which can be manufactured from a thin-wall graphite tube. The graphite heater shown in Figure 3 is a small version of those used in industry for silicon crystal growth in large Czochralski apparatuses.

Such resistance heater elements can easily be operated using commercial temperature controllers and thermocouples. Operating temperatures can be controlled at a very high accuracy of ± 0.1 K provided that the temperature sensor is in close thermal contact with the heating element.

Electromagnetic induction heating systems using RF coils are usually installed in commercial Czochralski pullers available for research purposes. For handling metallic materials, electromagnetic induction directly heats the melt rather than a conductive crucible which would be the case with non-conductive melts like oxides. RF heating has at least two interesting advantages: (i) Due to the interaction between the RF field and electrically conductive melts (Lorentz force), there is an additional contribution to convective stirring of the melt that is highly beneficial if crystal growth is done under off-stoichiometric conditions. (ii) When very sophisticated RF coils are used, melt levitation can be achieved. Such levitating melts are limited to a few grams of the whole charge. So-called *cold crucible* made from water-cooled copper segments have proved to be extremely useful in those cases where no crucible material is available that is inert with respect to the melt [6, 7].

Unfortunately, in RF-heated systems, temperature control can be more difficult due to heat induction into metallic thermocouples. In that case, optical pyrometers are used instead. Since they can be affected by changing surface conditions and emissivity on top of the melt, temperature accuracy is usually less when compared to resistance-heated systems.

Electric-arc heating has not so often been used for Czochralski growth but is quite popular in alloy preparation and homogenization of the elements. The material to be molten rests on a water-cooled metal hearth that acts as the anode. A sharp tungsten cathode is brought very close to the sample surface and almost touches it from the top, and an arc forms at the tip of the cathode thus heating the sample. The approximately spherical melt is hold by capillary forces and has almost no contact with the cold copper hearth. For single-crystal growth, more uniform heating is necessary and, at the minimum, tri-arc furnaces are used [8]. In quite sophisticated systems, the water-cooled crucible carrying the metallic sample can rotate in order to improve the homogeneity of the temperature field [9]. This type of heating device is most beneficial with highly reactive melts when no crucible material is available (see Chap. 9).

2.2.3 Atmosphere

With the only exception of a few noble metals, oxygen should be considered to be the most critical enemy of all metals at least at elevated temperatures. That is why crystal growth of intermetallic phases has to be done in closed vacuum chambers reducing the oxygen content as much as possible. Fully metal-sealed vacuum vessels are better suited than *Viton*®-sealed systems, but make the apparatuses as well as the operation more expensive. Depending on the vapor pressure of the intermetallics to be grown and the initial melts, very often inert gases like high-purity Ar or N_2, both in 5.0-grade quality, are used instead of a vacuum. Of course, inert gas filling does not change the vapor pressure of the molten metals, but it effectively reduces evaporation rates into the cold environment within a typical Czochralski growth chamber.

2.2.4 Example of a Czochralski apparatus

Commercially available Czochralski apparatuses hardly meet all the technical requirements that can be derived from specific intermetallic phases that should be grown. Moreover, usually they are oversized keeping in mind that crystals of a few cm^3 in size are regarded to be sufficient for basic research. Instead of adopting a commercial machine to the needs of single-crystal growth of intermetallics, the author of this chapter decided to construct his own Czochralski apparatuses that, as much as possible, follow the specific requirements of intermetallics (see Figure 4). Having in mind that extremely low pulling rates result in individual growth experiments that may last some weeks instead of hours or a few days, a very tight vacuum chamber was considered to be the most important prerequisite. Thus, a fully metal-sealed system was constructed with ultra-high-vacuum (UHV) components only, including all electrical feedthroughs and those for crystal pulling and the two rotations of crystal and crucible. This decision was simply made in order to reduce the leakage of the vacuum

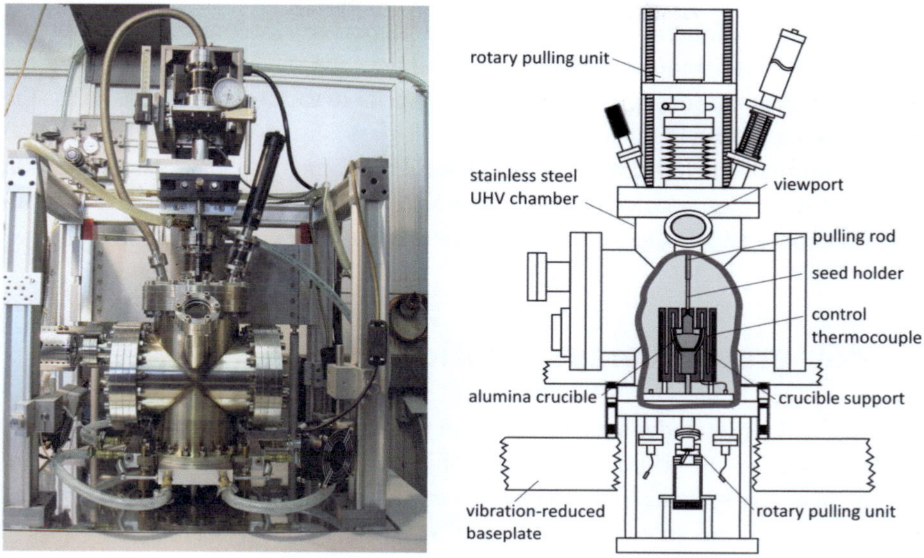

Figure 4: Photograph and scale sketch of a Czochralski apparatus used by the author for growing intermetallics in a fully metal-sealed vacuum chamber.

system to the lowest possible rate. It is not the ultra-high vacuum that is needed since crystal growth occurs under an argon atmosphere at approximately ambient pressure. But, the absence of any traces of oxygen is decisive in intermetallic growth, which is even more important in Czochralski growth where seeding occurs at the surface of the melt instead of at the bottom of the crucible such as, e. g., in Bridgman growth. All traces of oxides form a thin layer or sometimes isolated islands on top of the melt or solution and may, in severe cases, hinder seeding or even make it impossible.

The decision concerning the heating system was not as easy. Advantages and disadvantages of the various types of heating have been discussed already in paragraph 2.2.2. Since the chance of choosing extremely constant temperatures, as well as very low cooling rates in the order of a few 0.01 K/h was considered necessary, a resistance heater made from meander-shaped high-purity graphite was installed that can be easily operated using suitable temperature controllers. The control thermocouple being in close thermal contact with the heating element is important for the stability of temperature, but the exact temperature of the melt is not measured. Instead, a certain off-set between the temperature at the growth interface and that indicated by the temperature controller has to be considered.

With the exception only of the graphite heater, there is no porous material within the vacuum chamber in order to avoid that traces of oxygen can be trapped in open pores. This is the reason why thermal insulation is done by a system of radiation shields made from molybdenum instead of using porous carbon-based insulating materials. The 10-mL alumina crucible is held by a support crucible made from nickel that is connected to the rotation feedthrough. Heating of the vacuum chamber under dynamic vacuum for a couple of days is necessary prior to each experiment as this is the usual procedure with UHV technique.

Crystal and crucible rotations are driven by DC motors. For a smooth crystal pulling that enables rates sometimes as low as a few dozens of μm/hour, a reduction-gear unit operated by a step motor is extremely important.

3 Melt growth versus solution growth

In Chap. 2, the thermodynamic background of a crystal growth experiment was explained, and the typical phase diagrams of intermetallic systems were derived. Focusing on binary systems that can be more easily visualized than higher-component phase diagrams, two qualitatively different types of solid/liquid equilibria have been compared, which are once again summarized in the schematic diagram shown in Figure 5. Apart from the almost pure components A and B, i. e., phases α and β, there are two intermetallic compounds, γ and δ, showing a rather narrow existence region with respect to their exact stoichiometry AB and A_2B_3, respectively. Whereas phase γ can be crystallized either from a congruent melt or under incongruent conditions, i. e., from an A- or B-rich solution with tunable liquidus temperatures, the δ-phase can only be grown from a B-rich solution at temperatures lower than its peritectic temperature T_p. Since the different conditions of congruently or incongruently crystallizing liquids cause extremely different problems, they should be treated separately. Addi-

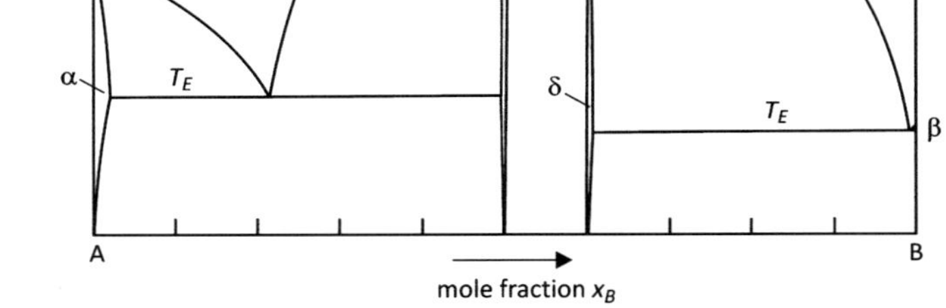

Figure 5: Schematic binary phase diagram of an intermetallic A–B system with two intermediate phases γ and δ (T_E: eutectic temperatures, T_P: peritectic temperature).

tionally, they even require sometimes different experimental parameters, which were already mentioned in paragraph 2.2.

Independent of these differences, crystal growth with the technique described here always occurs from a liquid phase that wets the growing crystal and, thus, is elevated with respect to the average level of the liquid that might be called a melt or a high-temperature solution. What is common in these two cases is that crystallization proceeds at the three-phase equilibrium line (solid/liquid/vapor) where the actual diameter of the growing crystal is determined by capillary forces. As we have defined this to be the decisive feature of the specific method, this is clearly the Czochralski method—independent of whether the liquid is a congruent melt or a high-temperature solution.

3.1 Growth from congruent melts

If thermodynamics allows congruent crystallization from a melt, this is always the easiest way to run the crystal growth experiment. Apart from the problem that it might be difficult to exactly establish congruent growth conditions, in principle we can assume that the melt simply has to solidify. This needs to take into account well-defined thermal conditions. But, with the only exception of re-arranging the atoms on an atomic scale, there is almost no materials transport problem and, accordingly, no such requirement that has to be fulfilled. Having a closer look at the congruent melting point in the phase diagram shown in Figure 6, it seems to be realistic that we will never *exactly* meet the composition of congruent crystallization.

Additionally, we have to distinguish between the congruent melting point that coincides with the local maximum of the liquidus curve, T_{MAX}, the melting temperature of the stoichiometric solid, T_L^{ST}, and the actual liquidus temperature of the melt, T_L, which usually deviates to some extent, e. g., T_L^1 and T_L^2 as shown in Figure 6. Among the many reasons for any deviation are: (i) inaccurate knowledge of the thermodynamic data; (ii) deviation in preparation of the starting melt; and (iii) later changes due to, e. g., evaporation of one of the components. Very often it is preferred to start an experiment from a melt composition that slightly deviates from exact stoichiometry in a well-defined manner instead of trying to meet exact stoichiometry that may result in unknown deviations that can either be A-rich or B-rich. Anyway, under near-congruent conditions, materials transport in the melt adjacent to the interface of the growing crystal does not play that dominant a role like in solution growth experiments which will be treated in more detail in the next paragraph. Liquid and solid phases under more or less congruent conditions of crystal growth have very similar compositions and only a very small amount of an excess component is rejected from the growing interface. Limits with respect to the growth rate in these cases come from thermal balance conditions and limited rates of crystal growth kinetics rather than

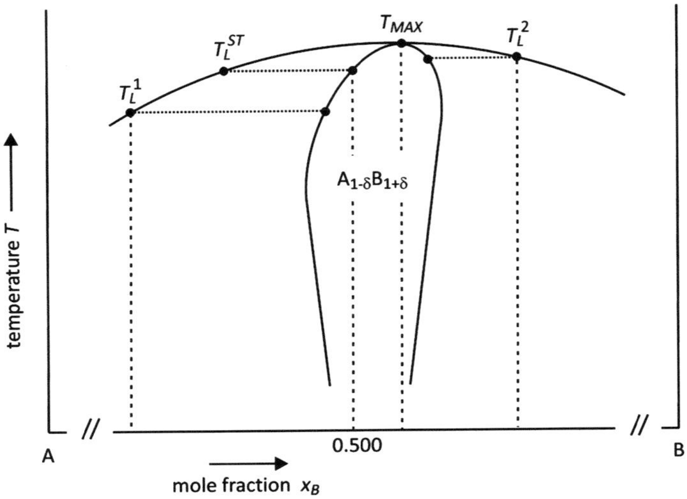

Figure 6: Schematic binary phase diagram of an intermetallic A–B system with an AB compound allowing slight deviations $\pm\delta$ from exact 1:1 stoichiometric composition (T_{MAX}: congruent melting point; T_L: various liquidus temperatures).

from materials transport. Thus, crystal growth rates, i. e., pulling rates with the near-congruent Czochralski method, can be much higher compared to growth conditions from high-temperature solution.

3.2 Growth from high-temperature solutions

The ultimate goal of any solution growth experiment is a remarkable decrease of the liquidus temperature T_L. Thus, not only slight deviations from the congruent composition as sketched in Figure 6, but a high amount of an excess component that considerably decreases the growth temperature is in the focus of this paragraph. Reasons for doing so may vary: (i) too high temperatures at congruent composition with respect to the limits of the crystal growth apparatus; (ii) too high vapor pressure of a component at the congruent melting point; and (iii) very often in intermetallic systems, no congruent melting behavior at all, i. e., decomposition of the compound to be grown at its peritectic temperature (see, e. g., δ-phase in Figure 5).

Independent of the specific reasons for high-temperature solution growth, materials transport conditions are much different from those in near-congruent growth. The composition of the high-temperature solution now strongly deviates from that of the solid in equilibrium. Very often the total amount of the excess component that plays the role of the solvent is comparable to the quantity of the solute to be crystallized.

The differences in the compositions of the solid and liquid phases being in equilibrium are determined by thermodynamics and occur as the jump (C_L–C_S) at the in-

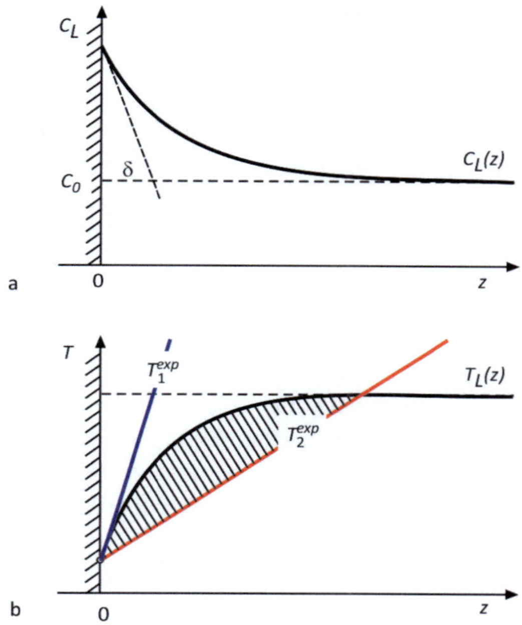

Figure 7: Schematic of the axial concentration profile in the liquid in front of a growth interface (**a**) with C_L being the concentration of the excess component, and the corresponding liquidus temperatures T_L according to the phase diagram (**b**). Two different experimental temperature gradients are plotted that are critical (red) or noncritical (blue) with respect to constitutional supercooling.

terface. If we allow the interface at position $z = 0$ in Figure 7 to move with the linear crystallization rate v, an excess of the rejected component (solvent) is accumulated in front of the phase boundary and has to be redistributed into the bulk of the liquid phase. But, next to the interface, there is no other transport mechanism than diffusion that leads to a so-called diffusion boundary layer of thickness δ (see Figure 7(a)). It is not a rigid layer of thickness δ, but a smooth transition between regions where either slow diffusion or the much more effective convection dominate the materials transport. This explains why solution growth is a very slow process limited by diffusion. Thus, at the solid/liquid interface, the continuity equation (1st Fick's law) holds:

$$(C_L - C_S)v = -D\left(\frac{dC_L}{dz}\right)_{z=0} \tag{1}$$

with D being the diffusion coefficient and z the distance from the solid/liquid interface that moves with velocity v. Already at this point, it should be noticed that according to eq. (1), the axial gradient of concentration at the interface is always proportional to the crystal growth rate. It is this concentration gradient that causes the effect of constitutional supercooling which has been described by Tiller et al. [10]. Due to limited materials transport next to the moving solid/liquid interface, there can be a region where the liquid is *constitutionally* supercooled, although the absolute temperature is higher than at the interface. This means the actual temperature is too low compared to the liquidus temperature that changes with the changing composition and can be calculated from the slope of the liquidus curve. In order to prevent the liquid adjacent to the growth interface from being constitutionally supercooled, the experimental temperature gradient has to be as high as the slope of the $T_L(z)$ curve at the interface (blue line in Figure 7(b)). With a too low temperature gradient (red line in Figure 7(b)), the entire region that intersects with the liquid phase has a temperature lower than its liquidus T_L and could crystallize, at least in principle. Unfortunately, due to relatively high thermal conductivities of metallic melts or solutions, low temperature gradients in front of the growing crystal are the usual condition. Violating the constitutional supercooling criterion to a large extent causes the crystal to grow into the supersaturated region and traps droplets of the liquid, thus forming fluid inclusions. Indeed, liquid inclusion formation is regarded as the most severe problem in crystal growth from high-temperature solutions. Going back to the reasons for the strange $T_L(z)$ curve next to the crystal (Figure 7), it is the rate of the moving interface v that determines the amount of the rejected excess component and the corresponding concentration gradient. Therefore, the pulling rate of a Czochralski growth experiment from a high-temperature solution has to be much lower than under near-congruent conditions and is typically limited to a few mm/day.

3.3 Role of hydrodynamics

In Czochralski growth, good mixing of the liquid phase would always be the best case that is, to some extent, achieved by heating the crucible from a surrounding heater and sometimes additionally from the bottom. Thus, radial temperature gradients are always present and allow buoyancy-driven convection without any threshold. Unfortunately, in metallic melts or solutions the thermal conductivity is relatively high, making temperature gradients rather low and natural convection weak. In Czochralski growth from congruent melts, this is not a severe problem, but, with high-temperature solution growth, materials transport becomes a decisive factor as explained in paragraph 3.2.

Dimensionless numbers have often been used in order to characterize hydrodynamic problems. The Prandtl number is defined as $\mathrm{Pr} = \nu/\kappa$ and describes the amount of heat transport driven by convection or conduction, with ν being the kinematic viscosity of the melt and κ its thermal diffusivity. Due to the high thermal conductivity of all metallic melts and solutions, there are always Prandtl numbers $\mathrm{Pr} \ll 1$, and the temperature distribution in the liquid is barely affected by convective motion in Czochralski growth of intermetallics.

Since buoyancy-driven convection is weak, forced convection using other driving forces than density gradients can play an important role. To some extent, already the counter rotations of the growing crystal and the crucible do assist mixing of the melt, especially next to the interface. As shown in Figure 7(a) and given by eq. (1), it is the concentration gradient and the δ-layer thickness that cause the constitutional supercooling problem and, thus, liquid inclusion formation. A better mixing by some kind of forced convection would reduce the thickness of the diffusion boundary layer δ and would minimize the risk to trap liquid droplets. For Czochralski growth of intermetallics, little is known about attempts to use techniques of forced convection that are otherwise quite common in crystal growth, e. g., accelerated crucible rotation technique (ACRT) [11] or rotating magnetic fields [12]. Among the earlier discussed various heating systems, RF heating has a very good side effect with respect to magnetic stirring. Even the otherwise not really preferred local heating by multiple arcs (see paragraph 2.2.2) causes, due to higher local temperature gradients, a stronger convection compared to that at the more uniform resistance heating principle.

Severe hydrodynamic problems may arise from mass density gradients that result from the crystallization process itself. With intermetallics, the growing crystal has occasionally a concavely curved solid/liquid interface. This is especially pronounced in the growth of semiconducting crystals from melts that have metallic properties, e. g., in $FeGa_3$ growth from a Ga-rich solution. If the excess component that is accumulated at the growing interface reduces the density of the solution compared to the average one, it will be trapped in the concave shape beneath the crystal. In these cases, a sufficient removal of the excess component is hard to achieve, and extremely slow pulling is sometimes the only way to avoid constitutional supercooling, i. e., liquid-inclusion formation. Forced convection using centrifugal forces resulting from very high rotation rates of the crystal (a few hundred rpm) have been studied [13].

4 Examples

The examples of successful growth of intermetallic alloys and compounds presented in the next paragraphs are those from our own projects during almost 20 years using the Czochralski method. Personal experience with a large number of various intermetallic phases is regarded as a useful prerequisite to describe the various problems and approaches required by the individual materials. Starting with decagonal quasicrystals from Al-rich solutions, we focused on intermetallics containing at least one low-melting element, such as Al, Ga, and In. This means we exclusively used the Czochralski technique from high-temperature solutions.

A much broader overview on crystal growth of intermetallic alloys and compounds by the Czochralski method is summarized in Table 1 which is far from complete.

4.1 Quasicrystals

The first thermodynamically stable quasicrystalline phases were discovered in the Al-based ternary systems Al–Cu–Li and Al–Pd–Mn (icosahedral quasicrystals) [48] and

Table 1: Intermetallic alloys and compounds grown by the Czochralski method.

Czochralski growth from *congruent* melt		
YB_{66}	Oliver et al. 1972	[14]
YB_4, TbB_4, ErB_4	Bressel et al. 1979	[15]
V_3Si	Wernick et al. 1979	[16]
$CePd_3$	Takke et al. 1980	[6]
$CeSi_2$, Ho_2Co_{17}, UGe_2, UPt_3	Menovsky et al. 1983	[17]
Mo_5Si_3	Nørlund Christiansen 1983	[18]
$MoSi_2$	Thomas et al. 1985	[19]
RAl_2 (R = Pr, Nd, Gd, Tb)	Abell et al. 1986	[20]
$Nd_2Fe_{14}B$	Tenaud et al. 1987	[21]
MnSi	Brown et al. 1990	[22]
R_3Ni, (R = Y, Gd, Tb, Dy, Ho, Er)	Talik et al. 1993	[23]
UPt_3, U_6Fe, $CeRu_2$, UPd_2Al_3, URu_2Si_2	Haga et al. 1998	[24]
Gd_7Rh_3, Gd_7Pd_3	Talik et al. 2005	[25]
Al_3Mg_2	Lipinska-Chwatek et al. 2007	[26]
$Gd_{7-x}Y_xPd_3$	Talik et al. 2008	[27]
$Mg_{32}(Al,Zn)_{49}$	Feuerbacher et al. 2008	[28]
$UNiSi_2$	Szlawska et al. 2013	[9]
Czochralski growth from *incongruent* melt/solution		
$Nd_2Fe_{14}B$	Swets 1986	[29]
Al_3Ti	Norby et al. 1986	[30]
i-$Al_{70}Pd_{20}Mn_{10}$	Yokoyama et al. 1992	[31]
i-$Al_{69}Pd_{23}Mn_8$	Boudard et al. 1995	[32]
TiAl	Bi et al. 1997	[33]
d-$Al_{71}Co_{18}Ni_{11}$	Gille et al. 1999	[34]
i-$Al_{64}Cu_{23}Fe_{13}$	Yokoyama et al. 2000	[35]
d-$Al_{68}Co_{19}Cu_{13}$	Gille et al. 2002	[36]
d-$Al_{72}Co_9Ni_{19}$	Bauer et al. 2006	[37]
d-$Al_{68}Co_{20}Cu_{12}$	Meisterernst et al. 2006	[38]
$Al_{13}Co_4$, $Al_{13}Fe_4$	Gille et al. 2008	[39]
GaPd	Gille et al. 2010	[40]
$Al_4(Cr,Fe)$, $Al_{13}(Fe,Cr)_4$	Bauer et al. 2011	[41]
$Al_{13}(Co,Ni)_4$, $Al_{13}Ru_4$	Gille et al. 2011	[42]
GaPd, $GaPd_2$, Ga_7Pd_3	Schwerin et al. 2014	[13]
InPd	Hahne et al. 2014	[43]
$FeGa_3$	Wagner-Reetz et al. 2014	[44]
$YbNi_4P_2$	Kliemt et al. 2016	[45]
Ga_3Ni_2	Wencka et al. 2016	[46]
$LaPd_2Al_2$	Doležal et al. 2017	[47]

Al–Co–Ni and Al–Co–Cu (decagonal quasicrystals) [49–51]. All these quasicrystals contain Al as the major component and can be grown under near-equilibrium conditions from an Al-rich liquid at temperatures of approx. 1000 °C or slightly higher. Single-crystal growth using other than trial-and-error approaches greatly benefitted from the excellent phase diagram studies published in the 1990s [52–54]. The first large single crystals were grown in the Al–Pd–Mn system by Boudard et al. [32] using the Czochralski method. These icosahedral quasicrystals were used for extended studies of the intrinsic physical properties of well-ordered solids that lack periodicity in all three spatial dimensions. Looking for unique physical effects, e. g., surface studies revealed special features with respect to surprisingly low surface energies. But, reports on the process of crystal growth did not contain much detail.

By far the best studied system containing a stable decagonal phase is the Al–Co–Ni system with an extended existence region of the quasicrystalline solid that corresponds approximately to $Al_{70}Co_{10+x}Ni_{20-x}$ ($0 \leq x \leq 15$). In Figure 8, a part of the ternary phase diagram according to Gödecke [53] is shown, including some tie-lines that mark the liquid-solid equilibria with Al-rich high-temperature solutions at liquidus temperatures of 1050 °C.

Based on these phase diagram studies, crystal growth by the Czochralski method was done using the above described fully metal-sealed growth chamber. Starting melt compositions according to the end point of a specific tie-line were synthesized *ex-situ* under an Ar atmosphere using bulk pieces of the pure elements in a RF-heated alumina crucible that was held in a graphite susceptor. During the synthesis, the high content

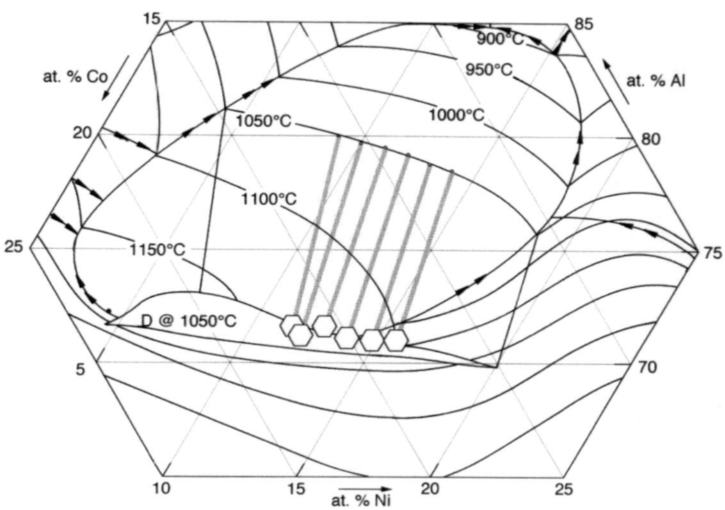

Figure 8: Part of the Al-rich region in the ternary Al–Co–Ni phase diagram showing liquidus projection and stability field of the decagonal AlCoNi phase at 1050 °C according to [53] and experimentally determined tie-lines from [55]. (Courtesy of Wiley-VCH.)

of Al served as the solvent, i. e., the high-melting point elements Co and Ni were dissolved in Al rather than molten. That is why there is no need to reach the individual melting points, but magnetic stirring due to RF heating is an important tool in order to sustain a homogeneous liquid at temperatures exceeding the liquidus temperature by not more than approximately 100 K for one hour. This minimized any evaporation loss. Once homogenization was considered to be complete, the high-temperature solution was quenched by simply switching off the heating power. Due to the small amount of material, crystallization occurs within less than a minute. Even the fast solidification follows the whole crystallization path according to the phase diagram and results in a multi-phase mixture. But, the fine-grained structure with very short distances between the individual phases can be easily homogenized in the next step after having transferred the ingot into the 10-mL alumina growth crucible and installed within the Czochralski apparatus. Performing the synthesis from the bulky elements right in the Czochralski growth chamber in-situ would require much more time to achieve a fully homogenized melt because of the poorer mixing conditions.

With all the fundamentally new phases to grow, a very special problem is how to start a Czochralski experiment without having an appropriate seed crystal. With AlCoNi quasicrystals, good results were obtained with spontaneous nucleation at a tapered alumina rod, as described in more detail for intermetallic approximants [39]. But, once a large enough single crystal of a new phase is available, a well-oriented seed crystal is by far the best choice to start a growth experiment. A full-ceramic seed holder that has been constructed for a 10° wedge-shaped single crystalline seed is shown in Figure 9.

Performing a quasicrystal growth experiment from an Al-rich high-temperature solution, there is nothing special to do compared to the growth of periodic crystals of intermetallics. What is a unique feature with AlCoNi and AlCoCu quasicrystals is

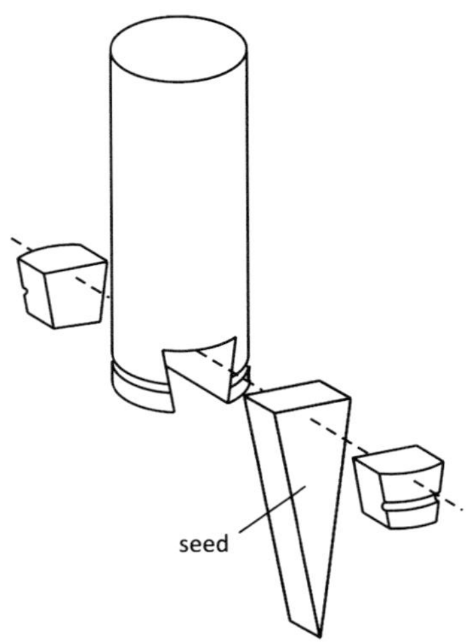

Figure 9: Full-scale sketch of the seed holder made from alumina ceramics. The nine-mm long intermetallic seed is hold in a dovetail and fixed within ceramic jaws using a wire. (Courtesy from Crystal Research & Technology.)

the pronounced anisotropy of the decagonal phases that are quasiperiodic (i. e., non-periodic) in two dimensions, whereas, perpendicular to the quasiperiodic layer, there is a periodic direction showing much faster growth rates. Thus, the natural morphology of a decagonal phase is a needle-like crystal with the long length parallel to the periodic direction which is the 10-fold axis. Using the Czochralski method, free growth along the periodic direction is hindered by the temperature field in the crucible, and the single crystal can be forced to grow into a well-defined shape that is determined by the three-phase equilibrium line at the melt meniscus. Figure 10 gives an impression how Czochralski-grown AlCoNi quasicrystals look. Their cross sections always reflect the symmetry of the crystal perpendicular to the pulling direction, i. e., being either almost circular if the crystal grows along its decagonal direction (a) or elliptic if pulled parallel to one of the two-fold axis (b).

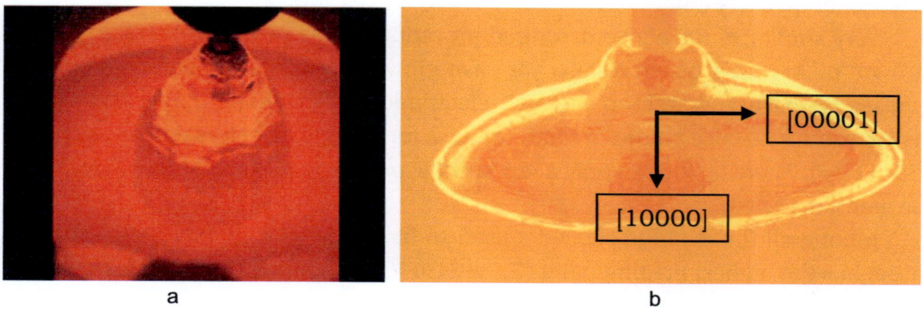

Figure 10: Photographs through the viewport during Czochralski growth of decagonal AlCoNi quasicrystals. Whereas pulling along the 10-fold [00001] axis results in an almost circular cross section of the growing crystal (**a**), crystal growth parallel to one of the 2-fold axes, i. e., perpendicular to [00001], occurs via an elliptical solid/liquid interface (**b**).

As in all high-temperature-solution growth experiments, crystallization is accompanied by a change of the composition of the remaining melt, i. e., an accumulation of the solvent Al that lowers the liquidus temperature. This has to be compensated by slowly decreasing the heating power. Low pulling rates have to be used for two different reasons: (i) to avoid liquid-phase inclusions caused by constitutional supercooling; and (ii) to meet the requirements of low growth kinetics along the non-periodic directions. The latter argument was even more decisive in crystal growth of decagonal AlCuCo. A very slow pulling along the two-fold directions was the only way to produce large single crystals [38].

As already indicated in the phase diagram shown in Figure 8, quasicrystalline phases often have a broad existence range, e. g., in the Al–Co–Ni system, they allow Co/Ni substitution between approx. 5–20 at.% Ni and accordingly 10–25 at.% Co. Thus, crystallization is not only accompanied by a change in the Al composition of the melt and the crystal, but also with respect to the Co/Ni content. For example, a growth experiment starting from a pre-synthesized $Al_{77}Co_6Ni_{17}$ melt resulted in a quasicrystal of an average composition $Al_{71}Co_{10}Ni_{19}$ with only weak axial segregation.

4.2 Approximants to quasicrystals

Crystalline phases that have similar local structures like quasicrystals but being periodic in all three dimensions are called approximants to quasicrystals. These approximant phases often occur in the compositional neighborhood of the quasicrystals and are characterized by very large unit cells, sometimes containing more than a thousand atoms. Table 2 gives some examples of periodic phases being approximants to the decagonal AlCoNi quasicrystal. The stacked-layer description refers to the structures' periodicity parallel to that direction that corresponds to the ten-fold axis of the decagonal quasicrystal.

Very similar to the above described procedure of the synthesis and Czochralski growth of Al-based quasicrystals, we have grown a number of these approximant phases for basic research on thermal, electronic and magnetic properties, as well as for surface science studies including those for the use as heterogeneous catalysts [59, 62–67]. There have been only a few specific problems and challenges that should be mentioned here.

Among all the metallic elements used so far in our growth experiments, ruthenium has the highest melting point (T_m = 2334 °C). In synthesizing the starting solution for $Al_{13}Ru_4$ growth, evaporation loss of Al at very high temperatures and, accordingly, inaccurate information about the exact composition of the starting charge have been frequently reported in the literature. But, melting temperatures of the individual components are not the decisive issue in solution growth. Instead, during the synthesis, Ru or any other high-melting point metal can be slowly dissolved in the solvent (here Al). The liquidus temperature of the final solution is the only important criterion and exceeding this limit by not much more than approx. 100 K in order to achieve homogenization within a reasonable time is good enough.

In all Cr-containing intermetallic phases, i. e., *o*-Al_4(Cr,Fe) [41], we faced severe problems resulting from the formation of native oxides. Any contact with ambient air in one of the necessary preparation steps (etching of the elements, weighing, synthesis and transfer to the crystal growth apparatus) leads to a stable film of Cr_2O_3, which can only be avoided or minimized if everything can be done under glove-box conditions. Not having this capability, we always started Czochralski growth with some thin islands of oxides on top of the melt and, more severely, with a native seed covered by a stable oxide layer that prevents wetting by the melt. The only solution to this problem that we found was to re-crystallize the surface of the melt and to carefully scratch with the rotating seed along this hard surface, thus forming tiny parts of fresh material at the tip of the seed. After re-melting the starting charge, a microscopic contact between the seed and the melt is enough for wetting, and after a short time the oxide cover of

Table 2: Structural data of some $Al_{13}TM_4$ phases being approximants to decagonal AlCoNi. Lattice constants of solid solutions refer to the specific composition of the investigated sample.

Phase	*m*-$Al_{13}(Co,Ni)_4$	*o*-$Al_{13}Co_4$	*m*-$Al_{13}Fe_4$	*m*-$Al_{13}Ru_4$	*o*-$Al_4(Cr,Fe)$
reference	Zhang et al. [56]	Grin et al. [57]	Grin et al. [58] Popčević et al. [59]	Edshammar et al. [60]	Deng et al. [61]
lattice constants	a = 1.7071(2) nm b = 0.40993(6) nm c = 0.74910(9) nm β = 116.17(1)°	a = 0.8158(1) nm b = 1.2342(1) nm c = 1.4452(2) nm	a = 1.5488(1) nm b = 0.80866(5) nm c = 1.24769(8) nm β = 107.669(4)°	a = 1.5862 nm b = 0.8188 nm c = 1.2736 nm β = 107.77°	a = 1.2500(6) nm b = 1.2617(2) nm c = 3.0651(8) nm
space group	$C2/m$ (No. 12)	$Pmn2_1$ (No. 31)	$C2/m$ (No. 12)	$C2/m$ (No. 12)	*Immm* (No. 71)
number of atoms per unit cell	32	102	102	102	366-59.56
stacked-layer description	2 layers along [010]	4 layers along [100]	4 layers along [010]	4 layers along [010]	6 layers along [100]

the seed is removed by undermining the thin layer. Once wetting and seeding have been successful, all the islands of oxides swim to the periphery of the melt surface due to the curvature of the melt meniscus. They accumulate at the rim of the melt, i. e., at the lowest surface level of the melt that is at the contact to the crucible wall, and do not affect the growth process any longer.

Many phase diagrams of intermetallic systems show a series of peritectically melting compounds, e. g., in the Al-rich region of the binary Al–Co system, see Figure 11. When growing o-$Al_{13}Co_4$ from an Al-rich solution, excess Al release at the growing interface leads to a permanently changing composition of the melt that has to be compensated by an appropriate temperature decrease. The primary crystallization of o-$Al_{13}Co_4$ ends with an interface temperature of 974 °C which is the peritectic temperature of the neighboring phase m-Al_9Co_2. In a classic discussion of the crystallization path in a peritectic system, one would argue that primary o-$Al_{13}Co_4$ reacts with the melt to form m-Al_9Co_2, which cannot be distinguished from polycrystalline m-Al_9Co_2 solidification onto o-$Al_{13}Co_4$. Any solid/solid interface of different lattices leads to high thermomechanical stress upon cooling that not only destroys the new phase but may also increase the defect density in a part of the primarily grown crystal. That is why crystal growth of the primary phase has to be stopped early enough by detaching the crystal from the rest of the solution prior to the onset of m-Al_9Co_2 crystallization.

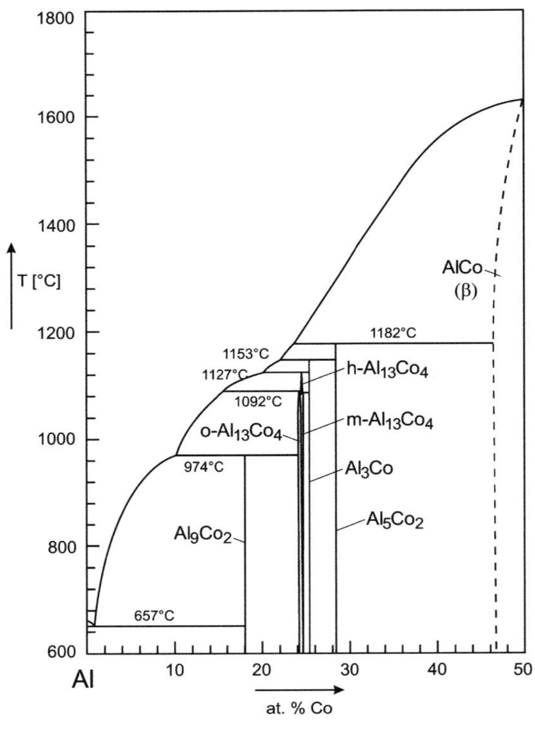

Figure 11: Al-rich part of the binary Al–Co phase diagram (after [68]).

4.3 Other intermetallic compounds

The above-mentioned intermetallics are characterized by quite complex crystal structures that are often measured in terms of number of atoms per unit cell. Quasicrystals, per definition, do not have a unit cell. Thus, it may be regarded as infinitely large, and some of their approximants have really large lattice constants containing hundreds, if not thousands, of atoms in a cell volume. But, unit cells are not at all an important issue in crystal growth kinetics. Assuming that individual atoms or, more realistically, preformed groups of atoms act as building units in the dynamic process of attachment and detachment kinetics, it is not surprising that intermetallic compounds with much simpler crystal structure do not behave differently in crystal growth when compared with complex metallic alloys and compounds. Again, thermodynamics defines the properties that are relevant for crystal growth. Some of the typical problems shall be treated here using examples of successfully grown intermetallic compounds of quite simple structures. Using the term “simple structure”, one should keep in mind that sometimes even these phases may have a rather complicated real structure given, e. g., by a slight off-stoichiometry, stacking faults, multi twinning etc. as has been discussed in Chap. 1.

Evaporation

Studying crystal growth in other than Al-based intermetallic systems, various low-melting point metals, e. g., Ga, In, and Sn, can serve as solvents for solution growth. In the usual Czochralski growth mode, i. e., without liquid encapsulation (LEC = liquid-encapsulated Czochralski [69]), the partial vapor pressures of the constituents define the limits. There are always “cold” walls in the vacuum chamber where vapor may condense. Inert-gas filling does not change the metals’ vapor pressure but drastically reduces the rate of evaporation. Nevertheless, even with rather “low-vapor pressure elements” of the 13th group, at high temperatures there is a considerably increasing material loss by evaporation when going from aluminum via gallium to indium. Whereas aluminum evaporation can almost completely be avoided, gallium as a solvent in the growth of $GaPd_2$ at temperatures exceeding 1200 °C [13] slowly evaporates and covers the viewport glass. Indium that has been used as a native solvent for InPd crystal growth [43] at these temperatures results in measureable material loss that has to be compensated by the temperature vs. time program.

Materials transport

Whereas a very high growth temperature required per the specific phase diagram may cause some trouble with respect to vapor pressures and radiation loss, really low temperatures are often detrimental for materials transport conditions. Keeping in mind the role of diffusional transport within the boundary layer, transport of the excess element that is rejected from the growth interface becomes the limiting parameter. A very difficult-to-solve problem was the growth of the Ga_7Pd_3 phase which decomposes at its peritectic temperature of $T_P = 460\,°C$ [70] and, therefore, has to be grown from a Ga-rich solution at $T < 460\,°C$ where diffusion is very slow. The only way to improve materials transport conditions seems to reduce the boundary layer thickness, i. e., to increase convective motion of the melt. Due to the low Prandtl numbers of all metallic melts, buoyancy-driven convection is too weak (see paragraph 3.3), and forced convection has been considered to be the only available tool. With respect to hydrodynamics, the growing crystal in a Czochralski growth experiment may be described as a rotating plate that is wetted by a liquid. For a flat plate, this model has been well-described long ago [71] and later on adapted to crystal growth problems in Czochralski geometry by Hurle [72]. In these models, the diffusive boundary-layer thickness δ is called the Ekman layer thickness and depends on the plate's rotation rate ω according to $\delta \sim (\nu/\omega)^{1/2}$ with ν being the kinematic viscosity of the melt. With unusually high crystal rotation rates of $\omega \geq 250$ rpm, we succeeded in growing inclusion-free Ga_7Pd_3 single crystals even at rather low growth temperatures of $T < 450\,°C$ [13].

Diameter control

In industrial crystal growth by the Czochralski technique, various methods of automatic diameter control are well-established. None of these tools can be easily adapted to the growth of small-scale single crystals of intermetallics using very low pulling rates. Because an exactly constant crystal diameter is not an important issue for basic research purposes, there is only the need to establish conditions that are stable in time, i. e., to avoid sudden changes of the diameter that would result in a too fast lateral growth and—the worst case—in touching the crucible wall by the counter-rotating crystal. Usually, it is sufficient to observe the current diameter of the crystal from time to time and accordingly to slightly change the temperature vs. time program to compensate for the decrease of the liquidus temperature. But, there are some specific phase diagrams that make it hard to achieve stable growth conditions in the stage of forming the crystal's shoulder, i. e., to achieve a slow diameter increase. One example is the Fe–Ga system with an extremely unusual liquidus line in the vicinity of the peritectic temperature of $FeGa_3$ which is 824 °C (see Figure 12).

When crystallizing $FeGa_3$, which is an interesting material for thermoelectric applications [44, 74], growth can be started from a Ga-rich solution with Ga contents higher than 81.5 at.%. In order to get a reasonable yield, the starting solution should be next to the peritectic point. It is this part of the liquidus curve that is almost horizontal, i. e., the usual stabilizing mechanism in solution growth does not work. Solidifying $FeGa_3$ changes the average composition of the rest of the melt, but, due to

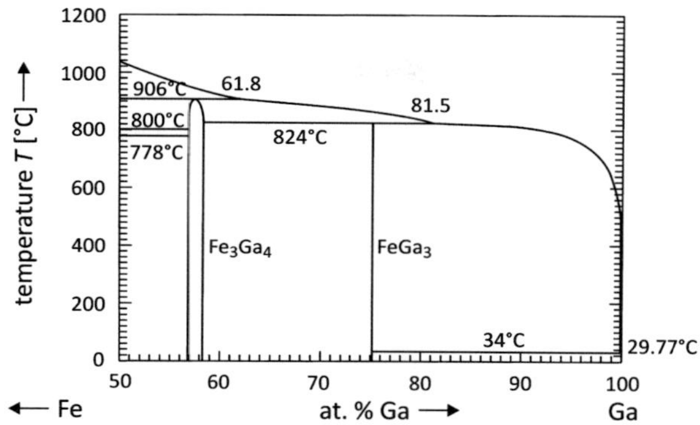

Figure 12: Ga-rich part of the binary Fe–Ga phase diagram (after [73]).

the almost horizontal slope of the liquidus, it does not reduce the equilibrium temperature to a remarkable extent. Thus, it does not avoid further crystallization that is assisted by the extension of the surface at the shoulder of the growing crystal that becomes larger and larger. Any additional surface promotes heat loss by radiation that again speeds-up crystallization. Figure 13 gives an impression how this self-amplifying process proceeds. Even with a nominally constant heat supply (constant set point of the temperature controller), lateral crystal growth does not have a cushioning effect. Instead, via the increased crystal surface and an accordingly higher heat loss, the temperature at the growth interface further decreases, resulting in further radial growth, etc. This worst-case scenario may happen within some tens of minutes and needs permanent observation by the operator. Since heat loss by radiation is proportional to T^4, a similar phase diagram situation in the Ru–In system next to the much higher peritectic temperature of $RuIn_3$ ($T_p \approx 1081$ °C [75]) is even more complicated to handle and could not satisfactory be solved in Czochralski growth so far.

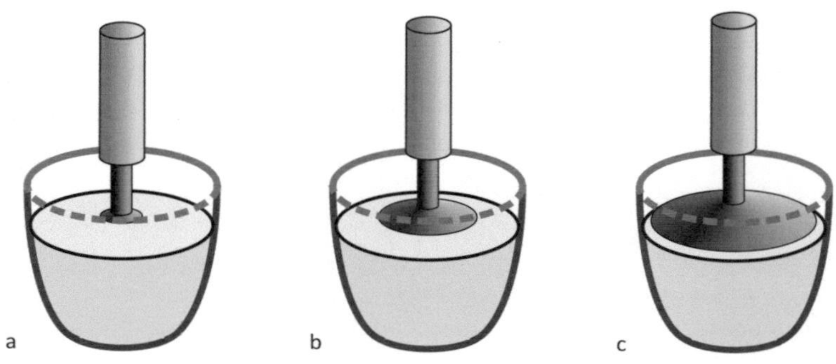

Figure 13: Schematic sketch of the early stage of the Czochralski process, i. e., formation of the crystal's shoulder. Starting diameter increase (**a**) yields a higher heat loss via radiation that further decreases interface temperature, resulting in a fast growth of the diameter (**b**). This self-amplifying process may end in a catastrophic failure when the rotating crystal touches the crucible (**c**).

Spontaneous nucleation

In the usual textbook descriptions of the Czochralski method, a native seed is assumed to start crystal growth. This is by far the best way if single-crystalline material from a preceding growth run is available. But, doing crystal growth of *new* intermetallic phases, i. e., of alloys or compounds that have never been grown before as single crystals large enough for seed preparation, one faces the chicken-and-egg problem. Unfortunately, there is no single trick that can be applied for all crystals; instead we can only suggest a few approaches that have yielded good results with various intermetallic phases.

a. In growing the first Al-based decagonal quasicrystals without having a suitable seed, we decided to use spontaneous nucleation on a tip of an alumina rod [34]. But, Al-rich melts do not wet alumina when such a tip touches the melt for the first time. After several hours of forced contact by deeply submerging the tip into the high-temperature melt, the alumina surface is modified in some way and, when slowly pulling the tip out of the melt, it is wetted and, thus, able to locally lift a tiny part of the liquid that forms a meniscus as schematically sketched in Figure 14(a)–(b). In the tiny region of the raised melt meniscus, conditions for heterogeneous nucleation at the sharp tip of the alumina rod are much preferred to all other sites at the surface of the melt. The alumina tip acts as a “cold” point, and heat loss by radiation into the colder environment is promoted as well. Now the set point of the temperature controller has to be slowly decreased until the first crystallization can be observed. Due to the sharp tip, it is very likely that only one nucleus forms making single-crystal growth possible. Unfortunately, in a few cases it happened that later on a too large crystal attached to the sharp tip had fallen into the melt. That is why we often preferred a compromise and used a slightly truncated tip of the alumina rods.

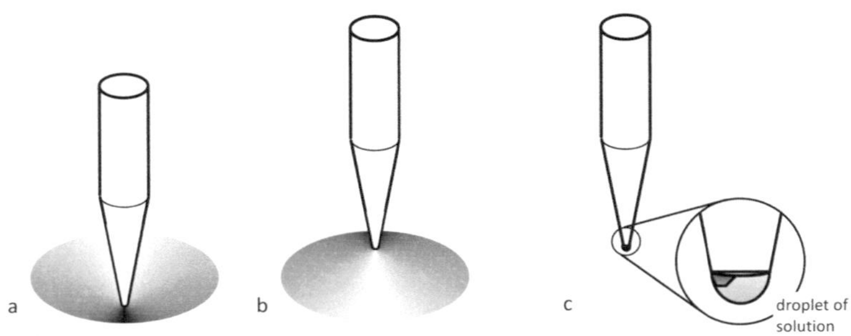

Figure 14: Surface depression due to capillary forces (**a**) and lifted surface after wetting the tapered alumina rod (**b**). After wetting the melt, spontaneous nucleation within a hanging droplet (**c**) may be used to get a native seed.

b. Occasionally, we observed that such a truncated tip of the alumina rod was detached from the melt after the above-mentioned wetting was successful. In these cases, after separation from the rest of the melt, a hanging droplet was still attached to the alumina tip (see Figure 14(c)). Having separated this tiny amount of the average composition of the solution, it can be slowly withdrawn. Pulling the alumina rod to lower temperatures according to a larger distance to the melt surface makes the solution droplet become partially crystallized. This cooling-down procedure is restricted to the tiny volume of the hanging droplet. Probably only one small crystal that is embedded in the remaining solution nucleates. When it again has been brought into contact with the melt reservoir at exact liquidus temperature, single-crystal growth starts from this preformed seed. This procedure has been successfully used with some Al-rich compounds [39].
c. Nucleation and first single-crystal growth in the Fe–Ga system was a severe problem. The Ga-rich solution with a composition next to the peritectic point of the $FeGa_3$ phase (see Figure 12) does not wet the tapered alumina rod. When spontaneous nucleation was tried by slowly cooling the bulk solution, a large number of small crystals always formed at the same time. Due to the tetragonal symmetry of $FeGa_3$, they appeared as tiny squares swimming at the surface of the melt. We never succeeded in dissolving all but one of the small crystals. Fortunately, we were able to catch one of them using the alumina tip. Once it had stuck at the end of the alumina rod, we could remove it from the melt surface and pull it to a safe position while overheating the starting melt. After carefully approaching the equilibrium temperature, the tip with the tiny seed was brought again into contact to the melt. This trick could be done thanks to the good observability through the viewport using a microscope, but was successful only after repeating the whole procedure several times. Of course, such a big effort cannot be implemented in a technology of single-crystal production, but one should keep in mind that it is a possibility if it succeeded once with a new phase.
d. Although ceramic materials are considered to be the best tools in contact with intermetallic phases and melts, wetting could not be achieved in individual case. For example, in crystal growth of InPd, which is an interesting material to be studied for heterogeneous catalysis, we failed with all the above-mentioned procedures of spontaneous nucleation. In the end, we prepared a wedge from polycrystalline refractory metal in the shape of the usual seed crystals to fit into the ceramic seed holder shown in Figure 9. In choosing tantalum as a foreign "seed", its high melting point and the absence of any low-melting composition in the binary In–Ta phase diagram [76] were important considerations. Using the tantalum tip, we wetted it immediately and, by cooling slowly, spontaneous nucleation at the slightly lifted tip could be initiated. With this strategy, we were able to grow our very first InPd single crystal, which was large enough to prepare oriented seeds from for further experiments [43].

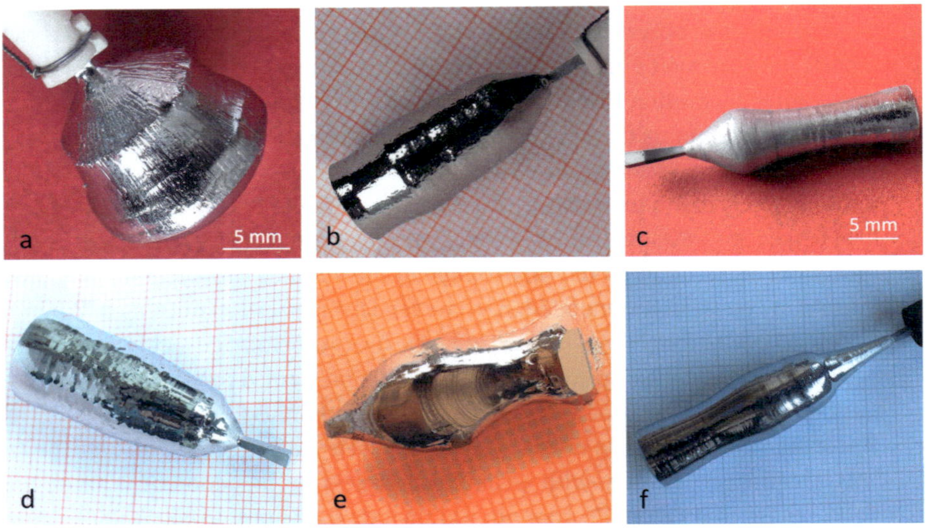

Figure 15: Some examples of Czochralski-grown intermetallic single crystals from the author's lab: tetragonal $FeGa_{3-x}Ge_x$ (**a**), trigonal Ga_3Ni_2 (**b**), orthorhombic $Al_{13}Co_4$ (**c**), cubic PdGa (**d**), monoclinic $Al_{13}(Fe,Cr)_4$ (**e**), and a decagonal $Al_{72}Co_9Ni_{19}$ quasicrystal (**f**).

5 Conclusions

The Czochralski method may be regarded as one of the most powerful techniques for single-crystal growth of intermetallic phases capable of producing well-defined samples for all kinds of basic research. Among the several advantages, easy observation during all stages of the growth process is considered to be the most important. Due to the various properties of specific intermetallics, there is no universal Czochralski apparatus that can meet the needs of all intermetallic systems with respect to accessible temperature ranges, vapor pressures, reactivity and hydrodynamic conditions. Thus, each Czochralski growth system should be well-adapted to the target group of specific intermetallic compounds or alloys. Figures 15(a)–(f) show a number of the intermetallic phases that were grown using the above-described Czochralski apparatus (see Figure 4).

Similar to these hardware conditions, there is no single recipe that can be applied to all materials, but single-crystal growth of a new phase will always be a process that requires a high degree of creativity. This has been explained, e. g., when discussing the question how to get the very first nucleus of a totally new compound. Nevertheless, a couple of problems have to be tackled in all Czochralski growth runs. These problems and possible solutions have been treated in this chapter in order to assist young colleagues in making their own first steps in the exciting field of Czochralski single-crystal growth of intermetallic phases. With any new crystal, one should be prepared to be surprised when looking through the viewport. This may be illustrated

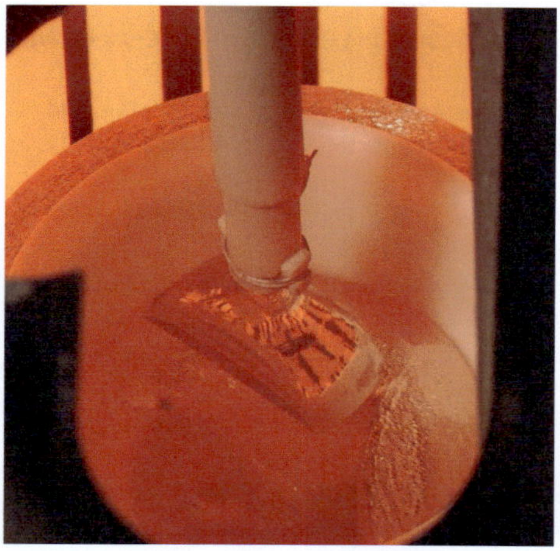

Figure 16: Photograph through the viewport during Czochralski growth of the orthorhombic Ga_3Pd_5 phase with morphologically stable (010) faces (pulling parallel to [001]).

by the last photograph in this chapter (see Figure 16) that shows a very recent single-crystal growth experiment of orthorhombic Ga_3Pd_5 under conditions that nominally ensure rotational symmetry.

Bibliography

[1] Czochralski, J. Ein neues Verfahren zur Messung der Kristallisationsgeschwindigkeit der Metalle. Z Phys Chem 1918, 92, 219–921.

[2] Teal GK, Little JB. Growth of germanium single crystals. Phys Rev 1950, 78, 647.

[3] Uecker R. The historical development of the Czochralski method. J Cryst Growth 2014, 401, 7–24.

[4] Scheel, HJ. Historical introduction. In: Hurle DTJ, ed. Handbook of Crystal Growth, Vol 1, Amsterdam, North-Holland Publishing Co, 1993, 1–42.

[5] Langsdorf A, Assmus W. Growth of large single grains of the icosahedral quasicrystal ZnMgY. J Cryst Growth 1998, 192, 152–6.

[6] Takke R, Assmus W. Growth and characterization of $CePd_3$ single crystals. J Cryst Growth 1980, 49, 97–101.

[7] Talik E, Oboz M. Czochralski method for crystal growth of reactive intermetallics. Acta Phys Pol 2013, 124, 340–3.

[8] Reed TB, Pollard ER. Tri-arc furnace for Czochralski growth with a cold crucible. J Cryst Growth 1968, 2, 143–7.

[9] Szlawska M, Kaczorowski D. Single-crystal growth of f-electron intermetallics in a tetra-arc Czochralski furnace. Acta Phys Pol 2013, 124, 336–9.

[10] Tiller WA, Jackson KA, Rutter RW, Chalmers B. The redistribution of solute atoms during the solidification of metals. Acta Metall 1953, 1, 428–37.

[11] Scheel HJ. Accelerated crucible rotation technique: a novel stirring technique in high-temperature solution growth. J Cryst Growth 1972, 13/14, 560–5.

[12] Gelfgat YuM. Rotating magnetic fields as a means to control the hydrodynamics and heat/mass transfer in the process of bulk single crystal growth. J Cryst Growth 1999, 198/199, 165–9.

[13] Schwerin J, Müller D, Kiese S, Gille P. Single crystal growth in the Ga-Pd system. J Cryst Growth 2014, 401, 613–6.
[14] Oliver DW, Brower GD, Horn FH. Cold metal crucible system for synthesis, zone refining, and Czochralski crystal growth of refractory metals and semiconductors. J Cryst Growth 1972, 12, 125–31.
[15] Bressel B, Chevalier B, Etourneau J, Hagenmuller P. Czochralski crystal growth of rare earth tetraborides. J Cryst Growth 1979, 47, 429–33.
[16] Wernick JH, Hull GW, Geballe TH, Bernardi JE, Buehler E. Czochralski growth of A15 structure intermetallic compounds. J Cryst Growth 1979, 47, 73–6.
[17] Menovsky A, Franse JJM. Crystal growth of some rare earth and uranium intermetallics from the melt. J Cryst Growth 1983, 65, 286–92.
[18] Nørlund Christiansen A. Preparation and characterization of Mo_3Si and Mo_5Si_3. Acta Chem Scand A 1983, 37, 519–22.
[19] Thomas O, Senateur JP, Mader R, Laborde O, Rosencher E. Molybdenum disilicide crystal growth, thermal expansion and resistivity. Solid State Commun 1985, 55, 629–32.
[20] Abell JS, Jones DW. Crystal growth and characterization of some RAl_2 intermetallic compounds ($R \equiv$ Pr, Nd, Gd, Tb). J Less-Common Met 1986, 115, 331–6.
[21] Tenaud P, Chamberod A, Vanoni F. Texture in Nd-Fe-B magnets analysed on basis of the determination of $Nd_2Fe_{14}B$ single crystals easy growth axis. Solid State Commun 1987, 63, 303–5.
[22] Brown SA, Howard BK, Brown SV, Julian SR. A new design for a UHV compatible Czochralski crystal growth system. Rev Sci Instrum 1990, 61, 2427–9.
[23] Talik E, Winiarska A, Winiarski A. Crystal growth of R_3Ni intermetallic compounds by Czochralski method from a levitated melt ($R =$ Y, Gd, Tb, Dy, Ho, Er). J Cryst Growth 1993, 126, 718–20.
[24] Haga Y, Yamamoto E, Kimura N, Hedo M, Ohkuni H, Õnuki Y. High-quality single crystal growth of uranium-based intermetallics. J Magn Magn Mater 1998, 177–181, 437–8.
[25] Talik E, Klimczak M, Winiarski A, Troć R. Gd_7T_3 ($T =$ Rh, Pd) intermetallics crystal growth. J Cryst Growth 2005, 283, 547–52.
[26] Lipińska-Chwałek M, Balenetskyy S, Thomas C, Roitsch S, Feuerbacher M. Single-crystal growth of the complex metallic alloy phase β-Al-Mg. Intermetallics 2007, 15, 1678–85.
[27] Talik E, Klimczak M, Winiarski A, Troć R. Crystal growth of $Gd_{7-x}Y_xPd_3$ intermetallics. J Cryst Growth 2008, 310, 1886–90.
[28] Feuerbacher M, Thomas C, Roitsch S. Single-crystal growth of the complex metallic alloy phase $Mg_{32}(Al,Zn)_{49}$. Intermetallics 2008, 16, 943–7.
[29] Swets DE. Growth of single crystals of $Nd_2Fe_{14}B$. J Cryst Growth 1986, 75, 277–80.
[30] Norby, P, Nørlund Christiansen A. Preparation and structure of Al_3Ti. Acta Chem Scand A 1986, 40, 157–9.
[31] Yokoyama Y, Miura T, Tsai AP, Inoue A, Masumoto T. Preparation of a large $Al_{70}Pd_{20}Mn_{10}$ single-quasicrystal by the Czochralski method and its electrical resistivity. Mater Trans, JIM 1992, 33, 97–101.
[32] Boudard M, Bourgeat-Lami E, de Boissieu M et al. Production and characterization of single quasicrystals of the Al-Pd-Mn icosaedral phase. Philos Mag Lett 1995, 71, 11–9.
[33] Bi YJ, Abell JS. Aluminum segregation of TiAl during single crystal growth. Scr Mater 1997, 37, 743–52.
[34] Gille P, Dreier P, Gräber M, Scholpp T. Large single-grain AlCoNi quasicrystals grown by the Czochralski method, J Cryst Growth 1999, 207, 95–101.
[35] Yokoyama Y, Note R, Fukaura K, Sunada H, Hiraga K, Inoue A. Growth of a single $Al_{64}Cu_{23}Fe_{13}$ icosahedral quasicrystal using the Czochralski method and annealing removal of strains. Mater

Trans, JIM 2000, 41, 1583–8.
[36] Gille P, Dreier P, Barz, RU. Growth of decagonal AlCoCu quasicrystals onto AlCoNi seeds. J Alloys Compd 2002, 342, 7–12.
[37] Bauer B, Meisterernst G, Härtwig J, Schenk T, Gille P. Czochralski growth and X-ray topographic characterization of decagonal AlCoNi quasicrystals. Philos Mag 2006, 86, 317–22.
[38] Meisterernst G, Zhang L, Dreier P, Gille P. Understanding Czochralski growth of decagonal AlCoCu. Philos Mag 2006, 86, 323–8.
[39] Gille P, Bauer B. Single crystal growth of $Al_{13}Co_4$ and $Al_{13}Fe_4$ from Al-rich solutions by the Czochralski method. Cryst Res Technol 2008, 43, 1161–7.
[40] Gille P, Ziemer T, Schmidt M, Kovnir K, Burkhardt U, Armbrüster M. Growth of large PdGa single crystals from the melt. Intermetallics 2010, 18, 1663–8.
[41] Bauer B, Gille P. Crystal growth of Al-rich complex metallic phases in the system Al-Cr-Fe using the Czochralski method. Z Anorg Allg Chem 2011, 637, 2052–8.
[42] Gille P, Bauer B, Hahne M, Smontara A, Dolinšek J. Single crystal growth of Al-based intermetallic phases being approximants to quasicrystals. J Cryst Growth 2011, 318, 1016–20.
[43] Hahne M, Gille P. Single crystal growth of the intermetallic compound InPd. J Cryst Growth 2014, 401, 622–6.
[44] Wagner-Reetz M, Kasinathan D, Schnelle W et al. Phonon-drag effect in $FeGa_3$. Phys Rev B 2014, 90, 195206.
[45] Kliemt K, Krellner C. Crystal growth by Bridgman and Czochralski method of the ferromagnetic quantum critical material $YbNi_4P_2$. J Cryst Growth 2016, 449, 129–33.
[46] Wencka M, Pillaca M, Gille P. Single crystal growth of Ga_3Ni_2 by the Czochralski method. J Cryst Growth 2016, 449, 114–8.
[47] Doležal P, Rudajevová, Vlášková K et al. Czochralski growth of $LaPd_2Al_2$ single crystals. J Cryst Growth 2017, 475, 10–20.
[48] Tsai AP, Inoue A, Yokoyama Y, Masumoto T. New icosahedral alloys with superlattice order in the Al-Pd-Mn system prepared by rapid solidification. Philos Mag Lett 1990, 61, 9–14.
[49] Tsai AP, Inoue A, Masumoto T. A stable decagonal quasicrystal in the Al-Cu-Co system. Mater Trans, JIM 1989, 30, 300–4.
[50] He LX, Wu YK, Meng XM, Kuo KH. Stable Al-Cu-Co decagonal quasicrystals with decaprismatic solidification morphology. Philos Mag Lett 1990, 61, 15–9.
[51] Tsai AP, Inoue A, Masumoto T. Stable decagonal Al–Co–Ni and Al–Co–Cu quasicrystals. Mater Trans, JIM 1989, 30, 463–73.
[52] Audier M, Durand-Charre M, de Boissieu M. AlPdMn phase diagram in the region of quasicrystalline phases. Philos Mag B 1993, 68, 607–18.
[53] Gödecke T. Schmelzfläche und Phasengleichgewichte mit Beteiligung der Schmelze im System Al-AlCo-AlNi. Z Metallkd 1997, 88, 557–69.
[54] Grushko B, Wittenberg R, Bickmann K, Freiburg C. The constitution of aluminium-cobalt alloys between Al_5Co_2 and Al_9Co_2. J Alloys Compd 1996, 233, 279–87.
[55] Gille P, Barz RU, Zhang LM. Growth of decagonal AlCoNi and AlCoCu quasicrystals by the Czochralski method. In: Trebin HR, ed. Quasicrystals. Structure and Physical Properties. Weinheim, Wiley VCH, 2003, 73–87.
[56] Zhang B, Gramlich V, Steurer W. $Al_{13-x}(Co_{1-y}Ni_y)_4$, a new approximant of the decagonal quasicrystal in the Al–Co–Ni system. Z Kristallogr 1995, 210, 498–503.
[57] Grin J, Burkhardt U, Ellner M, Peters K. Crystal structure of orthorhombic Co_4Al_{13}. J Alloys Compd 1994, 206, 243–7.
[58] Grin J, Burkhardt U, Ellner M, Peters K. Refinement of the Fe_4Al_{13} structure and its relationship to the quasihomological homeotypical structures. Z Kristallogr 1994, 209, 479–87.
[59] Popčević P, Smontara A, Ivkov J et al. Anisotropic physical properties of the $Al_{13}Fe_4$ complex

intermetallic and its ternary derivative $Al_{13}(Fe,Ni)_4$. Phys Rev B 2010, 81, 184203.
[60] Edshammar LE. The crystal structure of Ru_4Al_{13}. Acta Chem Scand 1965, 19, 2124–30.
[61] Deng DW, Mo ZM, Kuo KH. Crystal structure of the orthorhombic $Al_4(Cr,Fe)$ approximant of the Al-Cr-Fe decagonal quasicrystal. J Phys Condens Matter 2004, 16, 2283–96.
[62] Dolinšek J, Vrtnik S, Smontara A et al. Origin of anisotropic nonmetallic transport in the $Al_{80}Cr_{15}Fe_5$ decagonal approximant. Philos Mag 2008, 88, 2145–53.
[63] Smontara A, Smiljanić I, Ivkov J et al. Anisotropic magnetic, electrical and thermal transport properties of Y-Al-co-Ni decagonal approximant. Phys Rev B 2008, 78, 104204.
[64] Dolinšek J, Komelj M, Jeglič P et al. Anisotropic magnetic and transport properties of the orthorhombic Al13Co4. Phys Rev B 2009, 79, 184201.
[65] Alarcón Villaseca S, Ledieu J, Serkovic Loli LN et al. Structural investigation of the (001) surface of the Al_9Co_2 complex metallic alloy. J Phys Chem C 2011, 115, 14922–32.
[66] Ivkov J, Popčević P, Stanić D et al. Anisotropic Hall effect in $Al_{13}TM_4$ approximants. Philos Mag 2011, 91, 2739–45.
[67] Armbrüster M, Kovnir K, Friedrich M et al. Al13Fe4 as a low-cost alternative for palladium in heterogeneous hydrogenation. Nat Mater 2012, 11, 690–3.
[68] Gödecke T, Ellner M. Phase equilibria in the aluminum-rich portion of the binary system Co-Al and in the cobalt/aluminum-rich portion of the ternary system Co-Ni-Al. Z Metallkd 1996, 87, 854–64.
[69] Mullin JB. Liquid encapsulation and related technologies for the Czochralski growth of semiconductor compounds. In: Nishinaga T, ed. Handbook of Crystal Growth. 2nd ed. Amsterdam, Elsevier, 2015, Vol 2A, 105–30.
[70] Okamoto H. Ga–Pd (Gallium-Palladium). J Phase Equilib Diff 2008, 29, 466–7.
[71] von Kármán T. Über laminare und turbulante Reibung. Z Angew Math Mech 1921, 1, 233–52.
[72] Hurle DTJ. The dynamics and control of Czochralski growth. In: Dryburgh PM, Cockayne B, Barraclough KG, eds. Advanced Crystal Growth. New York, Prentice Hall Int, 1987, 97–123.
[73] Okamoto H. Fe–Ga (Iron-Gallium). J Phase Equilib Diff 2004, 25, 100.
[74] Häussermann U, Boström M, Viklund P, Rapp Ö, Björnängen T. $FeGa_3$ and $RuGa_3$: Semiconducting intermetallic compounds. J Solid State Chem 2002, 165, 94–9.
[75] Fritton M, Gille P. Unpublished result. 2013.
[76] Okamoto H. The In–Ta (Indium-Tantalum) system. Bull Alloy Phase Diagrams 1988, 9, 56–8.

Andrey Prokofiev

Floating zone growth of intermetallic compounds

1 Introduction

Floating zone (FZ) has its root idea in the zone-melting technique reported by W. G. Pfann in 1952 [1]. Germanium was heated in one end of a horizontal silica boat by radiation heating, and the narrow molten zone went through the boat to the other end. The idea of the method was a purification of germanium. Since the technique was not effective for the much more reactive silicon that was contaminated by the boat material, a crucible-free modification was designed. In this invention patented by H. C. Theuerer in 1952, the molten zone was suspended between two solid rods of the same composition as the melt [2]. Independently, the technique was invented by R. Emeis [3]. The first silicon crystals were produced by this technique by P. H. Keck and M. J. E. Golay [4]. Since that time the floating-zone technique has been established as one of the most effective methods for the growth of high-quality materials of various composition classes. This chapter is devoted to the floating-zone growth of intermetallic compounds. There are a few published reviews on the floating zone. The historical development of the method is considered by A. Mühlbauer [5]. Detailed overviews of the floating zone technique accentuated mainly on silicon are done by Lüdge et al. [6], Riemann and Lüdge [7] and by Muiznieks et al. [8]. Koohpayeh et al. [9] and Dabkowski et al. [10] reviewed this method with emphasis on oxide materials. Hermann et al. discussed flow processes in the liquid intermetallic zone heated by a high-frequency electromagnetic inductor of a special type [11]. Intermetallic compounds in relation to the FZ technique have their own specifics that are determined by a strong affinity to oxygen (and often to nitrogen), much higher electrical and substantially higher thermal conductivities, a higher density, and an absence of transparency, but a higher reflectivity of the melt surface.

2 Main advantages of the floating-zone technique

The main feature of the FZ is that it is a fully container-free technique. Neither hot-melt nor hot-crystal parts contact with a foreign material during the process. Therefore, it is the most universal technique for all materials. The absence of crucible material avoids any problem involving the growth atmosphere. Any type of atmosphere (oxidizing, inert, reducing atmosphere) compatible with the grown material can be applied, if necessary.

https://doi.org/10.1515/9783110496789-005

The temperature gradient on the crystal-melt interface is normally higher in the absence of a crucible since the latter contributes to the thermal flow. This can have both positive and negative effects. Sometimes a high-temperature gradient leads to thermal stress that in turn results in cracking of the grown crystal or the appearance of dislocations. This risk is, however, lower for intermetallic compounds due to their higher plasticity in comparison with oxide materials. On the other hand, the higher temperature gradient provides favorable conditions for incongruent or flux growth where the risk of unstable growth due to constitutional undercooling is high. The grown crystal experiences faster cooling due to the higher temperature gradient, which sometimes blocks metastable phases from transformations [12].

The crystal after the solidification is free standing, so mechanical stress due to different expansion coefficients of the crystal and the crucible material is naturally avoided.

For good control of the growth process, it is not necessary to know the melting point of the material since the melting process can be viewed directly by a video camera. This is an especially valuable technique for new phases whose thermal behavior has not been investigated.

An incongruent crystallization (e. g., flux growth or melt growth with a segregation-induced melt off-stoichiometry) usually causes additional problems when using techniques such as Bridgman or Czochralski because it is accompanied by a continuous variation of the melt composition and temperature. This in turn results in crystal inhomogeneity. A FZ growth can be performed as a steady-state process. Any zone-melting technique can be considered as transport of the material across the liquid zone regardless of the liquefaction mechanism (either melting or dissolution in a solvent). If the dissolving and the crystallizing phases are identical, the composition of the liquid zone and, hence, of the crystal (although different) are maintained in a constant state (Figure 1). For example, owing to this favorable circumstance, homogeneous single crystals of solid solutions or of phases with temperature-dependent existence range can be grown.

The technique has a few drawbacks. It is hard to grow materials with low surface tension in the liquid state. The melt zone is mechanically unstable, especially if the density of the material is high. Also, materials wetted well by their own melt cause problems because of the creep of the melt from the zone.

Finally, large crystals can hardly be grown since the mechanical stability of the zone imposes limitations on the radius of the crystal.

3 Experimental set-up of the FZ technique

The technique consists in the melting of a narrow zone in a vertically free-standing polycrystalline rod of material by a local heater and moving the molten zone along the rod (Figure 2). The zone is held by surface tension, and the main operation task is to

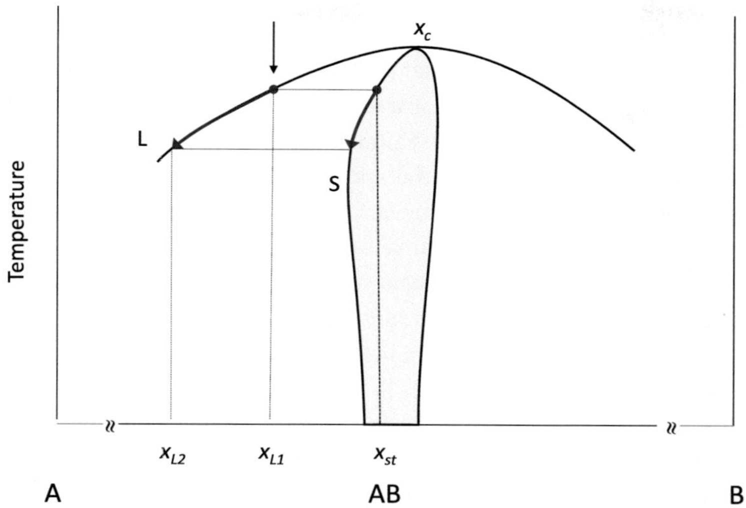

Figure 1: Comparison of incongruent crystallization in the FZ and in another technique (e. g., Bridgman, Chochraski). The congruent point x_c is shifted from the stoichiometric composition x_{st} of the compound AB. From the melt with composition x_{L1}, the stoichiometric phase first crystallizes. Bridgman or Chochraski technique: the segregation effect leads to the shift of the melt composition to x_{L2}. Correspondingly, the composition of the crystals shifts away from the stoichiometry x_{st} in course of crystallization. FZ: if the feed rod has the stoichiometric composition x_{st} and the liquid zone does x_{L1} the compositions of the zone and of the growing crystal remains constant during the whole process.

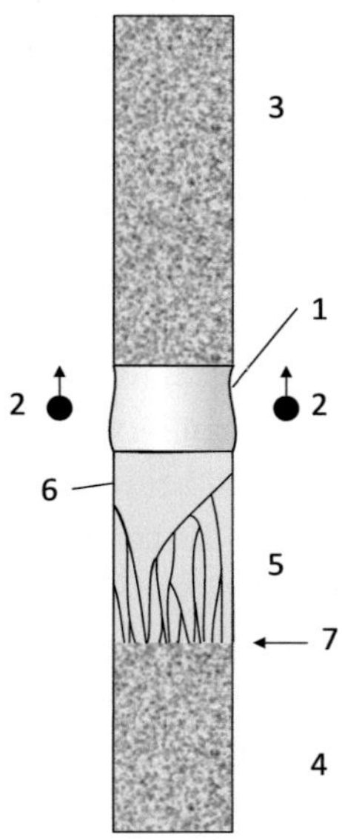

Figure 2: Crystal growth by floating-zone (FZ) technique. Heater 2 melts a liquid zone 1 which is held by surface tension between polycrystalline feed rod 3 and seed rod 4. The lower rod may be either single crystalline or polycrystalline. In the latter case, a grain selection stage is necessary (5) before the single crystal 6 begins to grow. 7 – starting point of the melting process.

maintain the zone mechanically stable against vibrations and down spilling. The rod held on its two ends is divided into two parts after the formation of the liquid zone. The part of the rod in the motion direction serves as a feed rod, whereas the other part is the seed rod whereon the single crystal is formed after an initial seed-selection stage. If the seed rod was *ab initio* a single crystal, then a single-crystal growth with the desired orientation begins immediately. The growth rate is controlled by the pulling velocity of the heater (or by the velocity of the rods relative to the fixed heater). The velocities of the seed and the feed rods may be different, and so a control of the diameter of the growing crystal becomes possible. Through this control, the necking is feasible on the starting stage of the growth in order to provide a better seed selection or reduction of the dislocation density.

The free surface of the melt is large relative to the zone volume; this may have both positive and negative implication on the crystal purity and composition. While the exposure of the intermetallic melt to oxygen is large, on the other hand, volatile impurities (including intrinsic ones) can be easily removed from the surface. The main control parameter to provide a stable zone is the heating power. Temperature control is difficult (see the next sections). In special cases, the temperature measurement of the liquid zone has been achieved using a pyrometer. The measurement is performed during short interruptions of the heating light beam by a shutter when using an optical heating system [13, 14]. However, a visual control of the form and the state of the surface of the zone is perfect, making the temperature control unnecessary for practical crystal growth.

For better homogenization of the composition in the liquid zone, the rotation of the upper and the lower rods is usual. However, in image furnaces, the concentration of the heating energy is high at the optical axis, and a small ex-centricity of the rotation leads to high temperature oscillation on the growth surface. This necessitates very high requirements for the rod centering and regular cylindrical shape of the rotating rods.

The cylindrically symmetric surface of the molten zone has the maximal curvature in the horizontal plane (being convex); a possible concavity in the vertical direction is much smaller. The dominantly convex surface implies an external (capillary) pressure that is reciprocally proportional to the zone radius. This pressure withstands the destructive hydrostatic pressure, which is proportional to the height of the zone. The balance between these two quantities mainly determines the stability of the FZ. This implies a limited size of the zone and hence of the crystal.

The maximal height of the FZ is linearly dependent on the zone radius for narrow rods, approaching, with an increase of the diameter, a constant value given by

$$h_{\max} = \lambda \cdot \sqrt{\frac{\sigma}{\rho g}}$$

where ρ is the density and σ is the surface tension of the melt, h is the height and R is the radius of the zone, g is the gravitational acceleration, and the dimensionless factor λ is ≈ 3 [15].

The surface tension of metallic melts is typically substantially greater than that of oxides and ionic salts [16]. However, some intermetallic compounds, especially of platinum group elements, have rather high densities, which makes (in combination with low surface tension) the molten zone especially unstable [17].

For the purpose of growth of large crystals, the needle-eye technique has been developed: the molten-zone radius remains small in the upper part whereas the seed and the feed rods have substantially larger radii (Figure 3). This technique is used in industry for the growth of oxygen- and dislocation-free silicon. However, this is possible due to a favorable confluence of the high surface tension and the low density of silicon. A special form of a one-turn inductor having a bowl shape was developed. The wetting of the seed rod by the melt is also an important factor: a very small contact angle may result in creeping of the melt along the solid rod near the melting zone and thereby in slow leakage of the melt from the hot zone, which is another type of the zone instability.

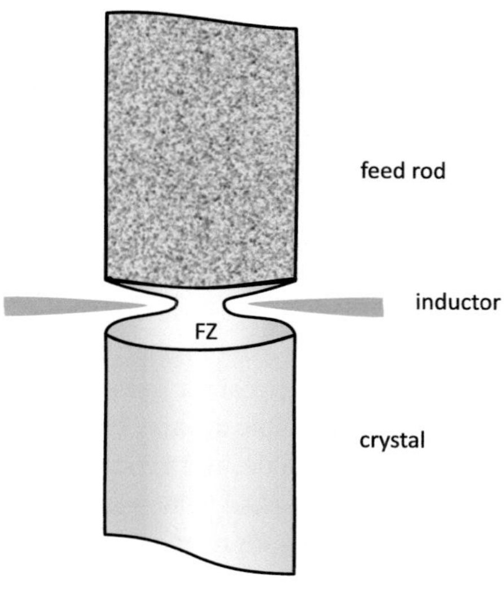

Figure 3: Needle-eye technique enables growth of large crystals of materials with low density and high surface tension of the melt. The diameter of the FZ is substantially smaller than that of the seed and feed rods.

4 Kinds of zone heating

The main advantage of FZ is the absence of any contact of the melt with other materials. This also makes compulsory a contactless heat transfer to the melt. In the case of non-conducting materials (e. g., a majority of oxides), only optical heating is a possibility, whereas for intermetallics quite a number of heating possibilities exist.

Radiofrequency (RF) heating

For the RF heating, a narrow (mainly a single turn) inductor is used (Figure 4). The electromagnetic (EM) field should couple directly to the material. In order to improve the coupling and to increase the density of the EM field, especially for melting of refractory materials, a concentrator of the EM power may be used [18].

Figure 4: High-frequency one-turn inductor for FZ melting of conducting materials. Adopted from Ref. Lüdge A. et al. [6].

Electron beam (EBFZ)

This method of heating is mainly applied to refractory metals like tungsten or molybdenum and their alloys. The electron gun is a system of electrostatic lenses. An arrangement of the electrodes of the electron gun makes it possible to vary the electron beam focusing from diffuse to sharp. The feed rod serves as anode, and a circular tungsten thermionic filament is usually used as a cathode. The applied voltage is of about 20–30 kV [19]. The advantage of this method is the possibility for many-channel control of the e-beam as to intensity and direction, which makes it feasible to form the desired temperature gradient in the hot zone. As a disadvantage, the necessity of vacuum and not the more beneficial case of an inert atmosphere in the growth chamber can be identified. Thus, this technique is not applicable for melts with relatively high vapor pressure.

Since an electron beam is only possible in a vacuum, EBFZ is applied to elementary refractory metals with very low vapor pressure (Zr, Nb, Ta, Mo, W) and their alloys. The vacuum provides, however, in this case a possibility of an additional purification through evaporation of volatile impurities.

The crystallographic perfection of the grown metal crystals is reported to be not high. A characteristic blocky structure and a high dislocation density make such crystals unsuitable, for example, for research of channeling high-energy particles or use in a number of other subtle physical experiments. However, among all other techniques for growth of refractory metallic crystals, the EBFZ provides crystals of the best quality [20].

Arc heating

Arc heating can suffice for intermetallic materials due to their high electric conductivity. However, this is a rather rarely used technique, and only a few attempts have been reported [21, 22]. An arc operates in an inert atmosphere, thus allowing the growth of materials with a perceptible vapor pressure. A substantial disadvantage is the contamination of the melt by the electrode material.

Optical heating

This is the most universal kind of heating suitable for all classes of materials. The essential parts of the equipment are power sources of light (lamps) and a focusing system. The latter consists of ellipsoidal or, rarely, paraboloid mirrors. The lamp is situated at one of the foci of the ellipsoid mirror, and the rod of the material is placed at the other focus. There are two types of setups—with vertical and horizontal optical axes (Figure 5). An essential part of the unit is a transparent tube forming the growth chamber that is filled with an inert gas for protection from oxidation (Fig. 7).

Horizontal axis (axes) furnace

Most of commercial image furnaces operate in the horizontal-optical-axis mode. Unlike the vertical-axis variant where the heating is all-sided, it is either two-sided (two-mirror furnace) or four-sided (four-mirror furnace). The sample rod stands perpendicular to the optical axis (axes) of the two (four) mirrors. The lamps are situated in one focus of the mirrors, and the other foci coincide. The sample is heated in this common focus of the optical system.

Vertical-axis furnace

In this variant, two ellipsoid mirrors are used with the material rod standing *along* the optical axis of both mirrors. The lamp is situated in the focus of one mirror and the sample in the focus of the other. The second foci of the two mirrors coincide. The light spot on the sample is all-sided, belt-like, having a circular symmetry, which makes the heating very homogeneous. This concept was designed or used for the growth of various kind of materials (e. g., [23–26]). This set-up provides sharper focusing and narrower light profiles on the crystal surface, resulting in very high efficiency of the

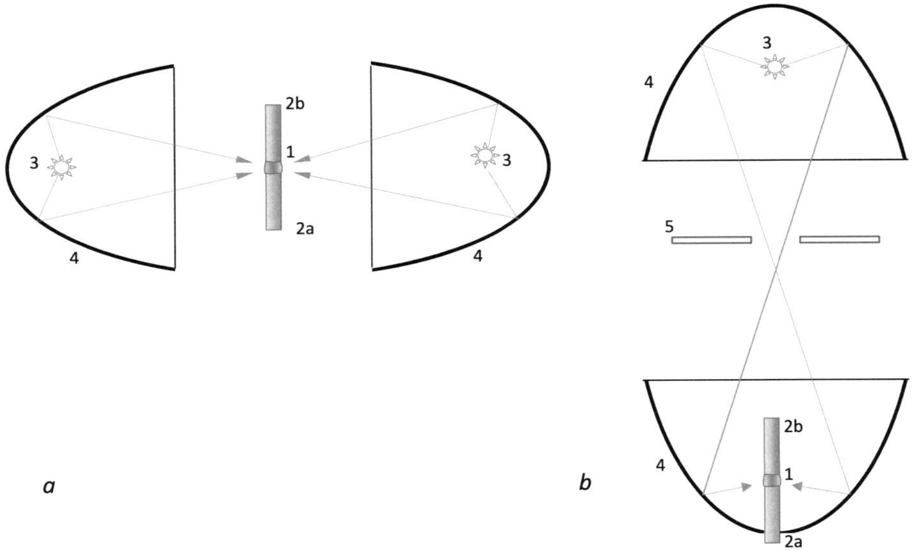

Figure 5: Two types of floating-zone furnaces with optical heating: *a*) with horizontal and *b*) with vertical optical axis. 1—FZ, 2a and 2b—seed and feed rods, 3—lamps, 4—elliptical mirrors, 5—light power shutter.

radiation flux. Thus, only one 5-kW xenon lamp is required for melting refractory oxides with melting temperatures about 2800 °C, whereas four 3-kW xenon lamps (12 kW in total) are necessary for achieving similar temperatures in four-mirror horizontal optical furnaces [27]. This practical result was interpreted in view of a more effective focusing of the light flux emitted by the arc lamp and a narrower light profile on the crystal surface. Another advantage is the possibility to use shorter silica glass or sapphire tubes for a growth chamber, which is very important for application of high gas pressures (up to 30 MPa) (Fig. 7) [28].

Tilted mirrors

In addition to the more widely used horizontal and (more rarely) vertical-optical axes, other axis orientations may be practicable. The aim of such tilted mirrors is the control of the shape of solid–liquid interface formed during the floating zone growth. It was revealed that the convexity of the interface can be controlled by adjusting the tilting angle. A less convex interface can more effectively stabilize the molten zone and can make it easier to grow larger crystals [29].

Halogen or xenon lamps are widely used as primary light sources for elliptical mirror systems. Lasers as light sources have more limited application. The bundled laser beam needs a different type of defocusing/focusing optics. The monochromatic radiation makes, however, pyrometrical temperature measurement much more feasible.

Even in white-light optical heating by a xenon or halogen lamp, a pyrometric measurement is possible. A patented technique permits temperature measurements on the rod and melt surface during the growth process [30]. The light flux is interrupted for some milliseconds, and a two-color pyrometer measures the unbiased radiation. Due to a moveable pyrometer head, temperature profiles can be recorded vertically (feed rod-melt-crystal) and horizontally (along the molten zone). Figure 6 shows a typical vertical temperature profile along the rod axis.

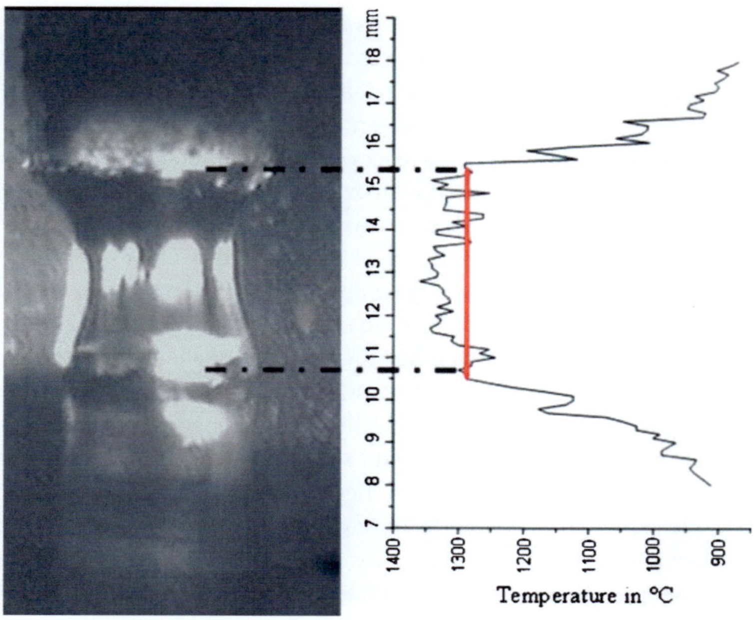

Figure 6: Pyrometric temperature measurement in an optical FZ furnace with a white light source. The measurement occurs during very short interruptions of the heating light beam. Right image—measured temperature profile along the FZ and in its vicinity. Adopted from D. Lindackers et al. Highlights 2008. Jahresbericht IFW Dresden (Leibniz Institute for Solid State and Materials Research).

Usually the FZ technique is characterized by a stronger concentration of heating energy and thus by a higher temperature gradient in the hot zone. Due to a higher thermal conductivity of intermetallic materials in comparison with oxides or similar phases, the temperature gradient is usually much lower. The high temperature gradient facilitates stable crystal growth. This is the well-known issue in respect to the constitutional undercooling. On the other hand, it leads often to crack formation, especially for brittle non-metallic crystal phases. Also a large energy concentration can result in larger temperature oscillations in case of a slight eccentricity of the rod rotation. Therefore special defocusing is done in image furnaces, or light sources with lower radiance are preferred (e. g., halogen lamps with large filaments instead of small-source arc lamps).

Due to the high surface-to-volume ratio of liquid zone two problems—melt oxidation and melt evaporation—may have an increased impact on the growth process and the quality of the grown crystals.

5 The problem of oxidation of melt

Metallic elements forming intermetallic compounds have very high affinity to two atmosphere gases, namely, oxygen and nitrogen. Because of the enhanced role of the melt surface, special provisions against the melt oxidation should be made, especially for materials containing elements with high oxygen affinity, such as alkaline, alkali earth and rare earth elements. These elements react with oxygen preferentially, which leads not only to contamination of the crystal with oxide-phase inclusions but also to a shift of the composition of the main phase because of the depletion of the highly active element. For many transition elements, an addition of hydrogen ($< 4\,\%$) to the growth atmosphere is helpful. However, this is completely useless in the case of the above-mentioned metals with the highest affinity to oxygen.

Therefore, a high vacuum or/and oxygen-free inert-gas atmosphere are prerequisites of high-quality intermetallic crystals. An upgrade of the standard image furnace FZ Crystal Systems Corp with ultra-high vacuum sealing has been reported [12, 31–33]. The standard Viton O-rings were replaced by metal sealing rings. The system can be baked under evacuation before being filled with inert gas. These measures enabled reaching a final gas pressure of 10^{-9} mbar. The inert-gas atmosphere was purified additionally before entering the growth chamber.

Once the growth environment is free from atmosphere gases, one faces a further problem. The starting elementary metals should be of a special high quality. However, elements certified by the supplier as high pure (> 99.9–$99.99\,\%$) are characterized "on metal basis". However, beyond this, they may contain comparatively high amounts of oxygen, hydrogen, or nitrogen. It is a particular problem for rare earth elements which have the highest affinity to oxygen. The commercially available highly pure elementary rare earth metals may contain 0.5–2.5 % oxygen, which is either dissolved in the lattice or encapsulated in the form of metal oxide inclusions [34, 35].

Self-purification of melts by reduction to volatile species

Table 1 summarizes the FZ experiments performed over the last 15 years. The majority of the crystals are silicides and borides. Melts of intermetallic compounds containing silicon, boron or carbon possess a mechanism for a "self-purification". It implies a formation of volatile low-valence oxides of the metalloid elements like SiO, CO and BO. Removing oxygen from the melt (and thus a purification of the melt) takes place due

Table 1: Intermetallic compounds grown by FZ in 2002–2018.

Silicides, Germanides		**Borides**		**Other compounds**	
UAu_2Si_2	[66]	SmB_6	[97–101]	**Carbides**	
$LaFe_{11}Si_2$	[67]	YB_{48}	[62]	Y_2C	[126]
Lu_2PdSi_3	[68]	Ru_7B_3	[103]		
$ErPd_2Si_2$	[69]	$Ce_{1-x}Pr_xB_6$	[104]	**Chalcogenides**	
RE–Si	[49]	LaB_6	[105, 110, 116]	FeSe	[127]
Nb–Si Alloys	[70]	CeB_6	[106, 111]		
Gd_2PdSi_3	[35, 71]	PrB_6	[112]	**Heusler comp.**	
Pr_2PdSi_3	[72]	$(La,Ce)B_6$	[108, 113]	NiMnSn	[128]
R_2PdSi_3	[35, 73, 74]	$(La,Pr)B_6$	[114]	$Co_2Cr_{1-x}Fe_xAl$	[129]
UCoGe	[75]	REB_6	[56, 115, 116]	Cu_2MnAl	[12]
Tb_5Si_3	[76]	LaB_6–MB_2	[107] (M = Zr),	Fe_2TiSn	[31]
Eu_2CuSi_3	[38]		[109] (M = Ti)	Fe_2VAl	[31]
$EuCu_2Si_2$	[39]	TiB_2	[117]	Co_2FeSi	[130]
Nd_2PdSi_3	[77]	GdB_6	[118]		
$CeCu_2Si_2$	[78]	LuB_{12}	[119]	**Clathrates**	
$CeNi_2Ge_2$	[32, 79]	CrB_2	[117, 120]	$Ba_8(CuGa)_5Ge_{41}$	[51]
Pr_5Si_3	[80, 81]	ZrB_2	[117, 120]	$Ba_8Au_5Si_{41}$	[51]
Nb–Si	[82]	ZrB_2–MeB_2	[122]	$(Ba,RE)_8Au_6Si_{40}$	[52]
$R_2Fe_3Si_5$	[83]	ReB_2	[123]		
Mo_3Si	[84]	Ru–Al–CrB_2	[124]	**Borocarbides,**	
$Ce_2Pd_xCo_{1-x}Si_3$	[85]	YbB_{66}:Nb	[125]	**borosilicides**	
Er_2PdSi_3	[86]			B_4C-MeB_2	[131, 132]
Cr_3Si	[87]			$YB_{41}C_{1.3}$	[102]
$CeSi_2$	[88]			$Y_{10}B_7C_{10}$	[133]
Ru_2Si_3	[89, 90]			R–T–B–C	[14]
RE–Si	[91]			YNi_2B_2C	[134]
Mn_3Si	[92, 32]			$RENi_2B_2C$	[135, 43, 136]
MnSi:Co, Fe	[93]			$YbB_{44}Si_2$	[137]
MnSi	[94, 48]				
$CePt_3Si$	[95, 96]			**Phosphides**	
$Ce_3Pd_{20}Si_6$	[55]			Ni_2P, Fe_2P	[138]
$CeRu_4Sn_6$	[17]				
CeAuGe	[17]			**Other**	
				CePtIn	[139]
				CoNiAl	[140]
				RuAl	[42, 47]
				TiAl	[47]
				TiNb	[47]
				$Al_{0.3}CoCrFeNi$	[141]
				GaSb	[142]
				$NbFe_2$	[143]
				YCo_5	[144]
				$GdCo_5$	[144]

R – rare earth elements, T – transition elements.

to evaporation of the monoxides and their condensation on cold parts of the growth chamber. The mechanism can reduce the oxygen concentration in intermetallic melts and grown crystals by 1–2 orders of magnitude [34], though this process occurs at temperatures above 1500 °C. However, the elimination of oxygen is accompanied by a loss of some amount of silicon (carbon, boron) from the melt. Therefore, this purification process may be detrimental if the growth chamber is not tight enough, and a concentration of oxygen remains at a certain constant level. In this case, the vaporization occurs during the whole growth experiment, and it causes a permanent transport of Si (C, B) from the melt to the cold parts. This can substantially change the stoichiometry of the melt and of the grown crystal.

6 The problem of evaporation of melt

Large surface-to-volume ratio of the FZ and its small volume in regard with the growth chamber volume makes evaporation a severe problem in FZ. In case of intermetallic compounds, some constituting elements may have a high vapor pressure at high melting temperatures. A loss of such an element from the melt results not only in a shift of the melt composition but also in condensation of the metal on the wall of the silica growth chamber, which becomes opaque. For optical heating, this means a loss of heating power or/and a corrosion of the silica tube. The change in melt composition may lead to destabilization of the single-crystal growth and crystallization of phases of a different composition.

In general, an evaporation can be thermodynamically suppressed by creating a counteracting atmosphere (for example, high-pressure oxygen atmosphere for suppression of the thermal decomposition of higher oxides [36, 37]). Since it is impossible to create an equilibrium metal atmosphere in a cool-wall growth chamber (like that of oxygen for growth of oxides); an atmosphere of an inert gas of a high pressure is used. Although it is thermodynamically neutral for the gas-melt equilibrium; it acts in the kinetic regime. The diffusion of the evaporated species from the melt is strongly inhibited in a high-pressure gas. As a result, a vapor layer is formed around the melt which reasonably suppresses further evaporation.

The commonly commercially available FZ furnaces operate with the gas atmosphere pressure up to 10–15 bar. For a high gas-pressure regime, floating-zone furnaces should be specially designed. For instance, the standard two-mirror furnace URN-2-ZM (MPEI, Moscow) has been upgraded by a high-pressure (up to 250 bar) growth chamber [13]. To resist the high pressure, the usual quartz silica tube of the wall thickness of 5 mm resisting the pressure of 10 bar was changed to a silica or sapphire tube of a 35-mm wall thickness (or even thicker for the pressure of 300 bar) (Figure 7).

For instance, Eu_2PdSi_3, $EuCu_2Si_2$, and Eu_2CuSi_3 have high Eu vapor pressure at elevated temperatures. Single crystals of these compounds have been successfully grown at a high argon pressure of 3.5 MPa using optical FZ technique [38, 39].

Figure 7: Transparent tubes forming the growth chamber of a mirror furnace. From left to right: single crystalline saphhire tubes for gas pressures up to 30 and 15 MPa, correspondingly; fused silica tubes for pressures up to 5, 2 and 1 MPa, correspondingly. Courtesy of Scientific Instruments Dresden GmbH (SciDre).

7 Convection in the liquid zone and the shape of the crystal-melt interface

High-quality single crystals can grow only in a highly controllable steady-state process. Understanding the transport processes near the crystallization front makes it feasible to organize such a process.

In the previous section, the material exchange with the growth atmosphere through the enhanced surface was discussed. Here the energy exchange through the surface and the thermal and material transport inside the molten phase will be discussed.

The impact of various factors—natural convection, Marangoni convection, rotation and electromagnetic forces by RF heating—in the flow pattern in FZ has been detailed in many publications (e. g., [6, 40]).

In a liquid phase, both diffusive and convective thermal-transfer mechanisms are possible. Their relative importance depends on the Prandtl number defined as the ratio between the kinematic viscosity ν and the thermal diffusivity λ, $\mathrm{Pr} = \frac{\nu}{\lambda}$.

At $\mathrm{Pr} \ll 1$, the thermal diffusivity dominates the convective heat transport. This is usually the case for intermetallic compounds that have a relatively high λ due to their electronic contribution to the total thermal conductivity. The loss of heat coming into the molten zone from the heater *via* the surface occurs due to back thermal radiation from the surface and to thermal conductance through the seed and the feed rods. The modified Biot number characterizes the importance of the heat dissipation through

Table 2: Some characteristic dimensionless numbers and their values for metallic melts.

Prandtl number	$\mathrm{Pr} = \frac{\nu}{\lambda}$	$2\text{–}3 \cdot 10^{-2}$ (for oxides > 1)	[6, 41, 64, 65]
Schmidt number	$\mathrm{Sc} = \frac{\nu}{D}$	10^2	[6, 64, 65]
Biot number	$\mathrm{Bi} = \frac{\varepsilon\sigma_B r}{\lambda} T_m^3$	$10^{-4}\text{–}0.2$ (for oxides > 1)	[6, 39]
Marangoni number	$\mathrm{Ma} = \frac{\frac{d\sigma}{dT}\Delta TL}{\nu\rho\lambda}$	100–300; $\mathrm{Ma}_{c2} = 2000 \cdot \mathrm{Pr}^{0.6}$	[65]

ν – kinematic viscosity, λ – thermal diffusivity, D – diffusion coefficient, σ_B – Stefan-Boltzmann constant, r – FZ radius, σ – surface tension, ΔT and L – temperature difference and the height of the FZ, T_m – melting point, ϵ – surface emissivity, ρ – melt density.

radiation by the zone surface related to the heat dissipation by thermal conduction. This value is much smaller for metallic melts in comparison with oxides, although it can approach one for some complex compositions [41]. Although the convection plays a negligible role in the heat transport, it is dominant for the material transport relative to the diffusion, which follows from the large Schmidt numbers. In Table 2, some of the characteristic numbers—Prandtl number, Schmidt number and Biot number—and their typical values for metallic melts are given.

Notwithstanding, the buoyancy convection plays a mostly secondary role in the thermal and material transport of intermetallic melts. Large free-liquid surface-to-volume ratio (the largest of all growth techniques) implies the importance of another convective mechanism—Marangoni (thermocapillary) convection [42]. The Marangoni convection occurs as a flow along the liquid surface due to temperature-dependent differences in surface tension. When applied to a molten zone heated from outside, the melt flows from the region of lower surface tension (i. e., from the higher temperature middle of the zone) to that of higher surface tension (the lower temperature crystal-melt interface). This leads to an enhanced surface melting of the rods, which facilitates a formation of convexity of the crystal-melt interface. Various kinds of flow patterns are shown in Figure 8 [43].

The most severe problem involving thermocapillary convection is related to the onset of oscillation, i. e., the transition from the laminar to the oscillatory convection. The onset of oscillatory convection is responsible for the appearance of striation in crystals so avoiding striation is important in controlling the crystal quality [44].

The Marangoni number, **Ma**, which characterizes the strength of the thermocapillary convection is reciprocally proportional to the thermal diffusivity of the melt. Therefore, such highly thermally conductive fluids as intermetallic melts have lower Marangoni numbers than other materials. Nevertheless, there are indications that thermocapillary convection is also important in metal-like melts for the formation of striations in grown crystals [42]. Unfortunately, due to the technical difficulties, there are limited test cases covering quantitative experiments of floating-zone convection for the cases of fluids with small Prandtl number, and the resulting data are dispersive [44].

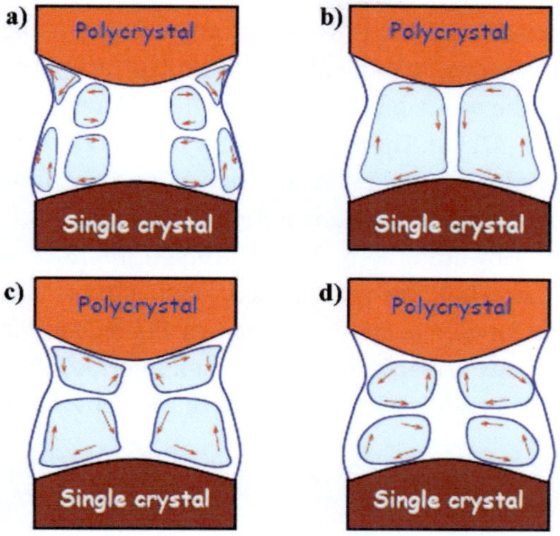

Figure 8: Flow patterns caused by various effects. a) thermocapillary (Marangoni) convection, b) natural (buoyancy) convection, c) flows due to rod rotation, d) electromagnetic field driven flows. Adopted from Souptel D, 2005 [43].

At high-enough Marangoni numbers (exceeding a *critical* Marangoni number, Ma_{c2}), the transition from laminar to oscillating convection flow starts. As a consequence, a reduction of the crystal quality because of striation takes place. Although metallic melts have lower Marangoni numbers than oxides, the critical **Ma**, Ma_{c2}, are also lower than for non-metallic melts (about 100–300 vs. 10^3–10^4). Over a wide range of Prandtl number $0.01 < Pr < 100$, the dependence of Ma_{c2} on the Prandtl number Pr was empirically derived: $Ma_{c2} = 2000 \cdot Pr^{0.6}$ [42].

7.1 Magnetic field control of flows in FZ and the shape of the crystal-melt surface

Since liquid intermetallics are highly conductive media external electromagnetic fields can strongly influence flows in the liquid zone. These flows arise sometimes spontaneously if RF heating is applied. However, with a special design of the magnetic fields, a useful control of the flow situation can be achieved for stabilization of the growth process and homogenization of the composition in the grown crystals.

The effect of high-frequency magnetic field on the Marangoni convection has been investigated numerically for the FZ crystal-growth process under space-experiment conditions. It was concluded from this study that the high-frequency magnetic field can be used to suppress Marangoni convection [45, 46].

The electromagnetic (EM) field of the RF single-turn heating inductor causes a melt flow radially inwards in the liquid zone with the maximal effect in the middle plane of the zone. In the center of the zone, the hotter flow splits in an ascending and a descending flow which melt the central parts of the solids of the upper and lower crystal-liquid surfaces [45]. This results in a concave crystal surface that is unfavorable for stable crystal growth. This instance can be, however, suppressed by using a composite inductor consisting also of a secondary coil along with the primary one wherein the magnetic field is induced by the primary field. Thus, the RF–EM field contains two components that differ by phase [47, 48]. The exploitation of the RF–EM field simultaneously for supporting the material in the molten state and for convection flow control makes independent fine control of both phenomena difficult [45].

In contrast to RF heating, a convex (towards the melt) crystal-melt interface is mostly observed in the growth with radiation heating by halogen or xenon lamps. This is because the Marangoni flow, dominant for optical heating, drives the hot melt along the surface from the middle of FZ toward the upper/lower crystal-melt interface [49]. A mixed approach combining the optical heating with RF control of the convection may be promising.

8 Flux floating-zone growth

Congruently melting intermetallic compounds can be grown by the FZ method straightaway. Usually a pulling rate of about a few millimeters per hour is enough for stable growth. Growth of incongruently melting phases is more demanding. Sometimes the congruent point is only slightly shifted from the stoichiometric crystal composition. In this case, only the pulling rate should be drastically reduced in order to avoid possible constitutional supercooling arising from the segregation effect. For peritectic phases, a high temperature solvent (flux) should be used, which provides primary crystallization of the desired phase below the peritectic temperature. The flux technique in relation to the FZ growth is usually called the travelling solvent floating-zone technique, TSFZ. Usually the polycrystalline rods have the stoichiometric (or the final desired) composition, and a small piece of solvent (flux) is fixed between the rods. After the initial melting of the flux and dissolution of the material therein, the heater is set in motion (usually upwards). The forward phase boundary thus becomes hotter, and here the polycrystalline material is dissolved. The crystallization from the solution takes place on the colder backward phase boundary. Thus, if the crystallized material has exactly the same composition as that of the feed rod, the composition of the floating zone remains constant during the growth. With this technique, large homogeneous crystals can be grown. Another modification of the flux technique may be applied for intentional growth of compositionally variable crystals, e. g., for investigation of the composition-property relation on small (quasi-homogeneous) crystals cut from the same growth batch. Sometimes a local probe measuring technique can be used for this purpose [50, 51]. In this case, the composition of the feed rod is off-stoichiometric (alloy of the growth phase and flux). In the course of crystallization, the FZ is permanently enriched with flux components, which may lead also to a crystal composition variation, if the solidus line has a temperature dependence.

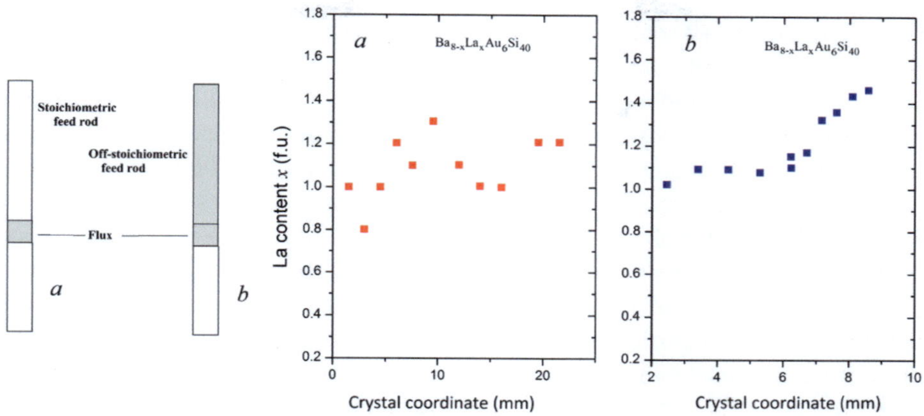

Figure 9: Two modes of the FZ self-flux growth and La concentration profile along the grown $Ba_{8-x}La_xAu_6Si_{40}$ clathrate crystals: a) FZ is off-stoichiometric, feed rod is stoichiometric (steady-state process), b) FZ and feed rod are off-stoichiometric (forced incorporation of La into clathrate cages).

This flux mode is especially effective for a particular purpose to force an incorporation of an element. For instance, rare earth elements can hardly be incorporated into intermetallic clathrate cages. The effectiveness of the latter growth mode was demonstrated by enhanced La incorporation into clathrates [52]. The composition profiles of crystals are compared for the two techniques—with a stoichiometric and with an off-stoichiometric feed rod—in Figure 9. The La content in crystals increases rapidly in the late stage of crystallization. Although crystals of the saturated composition grow, the solubility itself may increase with the progressive reduction of the growth temperate. Instrumentally, since the melting temperature changes constantly because of the variable zone composition, the heating power should be constantly adjusted during the entire process to keep the FZ stable.

9 Stoichiometry of FZ-grown intermetallic compounds

Principally, the TSFZ method makes it feasible to grow incongruently melting phases. Solvents may be either low-melting metals which do not form compounds or solid solutions with the solute phase. In, Bi or Pb are usually used for solution (flux) growth of intermetallic compounds. However, due to rather limited solving ability and high vapor pressure, they are more suitable for the slow-cooling crucible technique carried out in resistance-box furnaces. This technique also makes possible rather long growth time. However, due to the presence of foreign solvent atoms, crystal contamination is possible. For the FZ solution growth, using of solvents containing only the phase-related elements (self-flux) is more typical. They usually provide a higher solubility of the solute, which reduces the undesired segregation effect. Although the foreign atom contamination is absent in this case, the growth from an off-stoichiometric melt may cause severe problems for the quality of the grown crystals. This is because of the homogeneity range of the solute phase. Extended existence ranges are rather typical for many intermetallic phases. In this respect, related phases may differ strongly. For instance, different than the compound R_2PdSi_3, the phase $RPd_{2-\delta}Si_{2+\delta}$ exhibits a relatively extended homogeneity range ($\delta \leq 0.3$), which can give rise to a crucial dependence of physical properties on growth conditions [53]. Even the variation of the composition within a narrow homogeneity range can strongly influence the physical properties. A characteristic example is $Ce_3Pd_{20}Si_6$, a quantum critical system [54]. It can be grown from the stoichiometric melt. However, the congruent point is slightly shifted from the exact stoichiometry, and the crystals have a small Ce deficiency. The variance of the physical properties of $Ce_3Pd_{20}Si_6$ crystals obtained by the FZ technique with various modifications was investigated [17, 55]. Figure 10 shows the low-temperature resistivity which differs especially drastically. The polycrystalline highly stoichiometric sample (denoted as pc in Figure 10) served as a standard for the intrinsic behavior (established from a wide set of physical properties) in this study (in an incongruently crystallized polycrystal the complementary phases are mixed with the main phase on the microscopic level, and an annealing at lower temperature can restore the exact stoichiometrric composition). The segregation effect in single crystals is evident from the difference between the top and the bottom parts of crystals grown from the stoichiometric melt (compare e. g. sc1t and sc1b). The influence of crystal rotation is seen

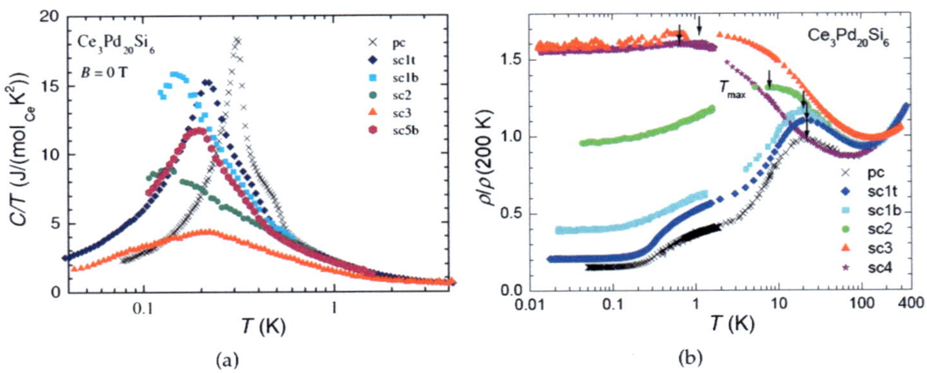

Figure 10: Low-temperature properties of $Ce_3Pd_{20}Si_6$ grown by FZ technique with variations. a) Specific heat in the vicinity of the AF phase transition, b) electrical resistivity normalized to the room temperature value. Samples: pc—high quality polycrystal, sc1b and sc1t—stoichiometric growth, bottom and top parts (difference through segregation effects), sc2—stoichiometric growth without rod rotation, sc3, sc4—flux grown crystals (self-flux Pd_5Si), sc5b—nearly stoichiometric growth (3 % Ce excess), bottom part. Adopted from Ref. [55].

from the difference between sc2 (without rotation) and sc1t,b (with rotation). The underlying reason of the crystal imperfection is the Ce deficiency. The study showed, that the deviation from the stoichiometry appeared to be more detrimental to the crystal quality than the reduced purity of the starting elementary metals (99.5 % Pd in sc1 and 99.95 % Pd in sc2).

FZ is the optimal technique for growth of nearly congruently melting rare earth hexaborides RB_6 (R = La to Sm). The congruent point is shifted from the perfect-lattice composition RB_6 (e. g., to $RB_{6.2}$ for R = La, Pr) [56]. Moreover, rare earth hexaborides exhibit a homogeneity range which extends from the stoichiometric RB_6 to $R_{0.7}B_6$ [57]. Therefore, the resulting crystals are boron-deficient if feed rods of the stoichiometric RB_6 starting composition were used. They have a compositional gradient. In particular, crystals of the composition $PrB_{5.87}$ result from the stoichiometric melt PrB_6. From the off-stoichiometric $LaB_{6.2}$ melt, $LaB_{6.09}$ crystallizes. On the other hand, crystals with the composition $CeB_{6.2}$ have been grown from the stoichiometric CeB_6 melt [58].

Deviation from the stoichiometric composition and systematical variation of the composition along the growth direction in single crystals of SmB_6 grown by stoichiometric FZ is small (< 1 % of Sm) [59]. However, they affect very strongly the physical properties of this system which became topical in recent years because of its properties as a correlated topological insulator [59, 60].

The phase on the base of YB_{66}, a potential thermoelectric material, has a wide homogeneity range between YB_{68} and YB_{58}, the congruent composition being YB_{62} [61, 62]. Hossain et al. succeeded recently in a FZ process further extending the phase composition to YB_{48}. This crystal showed a drastic enhancement of the thermoelectric efficiency in comparison with all previous yttrium borides [63].

Intermetallic silicides and borides cover a vast majority of intermetallic compounds grown by the FZ technique. Table 1 gives an overview of intermetallic materials (including composites) grown over the last 15 years with this method.

FZ has become an indispensable growth technique in basic materials research studies due to its universality and excellent visual control. With the progress in the development of physical property measuring systems which comply with small samples, the FZ technique has become a substantial source of single crystals for physical investigations, especially of newly discovered materials.

Bibliography

[1] Pfann WG. Principles of zone melting, Trans AIME, 1952, 94, 747–53.
[2] Theuerer HC. Method of processing semiconductive materials. Patent US 3060123 A 17. Dec. 1952.
[3] Emeis R. Tiegelfreies Ziehen von Silizium-Einkristallen. Z Naturforsch 1954, 9a, 67.
[4] Keck PH, Golay MJE. Crystallization of silicon from a floating liquid zone. Phys Rev 1953, 89, 1297.

[5] Mühlbauer A. History of Induction Heating and Melting. Vulkan-Verlag GmbH, 2008, 202 S.
[6] Lüdge A, Riemann H, Wünscher M, Behr G, Löser W, Muiznieks A, Cröll A. Floating zone crystal growth. In: Duffar T, ed. Crystal Growth Processes Based on Capillarity: Czochralski, Floating Zone, Shaping and Crucible Techniques. Chichester, UK, John Wiley & Sons, Ltd, 2010, 203–75.
[7] Riemann H, Lüdge A. Floating zone crystal growth. In: Nakajima K, Usami N, eds. Crystal Growth of Si for Solar Cells. Book Series: Adv. Mater. Res., 14, 2009, 41–53.
[8] Muiznieks A, Virbulis J, Lüdge A, et al. Floating zone growth of silicon. In: Rudolph P, ed. Handbook of crystal growth: Bulk crystal growth. Elsevier, 2015, 241–79.
[9] Koohpayeh SM, Fort D, Abell JS. The optical floating zone technique: a review of experimental procedures with special reference to oxides. Prog Cryst Growth Charact Mater 2008, 54, 121–37.
[10] Dabkowska HA, Dabkowski AB. Optical floating zone—complementary crystal growth technique for new classes of oxide materials. In: Handbook of Crystal Growth: Bulk Crystal Growth. Elsevier. 2015, 281–312.
[11] Hermann R, Priede J, Gerbeth G. Floating-zone single crystal growth of intermetallic compounds using a two-phase RF inductor. In: Handbook of Crystal Growth: Bulk Crystal Growth. Elsevier. 2015, 313–29.
[12] Neubauer A, Jonietz F, Meven M, et al. Optical floating zone growth of high-quality Cu_2MnAl single crystals. Nucl Instrum Methods Phys Res 2012, 688, 66–74.
[13] Behr G. et al. German Patent DE 10 2006 019 807.7 (21.04.2006) and PCT/EP2007/05157 (07.03.2007).
[14] Behr G, Löser W, Souptel D, et al. Crystal growth of rare earth-transition metal borocarbides and silicides. J Cryst Growth 2008, 310(7–9), 2268–76.
[15] Heywang W. Zur Stabilität senkrechter Schmelzzonen. Z Naturforsch 1956, 11a(3), 238–43.
[16] Waseda Y, Toguri JM. The Structure and Properties of Oxide Melts: Application of Basic Science to Metallurgical Processing. World Scientific, 1998, 236 S.
[17] Prokofiev A, Paschen S. Crystal growth and stoichiometry of strongly correlated intermetallic cerium compounds. In: Kolesnikov N, Borisenko E, eds. Modern Aspects of Bulk Crystal and Thin Film Preparation. Intech, 2012.
[18] Golubkov AV. Appliance for sealing of containers. Patent USSR 1551750. 1985.
[19] Glebovsky VG, Semenov VN, Lomeyko VV. J Less-Common Met, 1986, 117, 385–9.
[20] Glebovsky V. Some features of growing single crystals of refractory metals from the melt. In: Mastai Y, ed. Advanced Topics in Crystallization. InTech, 2015.
[21] Gerthsen P. Ein neues Verfahren zur Kristallzüchtung von hochschmelzenden Stoffen. Z Angew Phys 1963, 15, 301.
[22] Verhoeven JD, Gibson ED, Noack MA, Conzemius RJ. An arc floating zone technique for preparing single crystal lanthanum hexaboride. J Cryst Growth 1976, 36, 115–20.
[23] Balbashov AM, Egorov SK. Apparatus for growth of single crystals of oxide compounds by floating zone melting with radiation heating. J Cryst Growth, 1981, 52, 498–504.
[24] Prokofiev AV, Shelykh AI, Golubkov AV, Smirnov IA. Crystal growth and optical properties of rare earth sesquiselenides and sesquisulphides—new magneto-optic materials. J Alloys Compd, 1995, 219, 172.
[25] Kloc C, Cheong S-W, Matl P. Floating-zone crystal growth of perovskite manganites with colossal magnetoresistance. J Cryst Growth, 1998, 191, 294.
[26] Souptel D, Behr G, Balbashov AM. $SrZrO_3$ single crystal growth by floating zone technique with radiation heating. J Cryst Growth, 2002, 236, 583–8.
[27] Souptel D, Löser W, Behr G. Vertical optical floating zone furnace: principles of irradiation profile formation. J Cryst Growth 2007, 300(2), 538–50.

[28] Wizent N. Hochdruckkristallzüchtung ausgewählter Oxidverbindungen. PhD Thesis, TU Dresden, 2009 http://d-nb.info/1007715715/34.

[29] Sarker AR, Watauchi S, Nagao M, Watanabe T, Shindo I, Tanaka I. Effects of tilting mirrors on the solid–liquid interface during floating zone growth using tilting-mirror-type infrared-heating image furnace. J Cryst Growth 2010, 312(12–13), 2008–11.

[30] Lindackers D, Ziller S, Fischer F, Behr G. New Crystal Growth Furnace for Floating Zone Melting Under High Gas Pressure. Highlights 2008. Jahresbericht IFW, Leibniz Institute for Solid State and Materials Research, 2008.

[31] Neubauer A. Single crystal growth of intermetallic compounds with unusual low temperature properties. Ph.D. thesis, Technische Universität München, 2010.

[32] Neubauer A, Bœuf J, Bauer, et al. Ultra-high vacuum compatible image furnace. Rev Sci Instrum 2011, 82, 013902.

[33] Bauer A, Benka G, Regnat A, Franz C, Pfleiderer C. Ultra-high vacuum compatible preparation chain for intermetallic compounds. Rev Sci Instrum 2016, 87(11), 113902.

[34] Souptel D, Löser W, Gruner W, Behr G. Oxygen as impurity in crystal growth of intermetallics. J Cryst Growth 2007, 307, 410–20.

[35] Xu YK, Liu L, Löser W, Ge BM. Precipitates identification in R_2PdSi_3 (R = Pr, Tb and Gd) single crystal growth. Trans Nonferr Met Soc China 2011, 21(11), 2421.

[36] Balbashov AM, Egorov SK. Apparatus for growth of single crystals of oxide compounds by floating zone melting with radiation heating. J Cryst Growth 1981, 52, 498–504.

[37] Wizent N, Leps N, Behr G, Klingeler R, Büchner B, Löser W. The effect of process parameters on floating zone crystal growth of selected cuprates. J Cryst Growth 2014, 401, 596–600.

[38] Cao C, Löser W, Behr G, et al. Single crystal growth of Eu_2CuSi_3 intermetallic compound by the floating-zone method. J Cryst Growth 2011, 318(1), 1009–12.

[39] Cao C, Löser W, Behr G, et al. Self-flux growth of large $EuCu_2Si_2$ single crystals. J Cryst Growth 2011, 318(1), 1043–7.

[40] Giling IU. Crystal growth in technology. Elementary semiconductors: silicon. In: Arend H, Hulliger J, ed. Crystal Growth in Science and Technology. NATO ASI Series. Ser. B. Physics. 1989. 210 pages.

[41] Ganschow S. Melt Growth of Oxide Single Crystals in Controlled Oxygen Fugacity Atmosphere. Diss. Berlin, TU, 2015.

[42] Cröll A, Hibiya T, Shiratori S, Kakimoto K, Liu L. Marangoni convection. In: Duffar T, ed. Crystal Growth Processes Based on Capillarity: Czochralski, Floating Zone, Shaping and Crucible Techniques. Wiley. ISBN: 978-0-470-71244-3, 2010.

[43] Souptel D. Crystal Growth and Perfection of Selected Intermetallic and Oxide Compounds. PhD Thesis. Dresden, Leibniz Institute for Solid State and Materials Research, 2005.

[44] Hu WR, Tang ZM, Li K. Thermocapillary convection in floating zones. Appl Mech Rev 2008, 61, 010803-1.

[45] Munakata T, Someya S, Tanasawa I. Supression of Marangoni convection in the FZ melt by high-frequency magnetic field. J Cryst Growth 2002, 235(1–4), 167–72.

[46] de Pablo V, Rivas D. Effect of an axial magnetic field on the flow pattern in a cylindrical floating zone. Adv Space Res 2005, 36(1), 48–56.

[47] Hermann R, Vinzelberg H, Behr G, et al. Magnetic field controlled floating-zone crystal growth and properties of RuAl single crystal. J Cryst Growth 2008, 310(18), 4286–9.

[48] Hermann R, Gerbeth G, Priede J. Magnetic field controlled floating-zone single crystal growth of intermetallic compounds. Eur Phys J Spec Top 2013, 220, 227.

[49] Xu Y-K, Chen Y-N, Guo Y-J, et al. Experimental parameters during optical floating zone crystal growth of rare earth silicides. Rare Met 2014, 33(3), 343–7.

[50] Christensen M, Johnsen S, Søndergaard M, Overgaard J, Birkedal H, Iversen BB. Fast preparation and characterization of quarternary thermoelectric clathrates. Chem Mater 2009, 21(1), 122–7.
[51] Prokofiev A, Yan X, Ikeda M, et al. Crystal growth of intermetallic clathrates: floating zone process and ultra rapid crystallization. J Cryst Growth 2014, 401, 627–32.
[52] Prokofiev A, Svagera R, Waas M, Weil M, Bernardi J, Paschen S. Mechanism of rare earth incorporation and crystal growth of rare earth containing type-I clathrates. Cryst Growth Des 2016, 16, 25–33.
[53] Mazilu I, Teresiak A, Werner J, et al. Phase diagram studies on Er_2PdSi_3 and $ErPd_2Si_2$ intermetallic compounds. J Alloys Compd. 2008, 454(1–2), 221–7.
[54] Custers J, Lorenzer K-A, Müller M, et al. Destruction of the Kondo effect in the cubic heavy-fermion compound $Ce_3Pd_{20}Si_6$. Nat Mater 2012, 11, 189–94.
[55] Prokofiev A, Custers J, Kriegisch M, et al. Crystal growth and composition-property relationship of $Ce_3Pd_{20}Si_6$ single crystals. Phys Rev B 2009, 80, 235107.
[56] Gurin V, Konovalov M, Kuzanyan A, et al. Chemical composition topography in RE hexaboride crystals obtained by various solution-melt methods. Solid State Sci 2012, 14(11–12), 1705–9.
[57] Yajima S, Niihara K. In: 9th Rare Earth Res. Conf. 2, Blacksburg, 1971, 598–609.
[58] Hein H, Koeppel C, Vetter U, Warkentin E. Sc, Y, La-Lu Rare Earth Elements: Compounds with Boron. Springer Science & Business Media, 2013, 277 Seiten.
[59] Phelan WA, Koohpayeh SM, Cottingham P. et al. On the Chemistry and Physical Properties of Flux and Floating Zone Grown SmB_6 Single Crystals. 2015. arXiv:1510.07612v1.
[60] Kim DJ, Xia J, Fisk Z. Topological surface state in the Kondo insulator samarium hexaboride. Nat Mater 2014, 13, 466–70.
[61] Tanaka T, Sato A, Takenouchi S, et al. Floating-zone crystal growth of Nb-doped YB_{66} for soft X-ray monochromator use. J Cryst Growth 2005, 275(1–2), E1889–93.
[62] Slack GA, Oliver DW, Brower GD, Young JD. Properties of melt-grown single crystals of "YB_{68}". J Phys Chem Solids 1977, 38, 45.
[63] Hossain MA, Isao T, Takaho T, et al. YB_{48} the metal rich boundary of YB_{66}, crystal growth and thermoelectric properties. J Phys Chem Solids 2015, 87, 221–7.
[64] Chang CE, Wilcox WR. Inhomogenities due to the thermocapillary flow in floating zone melting. J Cryst Growth 1975, 28, 8–12.
[65] Yang YK, Kou S. Temperature oscillation in a tin liquid bridge and critical Marangoni number dependency on Prandtl number. J Cryst Growth 2001, 222, 135–43.
[66] Tabata C, Miura N, Uhlirova K, et al. Peculiar magnetism of UAu_2Si_2. Phys Rev B 2016, 94, 214414.
[67] Feng S, Fang Y, Zhai Q, et al. Magnetocaloric effect in a dual-phase coupled $LaFe_{11}Si_2$ crystal prepared by a modified high-pressure zone-melting technique. J Cryst Growth 2016, 451, 83–7.
[68] Cao C, Blum CGF, Löser W. Floating zone crystal growth of Lu_2PdSi_3 silicide. J Cryst Growth 2013, 401, 593–5.
[69] Cao C, Klingeler R, Leps N, et al. Single crystal growth of the $ErPd_2Si_2$ intermetallic compound. J Cryst Growth 2013, 401, 601–4.
[70] Guo H, Guo X, Zhao H. Research progress in directional solidification of Nb–Si based high temperature alloys. Rare Met Mater Eng 2014, 43(4), 1019–24.
[71] Xu Y-K, Löser W, Guo Y-J, et al. Crystal growth of Gd_2PdSi_3 intermetallic compound. Trans Nonferr Met Soc China 2014, 24(1), 115–9.
[72] Xu Y, Löser W, Behr G, et al. Crystal growth of the Pr_2PdSi_3 intermetallic compound. J Cryst Growth 2010, 312(12–13), 1992–6.

[73] Xu Y, Liu L, Löser W, et al. Vertical floating zone crystal growth of R_2PdSi_3 intermetallic compounds (R=Pr and Nd). In: Kim YH, Yarlagadda P, eds. Advanced Technologies in Manufacturing, Engineering and Materials, PTS 1-3. Book Series: Advanced Materials Research, 774–776, 2013, 720.

[74] Xu Y, Frontzek M, Mazilu I, et al. Floating zone crystal growth of selected R_2PdSi_3 ternary silicides. J Cryst Growth 2011, 318(1), 942–6.

[75] Pospisil J, Prokes K, Reehuis M, et al. Influence of sample preparation technology and treatment on magnetism and superconductivity of UCoGe. J Phys Soc Jpn 2011, 80(8), 084709.

[76] Xu Y, Löser W, Liu L, et al. Solidification and crystal growth of binary Tb_5Si_3 intermetallics. J Cryst Growth 2011, 321(1), 45–9.

[77] Xu Y, Löser W, Tang F, et al. Crystal growth of the intermetallic compound Nd_2PdSi_3. Cryst Res Technol 2011, 46(2), 135–9.

[78] Cao C, Deppe M, Behr G, et al. Single crystal growth of the $CeCu_2Si_2$ intermetallic compound by a vertical floating zone method. Cryst Growth Des 2011, 11(2), 431–5.

[79] Bergmann C, Jeevan HS, Schubert M, et al. Single crystal growth of $CeNi_2Ge_2$ using the floating zone technique. Phys Status Solidi B 2010, 247(3), 694–6.

[80] Souptel D, Leithe-Jasper A, Loser W, et al. Floating zone growth and characterization of Pr_5Si_3 single crystals. J Cryst Growth 2004, 273(1–2), 311–9.

[81] Legl S, Franz C, Neubauer A, et al. Pressure dependence of the magnetization in Pr_5Si_3. Physica B, Condens Matter 2009, 404(19), 2887–9.

[82] Guo JT, Tian YX, Cheng GM, et al. Microstructural characteristics and high temperature compressive properties at 1623 K of a directionally solidified Nb-silisides based in-situ composite. J Alloys Compd 2009, 470(1–2), 606–9.

[83] Tamegai T, Nakajima Y, Nakagawa T, et al. Two-gap superconductivity in $R_2Fe_3Si_5$ (R = Lu and Sc). In: 25th International Conference on Low Temperature Physics (LT25), Pt 5A: Superconductivity. Book Series: J. Physics Conference Series, 150, 2009, 052264.

[84] Rosales I. Synthesis and characterization of Mo_3Si single crystal. J Cryst Growth 2008, 310(16), 3833–6.

[85] Mazilu I, Löser W, Behr G, et al. Element segregation during crystal growth processes of $Ce_2Pd_xCo_{1-x}Si_3$ intermetallic compounds. J Cryst Growth 2005, 275(1–2), E109–14.

[86] Mazilu I, Frontzek M, Löser W, et al. Single crystal growth of the Er_2PdSi_3 intermetallic compound. J Cryst Growth 2005, 275(1–2), E103–7.

[87] Bei H, George EP, Pharr GM. Elastic constants of single crystal Cr_3Si and Cr-Cr_3Si lamellar eutectic composites: a comparison of ultrasonic and nanoindentation measurements. Scr Mater 2004, 51(9), 875–9.

[88] Souptel D, Behr G, Loser W, et al. $CeSi_2$-δ single crystals: growth features and properties. J Cryst Growth 2004, 269(2–4), 606–16.

[89] Ivanenko L, Filonov A, Shaposhnikov V, et al. Thermoelectric properties of Mn-doped Ru_2Si_3. In: 22nd International Conference on Thermoelectrics, Proceedings ICT'03, 2003, 157–60.

[90] Souptel D, Behr G, Ivanenko L, et al. Floating zone growth and characterization of semiconducting Ru_2Si_3 single crystals. J Cryst Growth 2002, 244(3–4), 296–304.

[91] Behr G, Loser W, Bitterlich H, et al. Single-crystal growth of binary and ternary rare earth silicides. J Cryst Growth 2002, 237, 1976–80.

[92] Hermann R, Wendrock H, Rodan S, et al. Single crystal growth of antiferromagnetic Mn_3Si by a two-phase RF floating-zone method. J Cryst Growth 2013, 363, 1–6.

[93] Bauer A, Neubauer A, Franz C, Münzer W, Garst M, Pfleiderer C. Quantum phase transitions in single-crystal $Mn_{1-x}Fe_xSi$ and $Mn_{1-x}Co_xSi$: Crystal growth, magnetization, ac susceptibility, and specific heat. Phys Rev B 2010, 82, 064404.

[94] Ritz R, Halder M, Wagner M, Franz C, Bauer A, Pfleiderer C. Formation of a topological non-Fermi liquid in MnSi. Nature 2013, 497, 231–4.
[95] Bauer E, Kaldarar H, Prokofiev, et al. Heavy fermion superconductivity and antiferromagnetic ordering in $CePt_3Si$ without inversion symmetry. J Phys Soc Jpn. 2007, 76, 051009.
[96] Fåk B, Raymond S, Braithwaite D, Lapertot G, Mignot J-M. Low-energy magnetic response of the noncentrosymmetric heavy-fermion superconductor $CePt_3Si$ studied via inelastic neutron scattering. Phys Rev B 2008, 78, 184518.
[97] Laurita NJ, Morris CM, Koohpayeh SM, et al. Anomalous three-dimensional bulk ac conduction within the Kondo gap of SmB_6 single crystals. Phys Rev B 2016, 94(16), 165154.
[98] Zhang X, Liang C, Zhang J, et al. Floating zone growth and electricity properties of topological Kondo insulator SmB_6. In: Egorov NV, ed. 10th International Vacuum Electron Sources Conference (IVESC), Saint Petersburg, RUSSIA, 30.6.–04.07.2014. 2014.
[99] Hatnean MC, Lees MR, Paul DMcK, Balakrishnan G. Large, high quality single-crystals of the new topological Kondo insulator, SmB_6. Sci Rep 2013, 3, 3071.
[100] Derr J, Knebel G, Braithwaite D, Salce B, Flouquet J, Flachbart K, Gabáni S, Shitsevalova N. From unconventional insulating behavior towards conventional magnetism in the intermediate-valence compound SmB_6. Phys Rev B 2008, 77, 193107.
[101] Liu H, Zhang X, Xiao Y, et al. The electronic structure and thermionic emission property of single crystal SmB_6. Vacuum. 2017, 145, 295–8.
[102] Hossain MA, Isao T, Takaho T, et al. Crystal growth and anisotropy of high temperature thermoelectric properties of yttrium borosilicide single crystals. J Solid State Chem 2016, 233, 1–7.
[103] Singh RP, Parzyk NA, Lees MR, et al. Crystal growth and properties of the non-centrosymmetric superconductor, Ru_7B_3. J Cryst Growth 2014, 395, 22–5.
[104] Zhang F-X, Zhang X, Zhang J-X, et al. Synthesis and properties of single crystal $Ce_{1-x}Pr_xB_6$ by floating zone melting. J Inorg Mater 2014, 29(10), 1073–6.
[105] Bao L-H, Zhang J-X, Zhou S-L, et al. Floating zone growth and emission properties of single crystal LaB_6 cathode. Acta Phys Sin 2011, 60(10), 106501.
[106] Petrosyan V, Vardanyan V, Kuzanyan V, et al. Thermoelectric properties and chemical composition of CeB_6 crystals obtained by various methods. Solid State Sci 2012, 14(11–12), 1653–5.
[107] Bogomol I, Nishimura T, Nesterenko Yu, et al. The bending strength temperature dependence of the directionally solidified eutectic LaB_6-ZrB_2 composite. J Alloys Compd 2011, 509(20), 6123–9.
[108] Bao LH, Tegus O, Zhang JX, et al. Large emission current density of $La_xCe_{1-x}B_6$ high quality single crystals grown by floating zone technique. J Alloys Compd 2013, 558, 39–43.
[109] Bogomol I, Nishimura T, Vasylkiv O, et al. High-temperature strength of directionally reinforced LaB_6-TiB_2 composite. J Alloys Compd 2010, 505(1), 130–4.
[110] Otani S, Aizawa T, Yajima Y. Floating zone growth of LaB_6 crystals from the CaB_6-added feed rods. J Cryst Growth 2002, 234(2–3), 431–4.
[111] Bao L-H, Zhang J-X, Zhou S-L, et al. Floating zone growth and thermionic emission property of single crystal CeB_6. Chin Phys Lett 2011, 28(8), 088101.
[112] Liu H, Zhang X, Xiao Y, et al. The thermionic and field emission properties of single crystal PrB_6 grown by floating zone method. Vacuum, 2018, 151, 76–9.
[113] Badalyan G, Kuzanyan A, Petrosyan V, et al. Investigation of the elemental composition of lanthanum-cerium hexaboride crystals. In: Papoyan AV, ed. International Conference on Laser Physics 2010. Book Series: Proceedings of SPIE-The International Society for Optical Engineering, 7998, 2011, 79980E.

[114] Ma R-G, Liu D-M, Zhou S-L, et al. Fabrication and emission property of polycrystalline $La_{0.4}Pr_{0.6}B_6$ bulk prepared by spark plasma sintering. J Inorg Mater 2010, 2(7), 743–7.
[115] Balakrishnan G, Lees MR, Paul DMK. Growth of large single crystals of rare earth hexaborides. J Cryst Growth 2003, 256(1–2), 206–9.
[116] Chen CH, Xuan Y, Otani S. Temperature and loading time dependence of hardness of LaB_6, YB_6 and TiC single crystals. J Alloys Compd 2003, 350(1–2), L4–6.
[117] Okamoto NL, Kusakari M, Tanaka K, et al. Anisotropic elastic constants and thermal expansivities in monocrystal CrB_2, TiB_2, and ZrB_2. Acta Mater 2010, 58(1), 76–84.
[118] Iwasa K, Igarashi R, Saito K, Laulhé C, Orihara T, et al. Motion of the guest ion as precursor to the first-order phase transition in the cage system GdB_6. Phys Rev B, 2011, 84, 214308.
[119] Dudka AP, Khrykina ON, Bolotina NB, et al. An exceptionally-high diffraction quality dodecaboride LuB_{12}: growth and single-crystal structure. J Alloys Compd 2017, 692, 535–44.
[120] Balakrishnan G, Majumdar S, Lees MR, et al. Single crystal growth of CrB_2 using a high-temperature image furnace. J Cryst Growth 2005, 274(1–2), 294–6.
[121] Hori K, Inoue S, Isogami M. Crystal growth of high-quality ZrB_2 single crystals using the floating-zone method. J Cryst Growth 2008, 310(1), 145–51.
[122] Otani S, Aizawa T, Kieda N. Solid solution ranges of zirconium diboride with other refractory diborides: HfB_2, TiB_2, TaB_2, NbB_2, VB_2 and CrB_2. J Alloys Compd 2009, 475(1–2), 273–5.
[123] Otani S, Korsukova MM, Aizawa T. High-temperature hardness of ReB_2 single crystals. J Alloys Compd 2009, 477(1–2), L28–9.
[124] Hashimoto Y, Okamoto NL, Acosta M, et al. Processing, microstructure, and thermal expansion measurements of high temperature Ru-Al-CrB_2 alloys. In: Palm M, Bewlay BP, He YH, et al., eds. Advanced Intermetallic-based Alloys for Extreme Environment and Energy Applications. Book Series: Materials Research Society Symposium Proceedings, 1128, 2009, 215–20.
[125] Mori T, Tanaka T. Effect of transition metal doping and carbon doping on thermoelectric properties of YB_{66} single crystals. J Solid State Chem 2006, 179, 2889–94.
[126] Otani S, Hirata K, Adachi Y, et al. Floating zone growth and magnetic properties of Y_2C two-dimensional electride. J Cryst Growth 2016, 454, 15–8.
[127] Ma M, Yuan D, Wu Y, et al. Flux-free growth of large superconducting crystal of FeSe by travelling-solvent floating-zone technique. Supercond Sci Technol 2014, 27, 122001.
[128] Yu J, Ren J, Li H, et al. A new approach to grow the Heusler Ni-Mn-Sn unidirectional crystal. J Cryst Growth 2014, 402, 147–50.
[129] Omar A, Dimitrakopoulou M, Blum CGF, et al. Phase dynamics and growth of $Co_2Cr_{1-x}Fe_xAl$ Heusler compounds: a key to understand their anomalous physical properties. Cryst Growth Des 2013, 13(9), 3925–34.
[130] Blum CGF, Jenkins CA, Barth J, et al. Highly ordered, half-metallic Co_2FeSi single crystals. Appl Phys Lett 2009, 95, 161903.
[131] Bogomol I, Nishimura T, Vasylkiv O, et al. Microstructure and high-temperature strength of B_4C-TiB_2 composite prepared by a crucibleless zone melting method. J Alloys Compd 2009, 485(1–2), 677–81.
[132] Bogomol I, Vasylkiv O, Sakka Y, et al. Mechanism of nucleation and growth of directionally crystallized alloys of the B_4C-MeB_2 system. J Alloys Compd 2010, 490(1–2), 557–61.
[133] Tanaka T, Sato A, Watanabe K, et al. Novel ternary Y–B–C compound: $Y_{10+x}B_7C_{10-x}$ (x approximate to 0.1). In: 16th International Symposium on Boron, Borides and Related Materials (ISBB 2008). Book Series: J. Physics. Conference Series, 176, 2009, 012006.
[134] Bud'ko SL, Schmiedeshoff GM, Lapertot G, et al. Anisotropic thermal expansion and magnetostriction of YNi_2B_2C single crystals. J Phys Condens Matter 2006, 18(35), 8353–65.
[135] Souptel D, Behr G, Löser W, et al. Crystal growth and perfection of $RENi_2B_2C$ (RE = Ho, Tb, Y). J Cryst Growth 2005, 275(1–2), E91–5.

[136] Souptel D, Behr G, Kreyssig A, et al. Growth features of $RENi_2B_2C$ (RE = Y, Ho, Tb) single crystals. J Cryst Growth 2005, 276(3–4), 652–62.
[137] Mori T, Tanaka T. Single crystal growth and physical properties of boron-rich ytterbium borosilicide. J Alloys Compd 2003, 348, 203.
[138] Otani S, Ohashi N. Preparation of Ni_2P and Fe_2P single crystals by the floating-zone method. J Ceram Soc Jpn 2013, 121(1412), 331–2.
[139] Klicpera M, Pospisil J, Rudajevova A, et al. Growth and characterization of CePtIn single crystal. J Cryst Growth 2014, 394, 61–6.
[140] Kopecek J, Jurek K, Jarosova M, et al. The preparation of oriented samples of ferromagnetic shape memory alloy CoNiAl. In: Proc. 11th European Workshop of the European-Microbeam-Analysis-Society (EMAS) on Modern Developments and Applications in Microbeam Analysis. Book Series: IOP Conference Series-Materials Science and Engineering UNSP, 2010, 012013.
[141] Yasuda HY, Shigeno K, Nagase T. Dynamic strain aging of $Al_{0.3}CoCrFeNi$ high entropy alloy single crystals. Scr Mater 2015, 108, 80–3.
[142] Danilewsky AN, Croell A, Tonn J, et al. Dislocations and dislocation reduction in space grown GaSb. Cryst Res Technol 2009, 44(10), 1109–14.
[143] Duncan WJ, Welzel OP, Moroni-Klementowicz D, et al. Quantum phase transitions in $NbFe_2$ and $Ca_3Ru_2O_7$. Phys Status Solidi B 2010, 247, 544.
[144] Patrick CE, Kumar S, Balakrishnan G, et al. Rare-earth/transition-metal magnetic interactions in pristine and (Ni,Fe)-doped YCo_5 and $GdCo_5$. Phys Rev Matter 2017, 1, 024411.

Marcus Schmidt and Michael Binnewies

Chemical vapor transport of intermetallics

1 Introduction – chemical transport reactions

Chemical transport reactions have played a firmly established role as a preparative method in solid-state chemistry for about 50 years [1]. With their help, single-phase products, in many cases made of monocrystalline material and of high purity, may be synthesized and/or crystallized with comparatively little effort. To date, several thousand publications have reported on such reactions, and the current state of knowledge in this area was summarized in a monograph a few years ago [2].

In the laboratory, chemical transport reactions are usually carried out in closed glass ampoules which are subjected to a temperature gradient for several days (see Figure 1). On one side of the ampoule, the so-called **source**, about 1 g of a solid—the **source material**—is placed with a much smaller amount of the so-called **transport agent**. This is a substance that reacts with the source material at the experimental temperature under the formation of one or several gaseous products.

$$iA(\mathrm{s}) + kB(\mathrm{g}) \quad \rightleftharpoons \quad jC(\mathrm{g}) + \cdots \tag{1}$$

For such a reaction to become an effective transport reaction, all reaction products have to be gaseous. Halogens or halogen compounds are most frequently used as transport additives.[1]

1 A transport additive is the substance added to the transport ampoule in order to transport the solids. In some cases, the transport additive may not be the actual transport agent according to eq. (1). The transport agent is formed in a simultaneously proceeding reaction (examples: 1. transport additive:

https://doi.org/10.1515/9783110496789-006

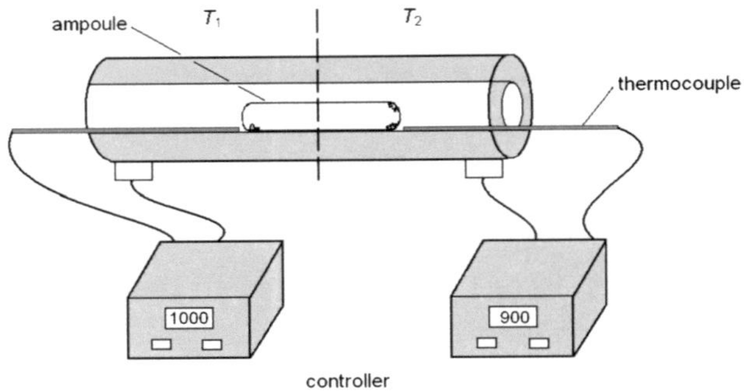

Figure 1: Experimental set-up for chemical transport in a dual zone furnace.

This technique makes use of the temperature dependence of the state of equilibrium of the introduced solids' reaction with the transport agent. The gaseous reaction products disperse within the entire transport ampoule starting from the source materials' surface and consequently also reaching the part of the ampoule kept at a different temperature, e. g., about 100 K lower. In the case of an endothermic reaction of the solids with the transport agent, the state of equilibrium will shift towards reformation of the source materials and release of the transport agent at the lower temperature, according to Le Chateliers' principle. The formation of a solid from the gas phase will occur in the so-called **sink**. The result of the process corresponds to a sublimation, although it is effected by a chemical reaction. In the case of an exothermic reaction, the back reaction will occur at the higher temperature. This aspect is also referred to as endothermic or exothermic transport reaction. The temperature dependence of a chemical reaction's equilibrium constant is described by the van't Hoff equation:

$$\ln K = -\frac{\Delta_r H_T^0}{R \cdot T} + \frac{\Delta_r S_T^0}{R} \tag{2}$$

K equilibrium constant
$\Delta_r H_T^0$ enthalpy of reaction/kJ·mol^{-1}
$\Delta_r S_T^0$ entropy of reaction/J·K^{-1}·mol^{-1}
R gas constant (8.314 J·K^{-1}·mol^{-1})
T temperature/K

It becomes apparent that the sign of the enthalpy of reaction alone decides whether the equilibrium constant increases or decreases with rising temperatures. It can be shown that chemical transport reactions are particularly efficient for values of the equilibrium constant as close as possible to unity [1, 2].

$PtCl_2$, transport agent: Cl_2 formed via decomposition, 2. transport additive: iodine, transport agent: a gaseous iodide formed with a component of the source material via simultaneous reaction).

1.1 Two examples

Chemical transport of nickel with iodine

Nickel migrates in the presence of a small amount of iodine within the temperature gradient of 860 → 1030 °C, indicating an exothermic reaction. Under the chosen experimental conditions [3], the transport agent iodine predominantly exists in atomic form. The transport reaction is essentially described by the following equation:

$$\mathrm{Ni(s) + 2I(g) \rightleftharpoons NiI_2(g)} \qquad (3)$$

$$\Delta_r H^0_{298} = -83\ \mathrm{kJ{\cdot}mol^{-1}}\ [4], \quad \Delta_r S^0_{298} = -57\ \mathrm{J{\cdot}K^{-1}{\cdot}mol^{-1}}\ [4]$$

Based on these values and eq. (2), disregarding the change in molar heat capacities, an equilibrium constant of 3.8 can be calculated for a mean transport temperature of 1218 K, coming rather close to the ideal value of one. This is a typical transport reaction for a metal. In principle, other halogens are also eligible to act as transport agents. To this end, let us consider the reaction of nickel with chlorine (chlorine existing as a diatomic gas under comparable conditions):

$$\mathrm{Ni(s) + Cl_2(g) \rightleftharpoons NiCl_2(g)} \qquad (4)$$

$$\Delta_r H^0_{298} = -70\ \mathrm{kJ{\cdot}mol^{-1}}\ [4], \quad \Delta_r S^0_{298} = 45\ \mathrm{J{\cdot}K^{-1}{\cdot}mol^{-1}}\ [4]$$

Once again disregarding the change in molar heat capacities, an equilibrium constant of $2.2 \cdot 10^5$ can be calculated on this basis for the just-mentioned temperature mentioned, showing that the equilibrium lies far to the side of the gaseous nickel chloride. Thus, the transport agent chlorine is almost completely consumed to form the very stable gaseous nickel chloride, which at 1030 °C does not perceptibly decompose into nickel and chlorine in terms of a back reaction. Much higher temperatures than 1030 °C are necessary for such a decomposition. Therefore, iodine is significantly better suited than chlorine as transport an agent for the transport of nickel. Nevertheless, transport reactions of metals with chlorine are also known, especially in the case of noble metals like gold or platinum. Here, the iodides are much too unstable to be formed at all in the gas phase. **The most commonly used transport agent for metals is iodine!**

Chemical transport of zinc oxide with chlorine

Zinc oxide can be endothermically transported with chlorine at temperatures around 900 °C. The following reaction takes place:

$$\mathrm{ZnO(s) + Cl_2(g) \rightleftharpoons ZnCl_2(g) + \frac{1}{2}O_2(g)} \qquad (5)$$

$$\Delta_r H^0_{298} = 83\ \mathrm{kJ{\cdot}mol^{-1}}\ [4], \quad \Delta_r S^0_{298} = 113\ \mathrm{J{\cdot}K^{-1}{\cdot}mol^{-1}}\ [4]$$

This is a typical transport equation for the transport of a salt-like compound containing metal and non-metal: The transport agent reacts with the metal atoms under formation of a gaseous compound, whereas the non-metal atoms do not react with the transport agent; they constitute the gaseous non-metal in molecular form, in the case of $O_2(g)$.

Based on the mentioned values, an equilibrium constant of 160 can be calculated for 1173 K, still being much larger than one but significantly smaller than the value of $2.2 \cdot 10^5$ discussed previously. Generally, it can be asserted that a noteworthy transport effect takes place for equilibrium constants in the range of about 10^{-4} to 10^4.

2 Chemical transport of intermetallic phases

The chemical transport of an intermetallic phase resembles that of metals, but differs fundamentally from the transport of salt-like compounds. It has been shown in the instance of the transport of zinc oxide with chlorine that the transport agent only reacts with the metal atoms. In the example, oxygen as non-metal atom forms gaseous elemental oxygen. In order to transport an intermetallic compound A_xB_y, however, the transport agent has to react with both the components *A* and *B*. *A* and *B* need to undergo the reaction with the transport agent simultaneously and, moreover, the temperature dependences of the equilibrium constants for the reactions of *A* and *B* with the transport agent need to show the same tendency and be of similar magnitude. These criteria apply to classical intermetallic phases where *A* and *B* are typical metals. They apply equally to compounds containing a typical metal and a semi-metal, e. g., silicon or germanium, which in elemental form cannot transform into the gas phase at the experimental conditions due to their much too low vapor pressures.

Section 3 in Ref. [2] concerning the chemical transport of metals mentions a total of 40 metals and semi-metals that have been successfully transported. Their element symbols are placed against a gray background in the following compilation.

H																	He
Li	**Be**											**B**	**C**	N	O	F	Ne
Na	Mg											**Al**	**Si**	P	S	Cl	Ar
K	Ca	Sc	**Ti**	**V**	**Cr**	Mn	**Fe**	**Co**	**Ni**	**Cu**	**Zn**	**Ga**	**Ge**	As	**Se**	Br	Xe
Rb	Sr	**Y**	**Zr**	**Nb**	**Mo**	Tc	**Ru**	**Rh**	**Pd**	**Ag**	**Cd**	**In**	**Sn**	**Sb**	**Te**	I	Kr
Cs	Ba	La	**Hf**	**Ta**	**W**	**Re**	**Os**	**Ir**	**Pt**	**Au**	Hg	Tl	Pb	**Bi**	Po	At	Rn

Ce	Pr	Nd	Pm	Sm	Eu	Gd	Tb	Dy	Ho	Er	Tm	Yb	Lu
Th	**Pa**	**U**	Np	Pu									

A calculation of how many binary intermetallic systems alone can arise from those 40 elements according to combinatorial principles yields an amount of 780 binary systems.

$$C_k^n = -\frac{n!}{(n-k)!\cdot k!} = \frac{40!}{(40-2)!\cdot 2!} = 780 \quad (6)$$

C_k^n number of k-combinations without repetition from a set of n elements ($n = 40$, $k = 2$).

Section 10 of the cited book about the transport of intermetallic systems gives examples for about 70 investigated systems, less than 10 % of the possible 780. Only a few of these 70 systems have been systematically explored concerning chemical transport. This demonstrates the great preparative potential that is still available. Moving on to ternary intermetallic phases, the analogous calculation results in 9880 different intermetallic systems.

$$C_k^n = -\frac{n!}{(n-k)!\cdot k!} = \frac{40!}{(40-3)!\cdot 3!} = 9880 \quad (7)$$

Out of this number, a mere six have been investigated concerning their transport behavior [2]! In particular, these are the systems Co–Cr–Ge [5], Co–Fe–Si [6], Co–Ta–Ge [5], Zr–Ge–Te [7], Zr–Sn–Te [7] and Cr–Ge–Si [8].

Regarding those binary intermetallic systems not yet explored concerning their transport behavior, there is the question which of those systems might promise preparative success. The basic requirements for a joint transport of both components have been stated. This makes it seem practically impossible to transport intermetallic compounds containing an alkali metal. The very reactive alkali metal will completely bind the transport agent, usually a halogen. Moreover, alkali metal halogenides are much too stable to decompose under deposition of the metal. The same applies to the alkaline earth metals (possibly with the exception of Be). Thus, all Zintl phases are practically excluded. According to the current state of knowledge, they are not accessible via transport reactions. The pre-conditions are particularly favorable in the case of chemically similar metals, especially when the constituent metals can be transported with the same transport agent under comparable conditions. In those cases, the chemical transport of intermetallic phases can also be expected to be possible.

The main features of the chemical transport of intermetallic phases are discussed in the following, using several examples.

2.1 Competition for the transport agent

Before carrying out the actual experiment, it is helpful to consider how the transport agent will presumably distribute among the elements involved. Since all metals included in the system compete for the transport agent, the ability of all solid components to form volatile compounds has to be guaranteed. This can be demonstrated in two examples: the system Cu–Ag and the system Pd–Ga. In both systems, intermetallic phases can be transported under addition of iodine [9, 10]. Looking at the expected distribution of the transport agent over both metals, the results are quite different in these two systems.

First, the state of equilibrium of the following reaction is calculated:

$$Cu(s) + AgI(g) \rightleftharpoons CuI(g) + Ag(s) \tag{8}$$

The state of equilibrium is depicted as a function of temperature in Figure 2.

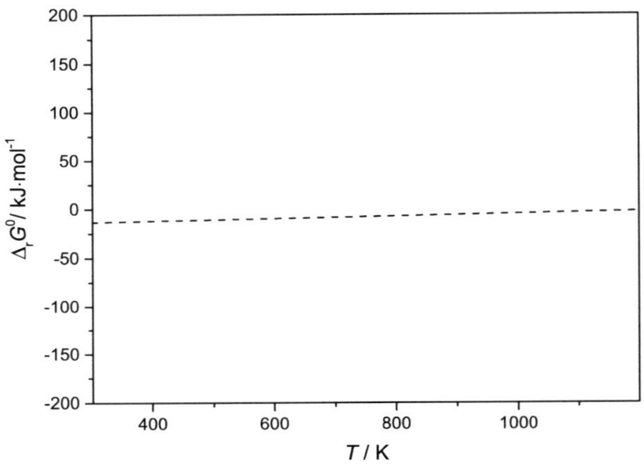

Figure 2: Gibbs energy of the reaction Cu(s) + AgI(g) $\rightleftharpoons$ CuI(g) + Ag(s) as a function of temperature.

The Gibbs free energy at a temperature of 1000 K is close to zero, thus the equilibrium constant is close to one. The transport agent iodine is distributed almost equally over the two components copper and silver. This case offers ideal conditions for a joint chemical transport of both metals.

The system Pd–Ga provides a very different scenario. The following reaction is to be regarded:

$$2GaI_3(g) + 3Pd(s) \rightleftharpoons 2Ga(l) + 3PdI_2(g) \tag{9}$$

A calculation of the state of equilibrium depending on temperature yields the correlation shown in Figure 3.

There is an extreme state of equilibrium at 1000 K, the transport agent is practically completely consumed in the formation of GaI_3. A joint transport of both metals appears to be impossible. Nevertheless, a chemical transport of both metals under formation of Pd_2Ga can be observed. This is explained by the formation of the gaseous compound $PdGa_2I_8$ which contains both elements but is not taken into account in this simple consideration. The formation of such gas complexes including both com-

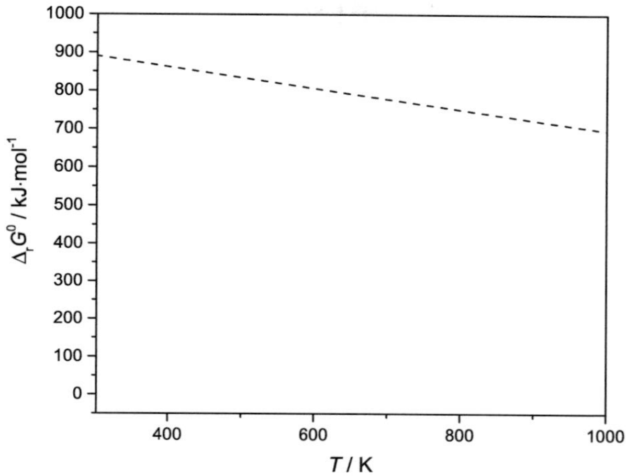

Figure 3: Gibbs energy of the reaction $2GaI_3(g) + 3Pd(s) \rightleftharpoons 2Ga(l) + 3PdI_2(g)$ as a function of temperature.

ponents of the source material provides favorable conditions for the transport of intermetallic phases. A survey of such gas complexes is given in [11]. The information compiled therein refers to gaseous chloride complexes. It is to be expected that bromine and iodine compounds will be formed in a similar manner. However, since scarcely anything is known about their stability, quantitative predictions about the transport behavior are hardly possible. Thermodynamic data of intermetallic phases are also insufficiently known, so that the solids also cannot be thermodynamically described in an adequate way. An exact prediction as to whether a particular intermetallic compound may be obtained through a chemical transport reaction is therefore not possible in the majority of cases. The rules given above in combination with conclusions based on already investigated analogous systems allow for certain predictions, though.

2.2 Examples

2.2.1 Binary systems with extensive regions of solid solutions

Intermetallic compounds are by no means always stoichiometrically composed; quite often they exhibit extensive homogeneity ranges. In the case of chemically similar elements with the same type of lattice and resembling lattice parameters, continuous series of solid solutions can be observed. This is particularly often true for homologous 4d and 5d elements. Within homogeneity ranges, the thermodynamic activities of the elements involved change with composition, in miscibility gaps, however, in miscibility gaps they are constant.

Molybdenum–tungsten

A typical example for a system with continuous formation of solid solutions is the system molybdenum–tungsten. The phase diagram is shown in Figure 4.

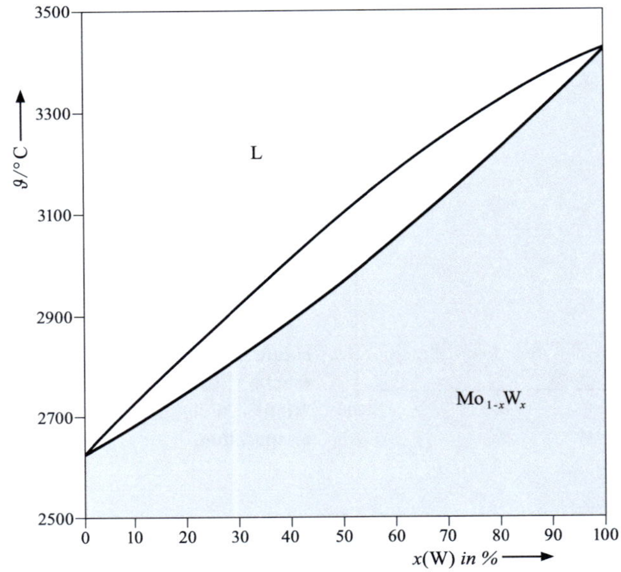

Figure 4: Phase diagram of the system molybdenum-tungsten according to [12].

This systems provides particularly good qualifications for a joint transport of both components, since molybdenum, as well as tungsten, can be transported with the same transport agent ($HgBr_2$) from 1000 °C to 900 °C [13, 14]. The transport is described by the following transport equilibria:

$$Mo(s) + 2HgBr_2(g) \rightleftharpoons MoBr_4(g) + 2Hg(g) \quad (10)$$

$$W(s) + 2HgBr_2(g) \rightleftharpoons WBr_4(g) + 2Hg(g) \quad (11)$$

and

$$W(s) + 2.5HgBr_2(g) \rightleftharpoons WBr_5(g) + 2.5Hg(g) \quad (12)$$

In the presence of water traces, oxidebromides are additionally formed. Based on the results of [13, 14], the transport of solid solutions in this system has been investigated regarding the composition of the source material, the transport temperatures and the amount of transport agent [15]. In this, as well as in many other systems, the composition of the source material proved to be the most significant determining factor regarding the composition of the sink material. A close examination of the deposited crystals yielded an unusual result: the deposited crystals possessed a molybdenum-rich core and a tungsten-richer shell, i. e., they exhibited an inhomogeneous composition. The reason for this lies in a rarely observed kinetic inhibition of the transport of tungsten that delays its onset for a while. Thus, molybdenum is initially transported more intensively [13].

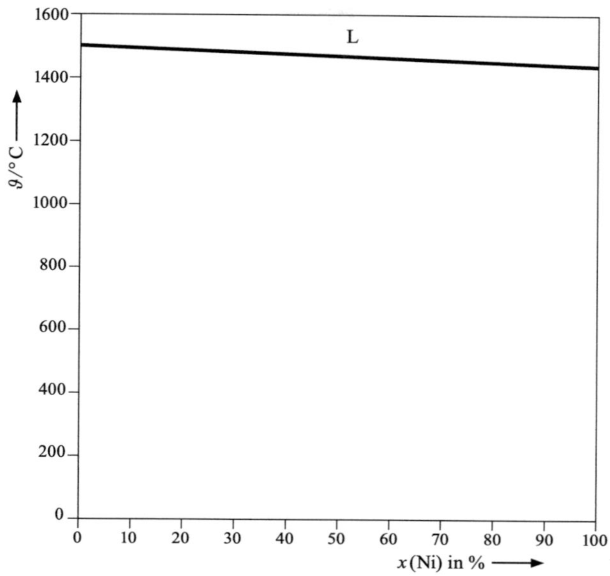

Figure 5: Phase diagram of the system cobalt-nickel according to [12].

Cobalt-nickel

Cobalt and nickel form a continuously miscible system in the solid state (see Figure 5).

The elements cobalt and nickel can be endothermically transported above 800 °C under addition of GaI_3 or $GaI_3 + I_2$ (e. g. 925 → 825 °C) [16]. In most instances, the composition of the source material is varied in order to specifically influence the composition of the deposited solid solutions. Another possibility, used in the cited paper [17], consists of choosing different temperatures for the two metals at the source. This requires a special working technique where cobalt and nickel are not intermixed in the source material, but are placed spatially separated in two separate source areas. A furnace with three temperature zones is necessary for this. Well-crystallized solid solutions that were homogenous regarding their composition in a range between 5 and 75 at.% nickel could be deposited in this way [17]. Contrary to the transport of tungsten, no reaction inhibition can be observed in the system cobalt–nickel.

2.2.2 Binary systems with formation of compounds

Especially well-researched on the basis of many examples is the transport of compounds comprising a transition metal and a group 14 element (Si, Ge, Sn). Such compounds are often attributed to the intermetallics composed of typical metals. Thus, the transport of silicides, germanides or stannides has been described for ***M*–Si** (M = Co, Cr, Cu, Fe, Ni, Mn, Mo, Nb, Re, Ta, Ti, U, V, W), ***M*–Ge** (M = Co, Cr, Cu, Fe, Ni, Nb, Ti) and ***M*–Sn** (M = Co, Cu, Fe, Nb, Ni, Ti) [2]. For some of these systems, extensive model calculations have been published, contributing significantly to an understanding of the chemical transport of intermetallic phases [18–21]. The majority of the compounds mentioned above can be deposited via gas phase with iodine as transport additive. In some cases, bromine (Cr–Si, Mo–Si, U–Si, W–Si) or hydrogen chloride (Nb–Ge, Nb–Sn) have turned out to be suitable. The phases in the systems Ni–Ge and Ni–Sn were obtained under the addition of Ga and I_2. The gallium/iodine ratio was chosen to enable GaI_3 to be formed in a preliminary reaction while maintaining a certain iodine excess. The available GaI_3 allows the formation of gas complexes, for example of the composition $NiGa_2I_8$. In this way, the nickel concentration in the gas phase is considerably increased.

Copper-germanium

As can be seen in the phase diagram of the system (Figure 6), copper and germanium form a number of compounds that in parts differ only slightly in their composition, are formed peritectically or are low-temperature phases with small regions of existence. Thus, their crystallization presents a particular experimental challenge.

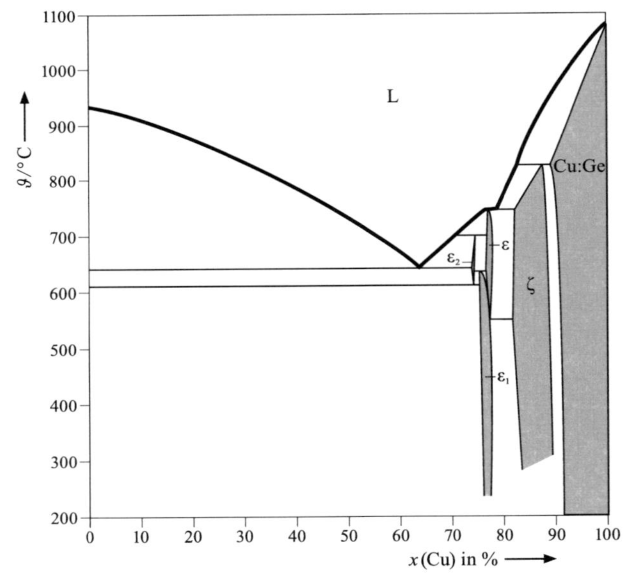

Figure 6: Phase diagram of the system copper–germanium according to [12].

Comparing the transport behavior of the elements copper and germanium, it becomes apparent that both elements can be transported with iodine in a broad temperature range, although with different transport directions: Copper is always deposited in the hotter zone, while germanium mostly in the less hot one. The transport of germanium towards the hotter zone has only been described for low iodine pressures and high temperatures (900 °C) [22]. Despite the elements' different transport directions, several intermetallic phases could be obtained via transport in this system.

The described transport experiments [23] were carried out in a temperature gradient of 570 °C to 630 °C under addition of iodine (initial pressure 1.5 bar) and varying source compositions ($Cu_{0.5}Ge_{0.5}$ ...$Cu_{0.95}Ge_{0.05}$).

Depending on the source composition, the phases Cu_3Ge (HT, RT), $Cu_{0.85}Ge_{0.15}$, Cu:Ge resp. Cu were deposited in the form of single crystals. Transport rates were in the range of 0.1 to 0.5 $mg{\cdot}h^{-1}$. To produce those phases as single crystals with other methods of preparation appears to be very difficult, if not impossible. The course of the reaction will be described in more detail in the following paragraphs. At first, there is the question which of the two elements constitutes the preferred reaction partner for the transport agent iodine. This question can be answered by regarding the state of equilibrium of the following reaction:

$$4Cu(s) + GeI_4(g) \rightleftharpoons 4CuI(s) + Ge(s) \quad (13)$$

Figure 7 shows the Gibbs free energy of this reaction as a function of temperature.

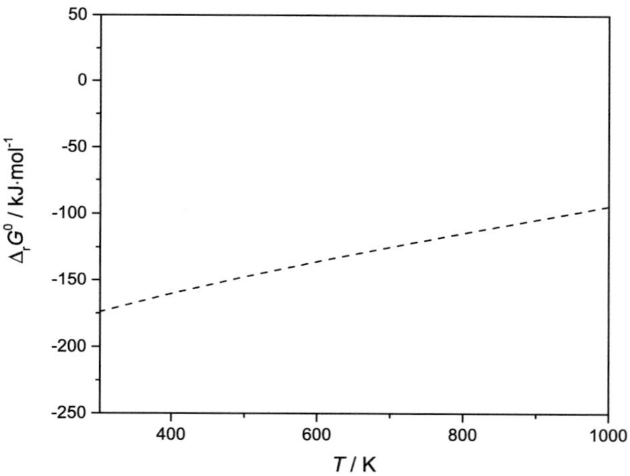

Figure 7: Gibbs energy of the reaction $4Cu(s) + GeI_4(g) \rightleftharpoons 4CuI(s) + Ge(s)$ as a function of temperature.

It becomes evident that the equilibrium lies far to the side of copper iodide which condenses under the transport conditions. Solid copper iodide is formed, binding most of the transport agent iodine. CuI has a sublimation pressure of about 10^{-4} bar at reaction temperature. The gas phase contains predominantly Cu_3I_3, which is thermodynamically not very stable and partially decomposes into the elements. This decomposition equilibrium determines the iodine pressure in the system. At this rather low pressure, iodine exists mostly in atomic form ($p(I) = 2{\cdot}10^{-5}$ bar). The atomic iodine effects the transport of germanium to the hotter zone according to the following equation:

$$Ge(s) + 2I(g) \rightleftharpoons GeI_2(g) \quad (14)$$

$$\Delta_r H^0_{298} = -147\ \mathrm{kJ{\cdot}mol^{-1}} \quad \Delta_r S^0_{298} = -75\ \mathrm{J{\cdot}K^{-1}{\cdot}mol^{-1}}$$

It thus becomes understandable why both copper and germanium are deposited in the hotter zone in the form of intermetallic phases. The different percentages of copper and germanium in the gas phase are determined by the different compositions of the source materials in the different experiments.

Iron–silicon

The phase diagram in Figure 8 shows that, in the binary system iron–silicon, the phases Fe_3Si, $Fe_{1.72}Si_{0.28}$, Fe_2Si, Fe_5Si_3, FeSi, $FeSi_2$ and $Fe_{0.92}Si_2$ are known. On the iron-rich side, a solid solution with up to 20 at.% silicon has been described. Examples for compounds with and without a noteworthy homogeneity range and for high temperature phases with small regions of existence can be found among the different iron silicides, as well as compounds that are formed peritectoidally.

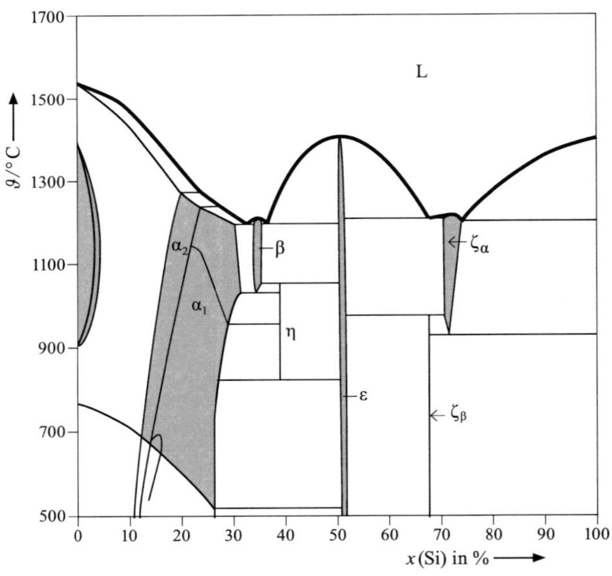

Figure 8: Phase diagram of the system iron–silicon according to [12].

This diversity and the fact that the system iron-silicon has been comprehensively investigated and described regarding chemical transport turns it into a typical example for the transport of a multitude of transition metal silicides [18, 24]. Due to the similarities between silicon and germanium concerning their transport behavior, the conclusions can be transferred to the transition metal germanides as well [25, 26].

The constituting elements can each be transported under addition of iodine. The transport of iron with iodine is described from 800 °C to 1000 °C [3]. For the reaction of iron with iodine in the given temperature range, the following equilibria are to be considered:

$$\mathrm{Fe(s) + I_2(g) \rightleftharpoons FeI_2(g)} \tag{15}$$
$$\Delta_r H^0_{298} = 24\,\mathrm{kJ{\cdot}mol^{-1}}\ [4] \quad \Delta_r S^0_{298} = 62\,\mathrm{J{\cdot}K^{-1}{\cdot}mol^{-1}}\ [4]$$

$$\mathrm{Fe(s) + 2I(g) \rightleftharpoons FeI_2(g)} \tag{16}$$
$$\Delta_r H^0_{298} = -128\,\mathrm{kJ{\cdot}mol^{-1}}\ [4] \quad \Delta_r S^0_{298} = -39\,\mathrm{J{\cdot}K^{-1}{\cdot}mol^{-1}}\ [4]$$

$$\mathrm{Fe(s) + 2I(g) \rightleftharpoons \frac{1}{2}Fe_2I_4(g)} \tag{17}$$
$$\Delta_r H^0_{298} = -209\,\mathrm{kJ{\cdot}mol^{-1}}\ [4] \quad \Delta_r S^0_{298} = -126\,\mathrm{J{\cdot}K^{-1}{\cdot}mol^{-1}}\ [4]$$

$$\mathrm{Fe(s) + 3I(g) \rightleftharpoons FeI_3(g)} \tag{18}$$
$$\Delta_r H^0_{298} = -308\,\mathrm{kJ{\cdot}mol^{-1}}\ [18] \quad \Delta_r S^0_{298} = -180\,\mathrm{J{\cdot}K^{-1}{\cdot}mol^{-1}}\ [18]$$

The first equilibrium is endothermic and can therefore not contribute to the transport of iron towards the hotter zone. Since iodine does not only exist molecularly in the gas phase at a mean temperature of 900 °C, but to a significant degree atomically as well, the second equilibrium is to be taken into account. It is exothermic and describes the observed transport behavior of iron with iodine. Besides the monomer, the dimeric Fe_2I_4 can also be found in the gas phase. Its formation is also exothermic. In addition to the formation of iron(II) iodides, that of gaseous iron(III) iodides is to be considered, too. Although this equally exothermic reaction does not contribute significantly towards the transport of iron with iodine, it is essential for the description of the transport of FeSi.

Like iron, silicon can be transported under addition of iodine. Depending on temperature and pressure, the transport can take place either endothermically, i. e., towards the less hot zone, or exothermically towards the hotter zone (1150 °C to 950 °C or 950 °C to 1150 °C) [27, 28].

The following equilibria can be formulated for the reaction of silicon with iodine:

$$\mathrm{Si(s) + 2I_2(g) \rightleftharpoons SiI_4(g)} \tag{19}$$
$$\Delta_r H^0_{298} = -235\,\mathrm{kJ{\cdot}mol^{-1}}\ [4] \quad \Delta_r S^0_{298} = -138\,\mathrm{J{\cdot}K^{-1}{\cdot}mol^{-1}}\ [4]$$

$$\mathrm{Si(s) + 4I(g) \rightleftharpoons SiI_4(g)} \tag{20}$$
$$\Delta_r H^0_{298} = -538\,\mathrm{kJ{\cdot}mol^{-1}}\ [4] \quad \Delta_r S^0_{298} = -325\,\mathrm{J{\cdot}K^{-1}{\cdot}mol^{-1}}\ [4]$$

$$\mathrm{Si(s) + SiI_4(g) \rightleftharpoons 2SiI_2(g)} \tag{21}$$
$$\Delta_r H^0_{298} = 295\,\mathrm{kJ{\cdot}mol^{-1}} \quad \Delta_r S^0_{298} = 207\,\mathrm{J{\cdot}K^{-1}{\cdot}mol^{-1}}$$

The silicon-transmitting gaseous species are therefore SiI_4 and SiI_2. The transport direction is determined by the partial pressure ratio $p(SiI_4)/p(SiI_2)$, which itself depends upon the total pressure and the temperature [27].

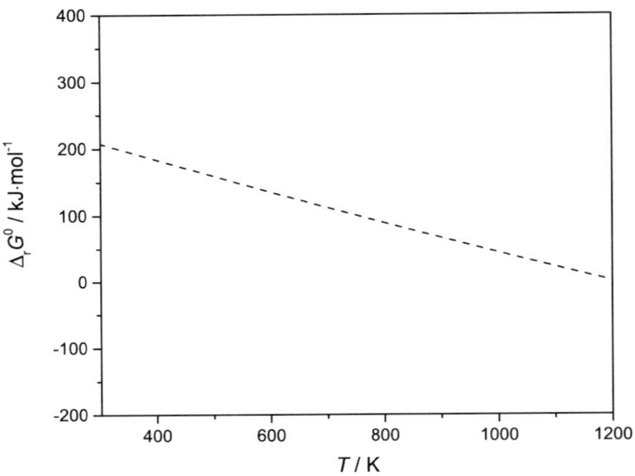

Figure 9: Gibbs energy of the reaction $2Fe(s) + SiI_4(g) \rightleftharpoons Si(s) + 2FeI_2(g)$ as a function of temperature.

Based on these observations, iodine could be a suitable transport agent for iron silicides. A comparison of the stabilities of the gaseous iodides SiI_4 and FeI_2 by means of eq. (22) shows a well-balanced state of equilibrium at 900 °C (see Figure 9).

$$2Fe(s) + SiI_4(g) \rightleftharpoons Si(s) + 2FeI_2(g) \tag{22}$$

The transport additive iodine is neither bound exclusively by iron nor by silicon and can form gaseous species with both components of the solids.

As supported by the literature, all known phases in the system iron–silicon are crystallizable by chemical transport [18]. On the iron-rich side of the binary system, starting with the iron-rich solid solution up to and including FeSi, the transport occurs from the less hot zone towards the hotter zone in accordance with that of elemental iron. The transport temperatures vary between 700 °C and 900 °C on the source side and 800 °C and 1000 °C on the sink side. The applied compositions of the source material are in the regions of the transported phase, as well as in the respective two-phase regions. Apparently, a depletion of silicon in comparison with the applied source occurs during the transport in the region of the iron-rich solid solution. Other transport experiments from the particular two-phase regions lead to the deposition of the respective co-existing silicon-rich phase in the hotter zone.

The transport of FeSi is exemplary for the exothermic chemical transport of iron silicides with iodine. As detailed thermodynamic model calculations show, it can in essence be described by the following equilibrium.

$$FeSi(s) + 3.5I_2(g) \rightleftharpoons FeI_3(g) + SiI_4(g) \tag{23}$$

Through the reaction with gaseous iodine in molecular and atomic form, the solid source material is transferred into the gas phase, forming gaseous FeI_2, Fe_2I_4, FeI_3, SiI_2, and SiI_4, with SiI_4 as the silicon transferring species. The gaseous species FeI_2

and Fe_2I_4, connected via the dimerization equilibrium, merely contribute to a small degree to the iron transfer. Therefore, the transport of iron under these conditions occurs predominantly via FeI_3.

On the silicon-rich side of the system, the compound $FeSi_2$ can be deposited. The transport in this case occurs endothermically from 1000 °C to 800 °C and thus follows the transport of elemental silicon at small total pressures [27]. The following transport equation can be formulated:

$$FeSi_2(s) + 2SiI_4(g) \rightleftharpoons FeI_2(g) + 3SiI_2(g) \tag{24}$$

The actual transport agent is thus not the added iodine, but the resultant SiI_4. It reacts with the source material under formation of SiI_2 and FeI_2 as transport effective species.

The two mentioned transport equations only roughly describe the events during the reaction. Generally, the transport process in such complex systems can only be covered accurately by a multitude of independent heterogeneous solid/gas equilibria and homogeneous gas equilibria. This is principally possible with the help of thermodynamic model calculations using the software packages TRAGMIN [29] or CVTRANS [30]. These calculations are based on the knowledge of all solid phases existing in the system, as well as the formed gas species and their respective thermodynamic data. The results are conclusions about the chemical composition of the expected sink and their transport rates and about the temperature dependences of the individual partial pressures. However, such a detailed analysis of transport behavior and precise predictions is not possible for many systems due to unavailable or inconsistent thermodynamical data.

Nickel–tantalum

The system nickel–tantalum, a system of two typical metals, has been closely examined regarding its transport behavior [25]. Figure 10 shows the phase diagram.

The transport took place with iodine as transport agent from 800 to 950 °C. The experiments were carried out with mixtures of both metals in varying ratios. The ratios were chosen in order to either match the composition of one of the intermetallic phases or that of one of the two-phase fields between two adjacent phases. Prior to the actual transport experiment, the samples were annealed in the presence of the transport agent at 840 °C to ensure that the solid reached equilibrium state. A transport could be observed in all experiments. In most cases, the transport was incongruent: the composition of the phase(s) deposited at 950 °C differed from that of the applied solid material. An enrichment of tantalum in the sink occurred in the majority of experiments. Ni_2Ta alone could be transported congruently. In the transport from the two-phase field between Ni_3Ta and Ni_2Ta, both these phases were detected in the sink, as well. When transporting from the two-phase field between the nickel-rich solid solution and Ni_8Ta, a reproducible enrichment of nickel took place and the nickel-rich solid solution was deposited. Moreover, the experiments revealed the apparent ex-

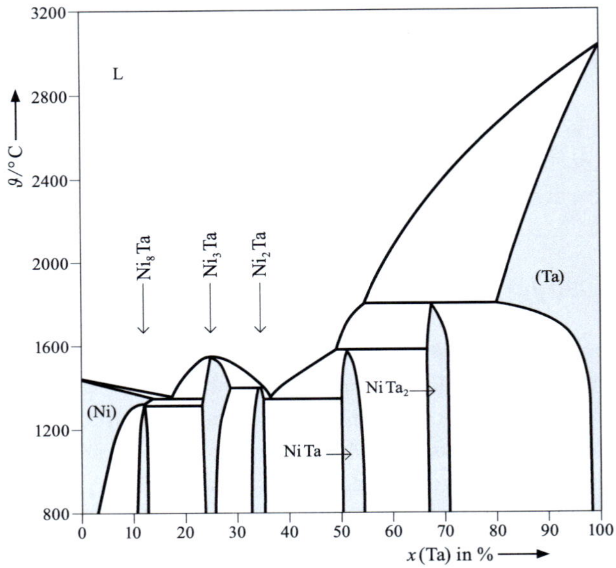

Figure 10: Phase diagram of the system nickel-tantalum according to [12].

istence of a so-far unknown phase of the composition Ni_7Ta_2 which is isostructural with Ag_7Y_2 [31]. A clear indication was furthermore provided of the existence of a phase with the analytically determined composition $NiTa_9$, which however could not be characterized by X-ray investigations. The correlation between the compositions of source and sink is depicted in Figure 11. The dots mark the composition of the source, arrowheads the composition in the sink. The three experiments that deviate from the others for reasons as yet unknown are symbolized by dashed lines.

When the direction of enrichment reverses at a certain composition of the source material in a binary system, this is also referred to as a *transport shed*.

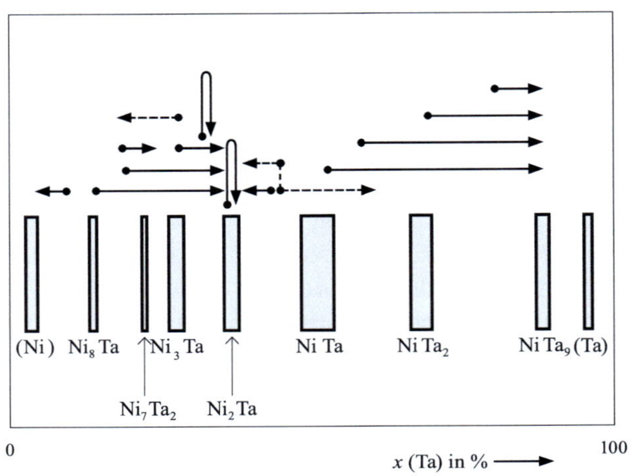

Figure 11: Transport behavior in the system nickel–tantalum.

Palladium–aluminum, gallium, indium

The chemical behavior of palladium differs fundamentally from that of group 13 metals of the periodic table. While group 13 metals form thermally very stable triiodides, the single known iodide of palladium, PdI_2, decomposes into the elements under heating. A transport of intermetallic phases composed of palladium and a group 13 element with iodine may thus hardly be expected to be feasible. Despite these prospects, however, the following intermetallic phases from these systems were successfully transported with iodine as the transport additive:

Pd–Al: Pd_2Al [32, 10], PdAl [10], Pd_5Al_3 [32], Pd:Al [32, 10]
Pd–Ga: Pd_2Ga [10, 33], PdGa [10], Pd:Ga [10]
Pd–In: Pd_3In [10], Pd_2In [10], PdIn [10], Pd:In [10]

Systems of this kind are apparently characterized by unusual properties that render the unexpected transport behavior under these conditions possible in the first place. This is consistent with the equally surprising observation that Pt–Al phases form at 1200 °C in a hydrogen atmosphere from platinum and aluminum oxide. In the absence of platinum, the reduction of aluminum oxide is not possible under these conditions [34].

An explanation of the unanticipated transport of the intermetallic phases just mentioned results from an inspection of the gas phase composition. This is to be illustrated with the example of the transport of Pd_2Ga [33].

Palladium cannot be transported with iodine, since the stability is too low for gaseous palladium iodides to be formed. Gallium, in contrast, forms thermodynamically very stable gaseous iodides (GaI, GaI_3, Ga_2I_4, Ga_2I_6). The energy of formation of gaseous PdI_2 has been estimated in a dissertation to be 230 $kJ{\cdot}mol^{-1}$ [10]. Gaseous GaI_3 ($\Delta H^0_{298} = -138\ kJ{\cdot}mol^{-1}$) is more stable by about 370 $kJ{\cdot}mol^{-1}$. The formation of GaI_3 should therefore consume the transport agent iodine almost completely. A simultaneous transport of palladium and gallium should thus not be expected according to this reasoning. The transport which can nevertheless be observed is to be ascribed to the formation of a gas complex of the composition $PdGa_2I_8$, which appears in the gas phase in noteworthy concentrations (see Figure 12). Merely by the formation of such gas complexes can palladium be transferred into the gas phase and thus enable the chemical transport of intermetallic phases in said systems.

2.2.3 Ternary systems

Hardly anything is known to date about the chemical transport in ternary intermetallic systems, an systematic investigations are particularly lacking. The few already published works show that ternary intermetallic phases are also accessible via chemical transport. However, it seems to be premature to draw general conclusions from the results published so far.

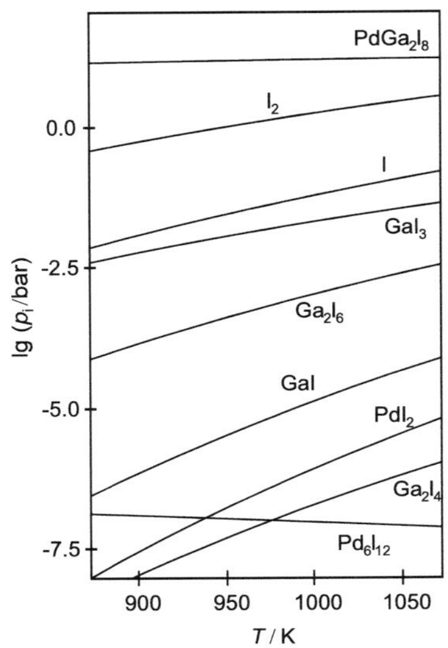

Figure 12: Calculated partial pressures for the chemical transport of Pd_2Ga with iodine according to [33].

Cobalt–tantalum–germanium [5]

The chemical transport in this system appeared promising since previous transport experiments in the binary systems Co–Ta, Co–Ge and Ta–Ge had led to the crystallization of intermetallic phases [2]. Transport in those three systems was carried out with iodine as transport additive at the following temperatures: Co–Ta (800 → 950 °C), Co–Ge (900 → 700 °C) and Ta–Ge (800 → 950), hence exothermic in two systems and endothermic in one system. Despite these different transport directions in the binary systems, the crystallization of ternary Co–Ta–Ge phases succeeded via transport towards the hotter zone. An incongruent transport was observed in all conducted experiments. The transport tends to result in an enrichment of tantalum. There have been reports that not all experiments in this system could be performed reproducibly, either. Figure 13 depicts the correlation between the compositions in source and sink.

The fact that reproducibility cannot always be granted in the incongruent transport of intermetallic phases becomes understandable in the following considerations. Let us presume a solid of the homogeneous composition A_xB_y to be at the source. When one of those two components, *A* in this example, is preferentially transferred towards the sink during the transport, an enrichment of *B* will necessarily occur at the source. In this solid/gas reaction, the enrichment will at first take place on the surface of the A_xB_y particles, so that gradients in the composition within single particles and possibly also the uncontrolled formation of new phases are to be expected. Such processes are hard to determine experimentally and an at least temporary state out of equilibrium may be reached which may result in limited reproducibility. The same is equivalently true for the transport of ternary phases. On this account alone, the thermodynamic modelling of an incongruent transport of binary or ternary inter-

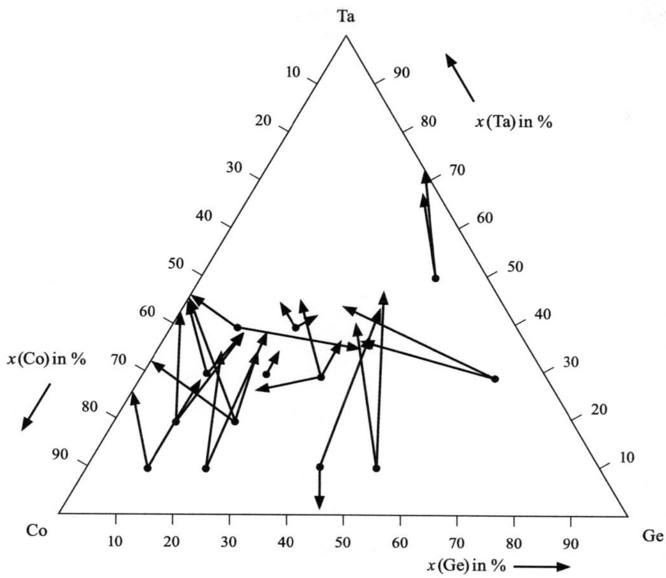

Figure 13: Transport behavior in the system cobalt–tantalum–germanium.

metallic phases does not seem very promising, even if the thermodynamic data of all participants in the reaction can be provided.

Chromium–germanium–silicon [8]

The chemical transport in this system is at present possibly the most exhaustively researched transport in a ternary intermetallic system. The transport is carried out with iodine in a temperature gradient from 800 to 1000 °C. The authors report on the formation of single crystals possessing homogeneous composition within individual crystallites. At the same time, they report on a very incongruent and time-dependent transport behavior. Attempts at thermodynamical modelling of the transport behavior proved unsatisfactory. For further investigations, the authors recommend a careful control of the solids in source and sink, not only on a few chosen crystals, as well as the application of solids without degrees of freedom. They advise to choose large amounts of substance at the source while keeping the duration of the transport short in order to guarantee conditions as constant as possible during the experiment.

3 Experimental approach

The experimental approach is comprehensively described in [2], so that the following section merely goes into the specifics of the chemical transport of intermetallic phases.

The most important control instrument for influencing the composition of the deposited sink materials is the composition of the source. Usually, a mixture of the elements in a specific ratio is applied as initial solids. In order to adjust reproducible and well-defined transport conditions, it makes sense to equilibrate the mixture of initial components in a pre-reaction. This is effected most simply by isothermically heating the prepared and closed transport ampoule containing initial solids and transport additive for a certain period of time. The duration of this pre-reaction depends on the nature of the applied metals (powder, foils, pieces, etc.) and the temperature. Finer material and high temperatures promote equilibration on the one hand, but on the other hand they increase the possible reaction rate with the ampoule material. Furthermore, the use of fine powders may be a source of unwanted water and/or oxygen in the system. The duration of such pre-reactions can be within the range of several days or weeks. If necessary, the progress of the reaction can be monitored via X-ray diffraction. For a systematic investigation of a binary system, it is recommended to vary the composition in steps of 10 at.% and conduct the series of experiments under otherwise identical conditions. Incongruent transport behavior can often be observed in the transport of intermetallic phases, thus gradually changing the composition of the source materials. In order to realize constant thermodynamic conditions, it is advisable to choose the composition of the source materials to lie within a two-phase region, if available. The chemical activities of all components are constant within a two-phase region. The use of high net weights of source materials also leads to quasi-stationary transport behavior for a while.

The above-mentioned reaction of the initial material with the ampoule material can be minimized by the use of suitable interior crucibles. However, a reaction of the gas phase with the ampoule material cannot be prevented this way. The amount of transport agent is generally chosen to achieve a pressure of about one bar in the ampoule at the mean transport temperature. In some cases, smaller amounts of transport agent, e. g., 0.25 to 0.5 $mg{\cdot}ml^{-1}$, are favorable in order to avoid the condensation of halogenides. Mostly, the concentration of the transport agent has only little influence on the composition of the sink material. It became apparent, however, that small total pressures and in particular small temperature gradients more often lead to congruent transports than high total pressures and large temperature gradients [19].

4 Summary

It was demonstrated that, for the crystallization of intermetallic phases, chemical transport reactions offer a great experimental potential which, however, is scarcely exploited yet. Even regarding binary systems, only a few have been investigated concerning their transport behavior. Less than ten examples have been published for ternary and more complex systems. Thus, we are far from tapping the full potential

provided by chemical transport reactions. The following characteristics can be recapped from the work published so far on chemical transport of intermetallic phases:
- The chemical transport of intermetallic phases follows that of the metals.
- The transport additive most commonly used is iodine.
- The preconditions for a joint transport of several elements under formation of intermetallic phases are particularly favorable in the case of chemically similar elements.
- In certain cases, the joint transport is successful even for chemically very different elements (Pd, Ga). In this instance, gas molecules containing several various metals (gas complexes) frequently play a role.
- Incongruent transport behavior can often be observed in the transport of intermetallic phases.
- The experimental and instrumental effort is yet meager.
- The products are not contaminated by impurities from, e. g., flux material.
- Generally, there is no contamination of the crystals by the transport agent.
- Well-developed individual crystals with readily identifiable crystallographic planes are frequently obtained.

Chemical transport reactions are particularly suitable as an alternative to other methods of synthesis or crystallization in the following cases:
- The phases exhibit very high melting temperatures (example Mo–W).
- The formation of the phases is peritectic or peritectoid (example Cu–Ge).
- The phases have very small regions of existence in the subsolidus region (example Cu–Sn [31]).
- The phases are not crystallizable from the melt (low-temperature modifications, e. g., FeGe [2]).
- The phases differ only slightly in their chemical composition (example Fe–Ge).

When considering a chemical transport reaction for the preparation of intermetallic phases in a system not yet examined to this effect, it appears to be reasonable to answer the following questions prior to the experiment:
- Can the elements involved be deposited via chemical transport?
- Can the transport be carried out for all elements involved with the same transport agent?
- Are the transport temperatures in the same range?
- Is the transport direction identical?
- Is the formation of gas complexes to be anticipated?

Should the answer to at least some of these questions be “yes”, then the preconditions for a joint transport are promising.

Bibliography

[1] Schäfer H. Chemische Transportreaktionen. Weinheim, Verlag Chemie, 1962.
[2] Binnewies M, Glaum R, Schmidt M, Schmidt P. Chemical Vapor Transport Reactions. Berlin, De Gruyter, 2013.
[3] Schäfer H, Jacob H, Etzel K. Die Verwendung der Zerfallsgleichgewichte der Eisen(II)- und Nickel(II)-halogenide zum Metalltransport im Temperaturgefälle. Z Anorg Allg Chem 1956, 286, 42–55.
[4] Binnewies M, Milke E. Thermochemical Data of Elements and Compounds, 2nd edition. Weinhein, Wiley-VCH, 2002.
[5] Wirringa J, Binnewies M. Der Chemische Transport ternärer intermetallischer Phasen in den Systemen Cr/Co/Ge und Co/Ta/Ge. Z Anorg Allg Chem 2000, 626, 1747–51.
[6] Boldt R, Reichelt W, Bosholm O, Oppermann H. Investigations on chemical vapour transport of intermetallic phases in the system Co/Fe/Si. Z Anorg Allg Chem 2003, 629, 1839–48.
[7] Wannek CH, Harbrecht B. Iodine-promoted systems of structurally ordered AlPd5. Z Anorg Allg Chem 2007, 633, 1397–402.
[8] Jandl I, Richter KW, Ipser H. Incongruent, time-dependent chemical vapour transport in multi-component systems: a case study in Cr–Ge–Si. Solid State Sci 2011, 13, 1108–14.
[9] Gerighausen S, Binnewies M. Der chemische Transport von Michkristallen im System Cu/Ag. Z Anorg Allg Chem 1995, 621, 936–40.
[10] Deichsel J. Dissertation, University of Hannover, 1998.
[11] Schäfer H. Gaseous Chloride complexes with halogen bridges – homo-complexes and hetero-complexes. Angew Chem, Int Ed 1976, 15, 713–78.
[12] Massalski TB, Okamoto H, Subramanian PR, Kacprzak L, eds. Binary Alloy Phase Diagrams, 2nd edition. ASM International, 1990.
[13] Lenz M, Gruehn R. Zum chemischen Transport von Molybdän mit $HgBr_2$ – Experimente und Modellrechnungen. Z Anorg Allg Chem 1993, 619, 731–40.
[14] Lenz M, Gruehn R. Zum chemischen Transport von Wolfram mit $HgBr_2$ – Experimente und Modellrechnungen. Z Anorg Allg Chem 1994, 620, 867–78.
[15] Neddermann R, Gerighausen S, Binnewies M. Der chemische Transport von Mischkristallen im System Mo/W. Z Anorg Allg Chem 1996, 622, 21–3.
[16] Schäfer H, Nowitzki J. GaX_3/Ga_2X_6 (X = Cl, J) als Transportmittel für die Metalle Fe, Co, Ni. Z Anorg Allg Chem 1977, 435, 49–55.
[17] Neddermann R, Binnewies M. Der chemische Transport von Mischkristallen im System Co/Ni. Z Anorg Allg Chem 1996, 622, 17–20.
[18] Bosholm O, Oppermann H, Däbritz S. Chemischer Transport intermetallischer Phasen, II: Das System Fe-Si. Z Naturforsch 2000, 55b, 614–26.
[19] Krausze R, Khristov M, Peshev P, Krabbes G. Crystal growth of chromium silicides by chemical vapour transport with halogens. Z Anorg Allg Chem 1990, 588, 123–32.
[20] Wirringa J, Binnewies M. Der chemische Transport von Titangermaniden. Z Anorg Allg Chem 2000, 626, 996–8.
[21] Krausze R, Khristov M, Peshev P, Krabbes G. Crystal growth of chromium by chemical vapour transport with halogens. Z Anorg Allg Chem 1989, 579, 231–9.
[22] Schäfer H, Odenbach H. Über den Einbau des Transportmittels Jod in Kristalle chemisch transportierter Stoffe. Z Anorg Allg Chem 1966, 346, 127–36.
[23] Wirringa J, Wartchow R, Binnewies M. Der chemische Transport von Kupfergermaniden und Kupfersiliciden. Z Anorg Allg Chem 2000, 626, 1473–7.

[24] Krausze R, Krabbes G, Khristov M. Estimation of the chemical transport behaviour of transition metal silicides $MeSi_x$ (Me = Cr, Ti, V, Mo, Fe, Mn) with X_2 (X = Cl, Br, I). Cryst Res Technol 1991, 26, 179–85.
[25] Neddermann R. Dissertation, University of Hannover, 1997.
[26] Bosholm O, Oppermann H, Däbritz S. Chemischer Transport intermetallischer Phasen IV: Das System Fe–Ge. Z Naturforsch 2001, 56b, 329–36.
[27] Schäfer H, Morcher B. Über den Transport von Silicium im Temperaturgefälle unter Mitwirkung der Silicium(II)-halogenide und über die Druckabhängigkeit der Transportrichtung. Z Anorg Allg Chem 1957, 290, 279–91.
[28] Lesser R, Erben E. Über die Reinigung von Silicium durch Transportreaktionen. Z Anorg Allg Chem 1961, 309, 297–303.
[29] Krabbes G, Bieger W, Sommer K-H, Söhnel T, Steiner U. Computerprogram TRAGMIN, Version 5.1, IFW Dresden, TU Dresden, HTW Dresden, 2014.
[30] Trappe O, Glaum R, Gruehn R. Computerprogram CVTRANS, University of Gießen, 1999.
[31] Cordier G, Henseleit R. Crystal structure of ytterbium silver (2/7), Yb_2Ag_7. Z Kristallogr 1991, 194, 146–7.
[32] Merker H, Schäfer H, Krebs B. Neue Pd_xAl_y-Phasen und die Verbindung Pd5Al12. Z Anorg Allg Chem 1980, 662, 49–56.
[33] Schmidt M, Kovnir K, Deichsel J, Binnewies M, Grin Y, Armbrüster M. Fest/Gas-Reaktion zur Darstellung von geträgerten intermetallischen Ga-Pd Katalysatormaterialien. Z Anorg Allg Chem 2015, 641, 1061–8.
[34] Bronger W, Müller P, Wrzesien K. Zur Struktur platinreicher Aluminium-Platin-Legierungen. Z Anorg Allg Chem 1997, 623, 362–8.

Michael Feuerbacher

Growth of high-entropy alloys

1 Introduction

High-entropy alloys (HEAs) represent a novel field in materials science, engaging with alloys that have a high number of components, typically five or more, none of which is present in a predominant proportion. The concept of these baseless multicomponent alloys was proposed in 2004 independently by Yeh et al. [1] and Cantor et al. [2]. The research was initially focused on equiatomic compositions, i. e., alloys, in which the molar ratios of all components are equal, and solidify as (near-) ideal solid solutions with a simple crystal structure. The pioneering works of Cantor and Yeh triggered intense worldwide research and led to a tremendous increase in publication activity. Recently a comprehensive book summarizing the state of the art was released by Gao et al. [3].

HEAs are conceptually different from the conventional alloys scientifically explored and technologically applied over the past 200 years—conventional alloys typically consist of one dominant primary element (e. g., Fe in steels; Al, Ti or Mg in light metals; Ni in high-temperature alloys) to which small amounts, typically less than 5 wt.% in sum, of empirically identified alloying elements are added to attain and optimize particular properties. Examples are the addition of C for the improvement of strength, Cr for corrosion resistance and Al for particle strengthening of nickel-based superalloys in high-temperature applications. In the history of metallurgy, there are only very few successful instances of alloys based on more than one element. Examples are Permendur, based on an FeCo alloy as soft-magnetic material with a 2 % addition of V ($Fe_{49}Co_{49}V_2$) for improved mechanical performance and machinability, Invar $Fe_{64}Ni_{36}$ as a material with a uniquely low coefficient of thermal expansion and $Fe_{50}Ni_{30}Cr_{20}$ as a material with a specifically high electric resistivity to be used as heating elements.

The conceptually new approach of HEAs opens up possibilities for novel physical properties. Indeed, HEAs display a variety of technologically appealing and tunable characteristics, such as high hardness, strength and ductility, oxidation and wear resistance, magnetism, etc., which render them attractive for applications (e. g., [4]). HEAs also show attractive characteristics, in conventional materials mutually excluding properties, such as, for example, low density *and* high hardness [5]. Several HEAs display exceptionally high strength: Yield stresses of about 2.5 GPa [6, 7] and fracture

Acknowledgement: The author thanks C. Thomas, E. M. Würtz, W. Sybertz (all Forschungszentrum Jülich) and M. Heidelmann (U. Duisburg-Essen) for continuing support and long-lasting cooperation. U. Glatzel (U. Bayreuth) and Y. Zhang (University of Science and Technology, Beijing) are acknowledged for sharing information on their experiments and for permission to use their figures in this chapter.

https://doi.org/10.1515/9783110496789-007

stresses approaching 4 GPa [8] at room temperature have been reported in alloys containing more than one phase. Single-phase bcc alloys can also be remarkably strong, the refractory single-phase HEA VNbMoTaW, e. g., shows a 0.2 % yield stress of 1246 MPa at room temperature [9]. Single phase FeCoCrMnNi was demonstrated to show fracture-toughness values above 200 MPa $m^{1/2}$ [10], which exceeds the toughness of virtually all alloys and pure metals. Technological potential of HEAs has been identified in the fabrication of machine components, dies and molds, corrosion-resistant parts, cutting tools, functional coatings, hard-facing, thin-film resistors, diffusion barriers and high-temperature structural components, as summarized by Yeh et al. [11]. Likewise, HEAs are highly interesting from a basic scientific point of view.

In the literature, there exist very few reports dedicated to single-crystal growth of HEAs. Ma at al. [12] succeeded in the growth of a single crystal of composition $CoCrFeNiAl_{0.3}$ by means of the Bridgman technique. The single crystal described in that paper, however, was fairly small, having a volume of about 0.35 cm^3. A larger single crystal was grown by Feuerbacher et al. [13] of equiatomic FeCoCrMnAl, also by a Bridgman approach. This single crystal of about 5 cm^3 consisted of a two-phase structure of coherent B2 inclusions in a bcc matrix. There are some experimental reports on single crystalline samples, which were in fact very small samples taken from coarse grained polycrystals, e. g., [14]. Otherwise, most experimental studies have been carried out on polycrystalline samples, often also on multiphase materials.

A huge body of work in the literature has to be considered with caution, since in many cases experiments have been carried out on ill-defined samples, the microstructural state of which has not been properly documented. In many published studies, the characterization is insufficient and no attention is paid to the thermal history that the samples have been subjected to. Nevertheless, frequently, conclusions are drawn that directly relate experimental results to the assumed disordered nature of the samples. This is particularly critical since, with the HEAs, a new field in materials science is being launched, and the basic knowledge required to discriminate between real effects and mere results of undetected deviant features in the investigated samples is not yet available. The community has recognized this issue, and accordingly alloy design, materials development and optimization, phase stability etc. are now increasingly being debated (e. g., [8, 15–18]). Definitely, there is a need for a well-defined basis of single phase or, preferably, single crystalline samples, to make possible reliable conclusions concerning structure-property relations, and eventually to develop an understanding of the salient physical mechanisms of HEAs.

One of the difficulties faced in materials development of HEAs is the non-availability of phase diagrams, which, in most intermetallic systems, form the basis of targeted materials development. This, of course, is due to the large parameter space of multicomponent systems, for which the systematic exploration of a single phase diagram would by far exceed laboratory timescales. Only selected isopleths have been determined experimentally [19] and by computational thermodynamics [20]. It has turned out, however, that the known HEA phases solidify congruently, so at least the melt composition for crystal growth is straightforward. The exploration of the extensions of HEA phase fields in composition and temperature, however, has to be done by trial and error. For the selection of HEA-forming alloy systems, one has to resort to empirical, HEA-specific formation criteria, which are described in Section 3.

2 What are high-entropy alloys?

HEAs are composed of a high number of metallic elements, typically around five [1] and occasionally up to 20 [2]. Their composition usually is chosen equiatomically or near equiatomically, i. e., none of the elements serves as a base for the alloy. It is commonly accepted for researchers to tolerate non-equiatomic compositions within certain limits that still allow the alloys to be considered baseless. For five-component alloys, this is the case for compositions in which all concentrations range between 5 and 35 at.% [21]. It is a particular feature of HEAs that they solidify as a single phase on a simple crystal lattice. A schematic illustration of a five-component HEA is shown in Figure 1.

Up to now, HEAs with fcc, bcc and hcp structures have been found, the unit cells of which contain between two and four atom positions. The number of symmetrically inequivalent lattice sites is thus lower than the number of alloy components, which is an obstacle against the formation of a chemically ordered crystal structure. This dilemma is solved in HEAs by random placement of the atoms onto the lattice sites: The structure of an ideal HEA consists of a perfectly random substitutional solid solution on a simple crystal lattice. HEAs thus simultaneously possess chemical disorder and topological order (Figure 1) and therewith constitute a third class of metallic materials be-

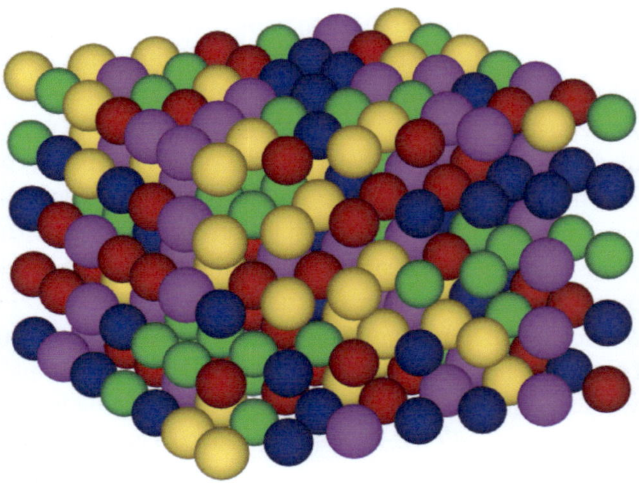

Figure 1: Schematic of a cubic five-component high-entropy alloy. Since the atoms are arranged randomly on the crystal lattice, the alloy possesses chemical disorder and topological order at the same time.

tween ordered crystals and metallic glasses. Materials belonging to the class of ordered crystals, which comprises conventional metallic crystals as well as quasicrystals and complex metallic alloys, possess both chemical and topological order (for which, in the case of quasicrystals, the lattice does not need to be periodic). The class of metallic glasses, on the other hand, comprises all materials possessing neither chemical nor topological order. Examples of ideal (or near ideal) HEAs are the equiatomic phases in the systems FeCoCrMnNi [2], ZrNbTiTaHf [22] and HoDyYGdTb [23], which solidify as a single phase with fcc, bcc and hcp structures, respectively. With their unique and salient features, HEAs open up a novel branch in materials science posing basic scientific challenges, ranging from the hardly explored physics of concentrated solid solutions to the implications of the presence of competing types of order. With their additional degrees of freedom in phase formation, HEAs also offer promising potential for technological applications.

In many alloy systems, it is observed that baseless multicomponent alloys solidify not as a single phase, but instead the formation of two or sometimes more phases is observed. This is still far below the number of phases allowed by Gibbs' phase rule [24], but excludes that these alloys are considered ideal HEAs. A typical example is the equiatomic two-phase alloy AlCoCrFeNi, which solidifies as an ordered B2 matrix with disordered bcc inclusions [25]. In order to discriminate this microstructurally different case, recently the term "compositionally complex alloys" (CCAs) was introduced (see, e. g., [26]). CCAs are thus baseless multicomponent alloys solidifying in multiphase form, while the term HEA should exclusively be used for such solidifying as a single-phase. Be aware, however, that in the literature multiphase alloys are often referred to as HEAs, particular in earlier publications.

Today, several alloy systems are known in which HEAs or CCAs form. Their constitutive elements are located in distinct regions of the periodic table: a first group of HEAs includes combinations of transition metals (TMs), ranging from the lighter TMs, e. g., the firstly discovered HEA FeCoCrMnNi [2], to the heavier TMs, e. g., the recently reported AuCuNiPdPt alloy [27]. These HEAs typically have fcc structure. A second group of HEAs is established by combining refractory metals. These typically form single-phase bcc HEAs, e. g., ZrNbTiTaHf [22] and NbMoTaWV [28]. A third group of hexagonal HEAs was discovered in rare-earth systems, e. g., HoDyYGdTb [23] or YGdTbDyLu [29]. Hexagonal HEAs in non-rare-earth systems were recently found in the four-component systems CoFeReRu [30] and HfScTiZr [31]. For a further group of well-investigated phases, transition metals are combined with aluminum. The inclusion of aluminum drives these systems towards bcc structures, and they are typically multiphase CCAs, since Al tends to the formation of intermetallic phases with many transition metals. Typical examples are equiatomic AlCoCrMnNi [1] and FeCoCrMnAl [32] both having bcc/B2 two-phase microstructures. It is however interesting to note that non-equiatomic alloys in the Al–Co–Cr–Fe–Ni system (below 0.3 molar fraction Al) are single-phase fcc HEAs [33].

Extensive listings of known HEAs and CCAs, summarizing alloy systems, compositions, microstructural states etc., are provided in diverse current review articles, see, e. g., [3, 34, 35].

3 Formation criteria

At constant temperature and pressure, the stability of a system is defined by its Gibbs free energy G, which is defined as

$$G = H - TS. \tag{1}$$

Here H is the enthalpy, a measure of the heat content, which in condensed systems is mainly determined by the total kinetic and potential energies of the atoms in the system. T is the absolute temperature and S the entropy, which is a measure of the randomness of the system. Generally, the system is stable at the lowest possible Gibbs free energy. Any phase transformation in the system is connected to a change in Gibbs free energy

$$\Delta G = G_2 - G_1, \tag{2}$$

where G_1 and G_2 are the Gibbs free energy of the initial and the final state, respectively. While in single-component systems the Gibbs free energy only depends on temperature and pressure, in alloys it additionally depends on the composition.

Alloying a number of elements involves two conceptually distinct steps that correspond to various changes in free energy. First, the composition of the alloy involves bringing together the various elements in their relative proportions. The free energy corresponding to this composition step, G_c, can straightforwardly be calculated as a weighted sum of the free energies of the individual components

$$G_c = \sum_i X_i G_i, \tag{3}$$

where X_i are the molar fractions of the i elements and G_i are their molar free energies at the given temperature and pressure. In a second step, the actual mixing of the atoms takes place. This will result in different bonding of the atoms, causing an enthalpy change, and to a different degree of randomness corresponding to an entropy change. As a result, the final Gibbs free energy after mixing G_f will be different and involve a change in Gibbs free energy given by

$$\Delta G_{\text{mix}} = G_f - G_c, \tag{4}$$

where for G_c and G_f eq. (1) individually holds, i. e., $G_c = H_c - TS_c$ and $G_f = H_f - TS_f$. Consequently, the Gibbs free energy of mixing, i. e., the change in Gibbs free energy due to mixing of all components of the alloy, is given by

$$\Delta G_{\text{mix}} = \Delta H_{\text{mix}} - T\Delta S_{\text{mix}}, \tag{5}$$

where $\Delta H_{\text{mix}} = H_f - H_c$ and $\Delta S_{\text{mix}} = S_f - S_c$. The mixing enthalpy ΔH_{mix} thus represents the difference in internal energy before and after mixing, and the mixing entropy ΔS_{mix} represents the difference in entropy as a result of the mixing step.

This reasoning shows us that the state of an alloy is determined by the Gibbs free energy of mixing. The mixing enthalpy can be positive or negative, depending on the bonding states of the atoms relative to the initial states in the individual components. It can be calculated in terms of a semi-empirical model of atom cohesion, the Miedema model [36], and values are available in tabulated form for binary systems [37]. The mixing entropy is always positive and will thus lead to a decrease of the Gibbs free energy of mixing.

According to statistical thermodynamics the entropy is related to randomness by the Boltzmann equation

$$S = k \ln w, \tag{6}$$

where k is Boltzmann's constant and w is the number of distinguishable ways to arrange the atoms in the solution. Neglecting thermal entropy, we can refer to this entropy as the configurational entropy. In the initial state before mixing, there is only one distinguishable way to arrange the atoms, i. e., $S_c = k \ln 1 = 0$, and therefore

$$\Delta S_{\text{mix}} = k \ln w. \tag{7}$$

The number w can be calculated by standard combinatorics. Applying Stirling's approximation, we end up with an expression for the mixing entropy

$$\Delta S_{\text{mix}} = -R \sum_{i=1}^{n} X_i \ln X_i, \tag{8}$$

where R is the universal gas constant. Since X_i are molar fractions, the sum is always negative and ΔS_{mix} always positive. For the special case of equiatomic alloys, ΔS_{mix} is maximum [38]. For all i, $X_i = 1/n$, and eq. (8) reduces to the simple form

$$\Delta S_{\text{mix}} = R \ln n. \tag{9}$$

In ideal solutions, $\Delta H_{\text{mix}} = 0$ holds, i. e. the mixing leads to no change in internal energy and the Gibbs free energy change is only due to the change in entropy. Therefore, the Gibbs free energy of mixing for ideal solutions is always negative, i. e., mixing always leads to a state of lower free energy. This is consistent with our experience that mixing of ideal systems, e. g., particles without any interaction, always leads to a randomly mixed final state and never to a separation of the individual components.

In real systems, on the other hand, $\Delta H_{mix} \neq 0$. Mixing is endothermic or exothermic, i. e., leads to absorption or a release of heat. In a quasi-chemical approach, assuming that the heat changes are only due to the changes in bond energy of adjacent atoms, the mixing enthalpy can be included in a quasi-ideal solution model: for a two-component system of components A and B, it can be shown that the mixing enthalpy is proportional to the difference ε between the energy of unlike bonds ε_{AB} and the average of like bonds ε_{AA} and ε_{BB}, $\varepsilon = \varepsilon_{AB} - \frac{1}{2}(\varepsilon_{AA} + \varepsilon_{BB})$ [39]. The mixing enthalpy is then given by

$$\Delta H_{mix} = N_a \varepsilon z X_A X_B, \tag{10}$$

where N_a is Avogadro's number and z is the number of bonds per atom. For an ideal solution, $\varepsilon = 0$ and hence $\Delta H_{mix} = 0$, consistent with the case discussed previously. In that case, the arrangement of the atoms is completely random. If $\varepsilon < 0$ the atoms in the system prefer to be surrounded with unlike atoms, if $\varepsilon > 0$, they prefer to be surrounded by their own species.

For multicomponent alloys commonly the relationship

$$\Delta H_{mix} = 4 \sum_{i<j} c_i c_j \Delta H_{ij} \tag{11}$$

is used [38], where c_i and c_j are the concentrations of the elements i and j, respectively, and ΔH_{ij} is the binary mixing enthalpy of elements i and j. The mixing enthalpy of the alloy can then be calculated using, e. g., the values for ΔH_{ij} provided in [37].

The Gibbs free energy of mixing for non-ideal multicomponent alloys thus reads

$$\Delta G_{mix} = 4 \sum_{i<j} c_i c_j \Delta H_{ij} + RT \sum_i X_i \ln X_i. \tag{12}$$

On the right side of the equation, the first term corresponds to the mixing enthalpy and the second term to the mixing entropy. As in the case of ideal solutions, the entropy term always leads to a decrease of free energy upon mixing. In the non-ideal case, the entropy term balances against the enthalpy term, which can be positive or negative.

A formation criterion for HEAs has been formulated along these lines: Yang and Zhang [40] introduced the parameter

$$\Omega = \frac{T_m \Delta S_{mix}}{|\Delta H_{mix}|}, \tag{13}$$

where T_m is the absolute melting temperature of the alloy. For ideal solutions, the melting temperature can be approximately calculated as the weighted mean of the melting points of the constituting elements, i. e.,

$$T_m = \sum_i T_{m,i}, \tag{14}$$

where $T_{m,i}$ is the melting temperature of element i of the alloy.

If $\Omega > 1$, the entropy term dominates the Gibbs free energy of mixing when the alloy solidifies and a solid solution can form. The larger Ω, the higher are the changes that a solid-solution phase forms. If, on the other hand, $\Omega \leq 1$ the mixing enthalpy is predominant at solidification, and either an ordered intermetallic compound will form (when ΔH_{mix} is negative), or the components will segregate (when ΔH_{mix} is positive).

Other thermodynamics-based criteria have been proposed in the literature. In their simplest form, only ΔS_{mix} is considered, and HEAs are expected to form if ΔS_{mix} is large [1]. This also illuminates why this class of alloys was referred to as high-entropy alloys by Yeh. Later on, Zhang and Zhou [38] took the competition between ΔS_{mix} and ΔH_{mix}, which is implied in eq. (5), into account. They formulated a HEA criterion based on ΔH_{mix}, which implies the assumption that ΔS_{mix} is large anyway, because of the high number of components of the alloy. More elaborately, thermodynamics-based criteria were formulated by Ye et al. [41], relating the Ω parameter to the ratio of the total entropy of mixing and the purely configurational entropy of mixing (which implies taking not only configurational effects, but also coupling effects of atom size, atomic packing fraction and the number of elements into account). Senkov and Miracle [42] explicitly addressed the competition between the formation of intermetallic phases and solid-solutions by comparing their respective formation enthalpies for a given annealing temperature.

Purely thermodynamic criteria, however, do not suffice to predict whether or not a solid-solution phase will form, but other critical factors have to be accounted for. Highly important, as already recognized by Hume-Rothery, is the relative size of the atomic species in the alloy [43]. If the size difference is too large, substitutional solid solutions cannot form. As a corresponding criterion, the radius difference

$$\delta r = \sqrt{\sum_i c_i (1 - r_i/\bar{r})^2} \tag{15}$$

is considered, where r_i are the radii of the components and $\bar{r}$ is the weighted average radius calculated according to

$$\bar{r} = \sum_i c_i r_i . \tag{16}$$

Taking both thermodynamic and radius-difference criteria into account, limits for HEA formation can be formulated by statistical evaluation of solidification data. For as-cast samples, several references provide more or less consistent values: Wang et al. [44] find that $-15\,\text{kJ/mol} \leq \Delta H_{\text{mix}} \leq 5\,\text{kJ/mol}$ and $\delta r < 6.6\,\%$ are required for the formation of HEAs. This is comparable to the values provided in [45] ($-12\,\text{kJ/mol} \leq \Delta H_{\text{mix}} \leq 5\,\text{kJ/mol}$ and $\delta r < 6.2\,\%$) and [46] ($-20\,\text{kJ/mol} \leq \Delta H_{\text{mix}} \leq 5\,\text{kJ/mol}$ and $\delta r < 8\,\%$) and, in terms of the Ω parameter, $\Omega \geq 1.1$ and $\delta r < 6.6\,\%$ [40]. All authors agree that occasionally intermetallic phases occur within these limits. For a more likely formation of pure single-phase HEAs the limits should be narrowed down to about $-5\,\text{kJ/mol} \leq$

$\Delta H_{mix} \leq 5\,kJ/mol$ and $\delta r < 4\,\%$ [47], or, in terms of the Ω parameter, $\Omega \geq 10$ and $\delta r < 3.8\,\%$. This compares to the findings of Wang et al. [44], which report that annealing between 50 and 90 % of the melting temperature leads to narrower limits requiring $-7.5\,kJ/mol \leq \Delta H_{mix}$ and $\delta r < 3.3\,\%$ due to the formation of intermetallic phases [42].

Two further indicative criteria for HEA formation exist: first, it is favorable if the components of the alloy possess identical crystal structure. This criterion corresponds to one of the Hume-Rothery rules for substitutional solid solutions [43], requiring identical crystal structures of the solvent and the solute atoms. However, this is not a condition *sine qua non*, as e. g., demonstrated by the single-phase Cantor HEA FeCoCrMnNi [2], which solidifies in an fcc structure, but contains bcc (Fe, Cr and Mn) fcc (Ni) and hcp (Co) elements. Second, it is favorable if the binary phase diagrams of the alloy components show a high mutual solubility. Again, this condition is not inevitable as displayed by the single-phase HEA ZrNbTiTaHf [22], which forms despite the fact that, e. g., Zn has a very low solubility in Ti and diverse intermetallic phases exist in the Ti–Zn system [48]. This condition also is not independent of the formation criterion considering the enthalpy of mixing: if the binary mixing enthalpy of two components has a high negative value, they have a strong tendency to form intermetallic phases.

4 Growth methods

4.1 Polycrystals

HEAs have been produced in bulk form by solidification from the melt using various techniques and via a solid-state route by mechanical alloying and subsequent sintering. HEAs in the form of thin films were produced by physical vapor-deposition techniques such as arc deposition, pulse laser deposition or sputtering. In this chapter, we only consider bulk solidification from the melt; an overview of the other approaches is provided in [49].

4.1.1 Arc melting

The most commonly used technique for HEA polycrystals is arc melting, where the melt is directly exposed to an electric arc between a moveable water-cooled tungsten electrode, the “stinger”, and a water-cooled copper hearth, which acts as counter-electrode and contains one or more crucibles. The arc is formed by igniting a gas plasma at low voltage and high current. Laboratory-scale arc-melting devices typically have sample capacities in the range from a few grams to about 200 grams and reach temperatures of up to 3500 °C. Therefore, arc melting is essential for the production of samples with high melting points, e. g., refractory HEAs in the systems Zr–Nb–Ti–Ta–Hf or W–Nb–Mo–Ta–V.

Arc melters can be used to produce button-shaped samples if the melt is solidified in the crucibles, or as rods by pouring into a cylindrical copper mold. For the latter, the pouring process can be accelerated by vacuum in order to achieve higher cooling rates, a process referred to as suction casting.

The advantages of arc melting are the high temperatures, and the fast heating and cooling rates that can be reached. A downside of the fast cooling rates is that the solidification process cannot be controlled very well, and the solidification rate is not constant throughout the sample. As a result, the microstructure in arc-melted samples often changes from the surface to the center. Further typical artefacts are elemental segregation, suppression of equilibrium phases, pores and occasionally cracks, and tungsten contamination from the anode.

4.1.2 Induction melting

Induction melting is also a very widespread technique, particularly suited to alloys with not too high melting points, such as, e. g., FeCoCrMnNi or AlCoCrFeN. The material is placed inside a high-frequency coil and is heated via electromagnetic induction of eddy currents in its interior. The melt can be contained in a conventional crucible or levitated within the high-frequency coil. Levitation melting is particularly advantageous because contamination and reaction of the melt with a crucible can be avoided. The maximum temperature that can be reached depends on the available generator power, the geometry of the coil and the physical properties of the constituting elements.

Induction heating enables precise control of the heating and cooling rates and also achieves very good homogenization of the material in the molten state. Melting can be carried out over a wide pressure range, e. g., under a protective atmosphere or under vacuum. The technique can be combined with rod casting options, which allow for higher cooling rates, and can be extended to non-conductive elements by the use of a susceptor. Figure 2(a) displays depicts an equiatomic, 63-g FeCoCrMnAl ingot with a diameter of about 30 mm, produced by levitation melting in a water-cooled copper crucible.

4.1.3 Homogenization

HEA samples in the as-cast state generally cannot be considered homogeneous and in equilibrium. Typically, the microstructure of as-cast HEA samples displays a dendritic microstructure (Figure 2(b)). A subsequent annealing step of sufficient time at elevated temperature will reduce inhomogeneities, equilibrate stable phases and resid-

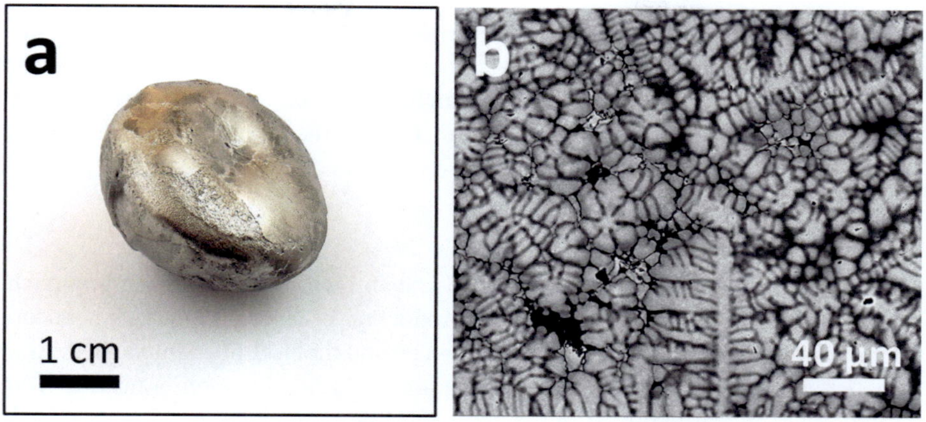

Figure 2: (a) Equiatomic FeCoCrMnAl ingot produced by levitation melting. (b) Backscattered-electron SEM micrograph of the dendritic microstructure of an equiatomic ZrNbTiTaHf HEA.

ual stresses. The required annealing conditions have to be adapted to the individual alloy: the annealing temperature usually is chosen close to the melting temperature, typically $0.8T_m$ or $0.9T_m$. Using a diffusion constant D of about $10^{-14}\ \mathrm{m}^2\ \mathrm{s}^{-1}$ [50], diffusion distances $\sqrt{2Dt}$ of 10 nm to 50 nm, which correspond to typical dendrite diameters in HEAs, require annealing times of about two to 50 h at $0.9T_m$.

4.2 Single-crystal growth techniques

4.2.1 Bridgman technique

The Bridgman technique is a classical approach to single-crystal growth. The melt is placed in a furnace, in most implementations vertically, under a well-defined temperature gradient. Solidification is induced by downward movement of the melt relative to the gradient. The geometry of the furnace and the crucible are chosen such that nucleation takes place in a small volume at the bottom of the crucible. This is frequently attained by using a pointed-tip shaped crucible in combination with a cold finger. The result of a Bridgman growth typically is a rod-shaped ingot with a diameter equal to the interior of the crucible.

An advantage of Bridgman technique is that the diameter of the crystal grown depends only on the diameter of the crucible used, which is in contrast to the Czochralski technique (see the following section). The growth temperature can be chosen freely such that optimum growth conditions according to the phase diagram are met. Furthermore, the cost of manpower applying this technique is low since the process does not have to be continuously monitored. A technical disadvantage of the method is that spontaneous nucleation may take place at the crucible walls and lead to unwanted secondary grains. Also, the ingot may stick rigidly to the crucible after cooling, so that its removal requires destruction of the (sometimes expensive) crucible and can lead to fracture of the crystals. Seeded growth can be carried out with the Bridgman technique, but is more difficult to implement than, e. g., for the Czochralski technique [51].

4.2.2 Czochralski technique

The Czochralski technique is another well-known single-crystal growth approach, which is particularly widespread for the production of large high-quality semiconductor wafers. This technique employs a seed crystal that is dipped into a melt. After establishing a connection between the melt and the seed, the latter is steadily withdrawn and a crystal is pulled out of the melt. The process temperature is chosen such that a stable meniscus forms at the solid-liquid interface and the crystal grows with the desired diameter. The crucible and the seed can be rotated during growth to achieve a higher homogeneity of the product.

In order to keep the number of grains in the crystal as low as possible, a region of very small diameter can be pulled in an early stage of the growth. This so called "thin neck" is a means for geometrical grain selection by decreasing the diameter so far that grain boundaries potentially present in the primarily solidified crystal run out of the crystal and that only one grain survives.

One should discerns between homogeneous and heterogeneous seeding, meaning that either a piece of crystal of the phase to be grown or another suitable material, respectively, is used as a seed. The first is of particular importance, since it allows for a deliberately oriented growth of a crystal. If a small oriented single crystal is used, the newly solidified material preferentially continues the predetermined orientation of the seed.

The Czochralski technique requires experience to master, but has a number of valuable advantages. First, the crystal is essentially unstrained during cooling, thus a high structural perfection can be obtained. Second, the crystal grows freely without contact to a crucible and thus there is less chance of contamination or spontaneous nucleation than for the Bridgman technique. Third, as mentioned above, the orientation of the grown crystals can be controlled via the seed crystal. It is furthermore advantageous in many cases that the growth of the crystal can be observed throughout all stages of the process.

4.2.3 Zone melting

Zone melting, also referred to as float-zone method, is a two-step process particularly suitable for materials that potentially react with available crucible materials, or which have a very high melting point. In a first step, a feed rod of suitable composition is produced, e. g., by arc melting and subsequent casting into a cylindrical mold. Second, the feed rod is placed in a heating device and locally heated, so that a narrow molten zone forms, which is then moved slowly and uniformly along the ingot. This growth technique is containerless, i. e., at no time during the growth does the melt come into contact with a crucible. Most commonly, heating is carried out inductively, by placing the feed rod vertically inside a high-frequency coil. Other widespread heating options are mirror furnaces using of high-power lamps as heaters and electron-beam heating for very high temperatures.

Once the zone is molten, the thickness of the crystal can be regulated by changing the relative velocity of the upper and lower feed-rod attachments. This makes it possible, by increasing the relative velocity, to attempt to grow a thin neck for grain selection similar to the Czochralski technique.

4.2.4 Further techniques

All single-crystal growth experiments on HEAs published to date were carried out by the three techniques described above or variations thereof. However, there are numerous further single-crystal growth techniques that may be applicable to HEAs. Examples are: flux growth, where a temperature transient instead of a temperature gradient as in the Bridgman approach is employed; the Kyropoulos technique, where a crystal grows from a cooled seed into the melt; horizontal versions of the Bridgman or zone-melting techniques, etc. For an overview, see, e. g., [51].

5 Single-crystal growth of HEAs

In the field of HEAs to date, not much work has been dedicated to the development of single-crystal growth routes. The major body of experimental investigations in the literature has been carried out on polycrystalline samples, produced by arc or induction melting as described previously. The so-produced as-cast materials for many (but not all) investigations were subsequently heat treated in order to bring them into a more equilibrated and homogenized form.

In the following, we will describe the single-crystal growth approaches that were published, were carried out by the author, or otherwise came to the attention of the author. To date, this comprises HEAs in only four different alloy systems.

5.1 FeCoCrMnNi

The HEA FeCoCrMnNi was one of the alloys Cantor investigated in his pioneering paper of 2004 [2] and the first single-phase HEA discovered. Single-crystal growth ap-

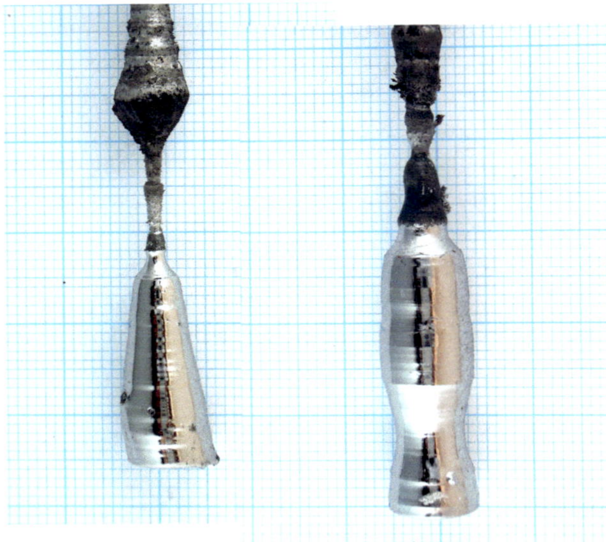

Figure 3: Crystals of equiatomic FeCoCrMnNi produced by Czochralski growth.

proaches were up to then carried out via the Czochralski technique and investment Bridgman growth, both using a melt of equiatomic composition.

Figure 3 displays two crystals grown by the Czochralski technique. Both were grown under a protective argon atmosphere from an equiatomic melt of high-purity elements using a coarse-grained homogeneous seed of about 3 mm in diameter, which was extracted from a previously produced Bridgman ingot. The seed was rotated at 20 rpm and withdrawn at 1 mm/h during the growth. After establishing a stable contact between the seed and the melt, the temperature was raised in order to decrease the diameter of the growing crystal, and a thin neck was pulled for grain selection. Subsequently, the temperature was decreased again, which lead to an increase in crystal diameter. The desired diameter in these growth attempts was about 10 mm, which was reached faster in the example on the right. The growth was stopped after having been pulled to about 20 to 25 mm.

The surface of the crystals is clean and shiny, and careful inspection reveals facets running over the surface from top to bottom. The volumes of the crystals shown are about 0.8 (left) and 1.6 cm^3 (right).

Back-reflection X-ray Laue diffraction of the grown crystals was used for a non-destructive check of the single-crystallinity of the grown crystals. Figure 4(a)–(c) display a set of Laue patterns of the fourfold (1 0 0) plane (a), the twofold (1 0 1) plane (b) and the threefold (1 1 1) plane (c) taken of the surface of the left crystal in Figure 3. These patterns are consistently found on the surface by rotating the crystal by the corresponding relative angles on a goniometer stage. The Laue reflections are clear and sharp, indicating the high structural quality of the material.

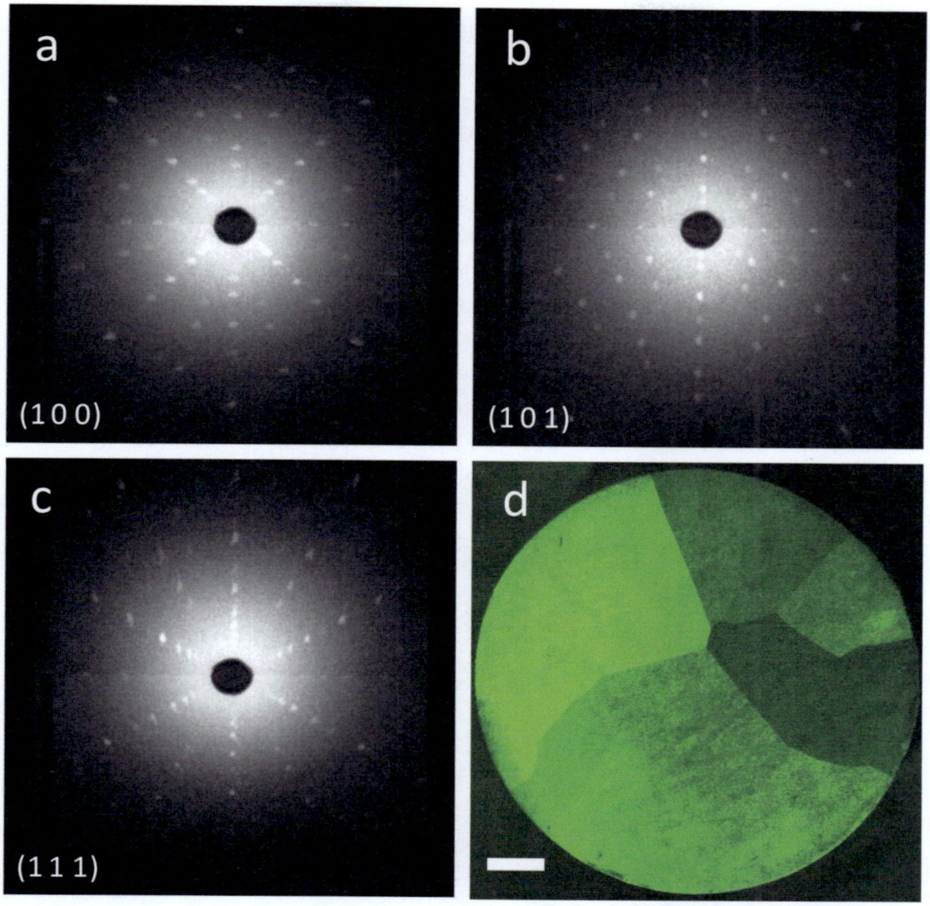

Figure 4: Back-reflection X-ray Laue patterns of the fourfold (1 0 0) plane (a), the twofold (1 0 1) plane (b), and the threefold (1 1 1) plane (c) from a Czochralski-grown FeCoCrMnNi single crystal. Optical micrograph of an etched slice of a coarse-grained FeCoCrMnNi crystal with five grains (d). The length of the scale bar is 1 mm.

Seeded Czochralski growth does not always reproducibly lead to single-crystalline results. Figure 4(d) shows an optical micrograph of a slice extracted from the crystal on the right in Figure 3. The surface was polished and etched in order to make the grain boundaries visible. The slice consists of five grains, the largest of which extends over about a quarter of the surface area. The grain boundaries are straight and separate regions of clearly different contrast. Laue investigation of this crystal shows consistent images only within the surface areas corresponding to a single grain, traversing a grain boundary leads to discernibly different patterns.

Even the single crystals displaying consistent Laue images, which are thus concluded to be single crystals, occasionally display the presence of subgrains, i. e., crystal areas tilted by relative angles of a few degrees, which are separated by low-angle grain boundaries. Figure 5(a) shows a backscattered-electron image taken in the scanning electron microscope (SEM) at a high probe current. The sample area consists of two grains, a triangular grain G1 in the lower part of the image and the rest G2. These grains are imaged with discernible contrast in the SEM and are separated by a grain boundary (black arrow), which is essentially straight. The larger dark grain G2 consists of four subgrains, imaged with much lower contrast. The subgrains are separated by subgrain boundaries, which often appear much wavier (white arrow) than the regular grain boundaries. Subgrain boundaries can also be detected in Laue diffraction: Figure 5(b) is a Laue pattern taken with the X-ray beam placed exactly on a subgrain boundary. All spots are doubled, and the pattern corresponds to two superposed similar Laue patterns belonging to the two subgrain orientations. The relative angle of the subgrains can be determined by measuring the distance between two spots (e. g., using a Wulff net) and in this example amounts to about 2°.

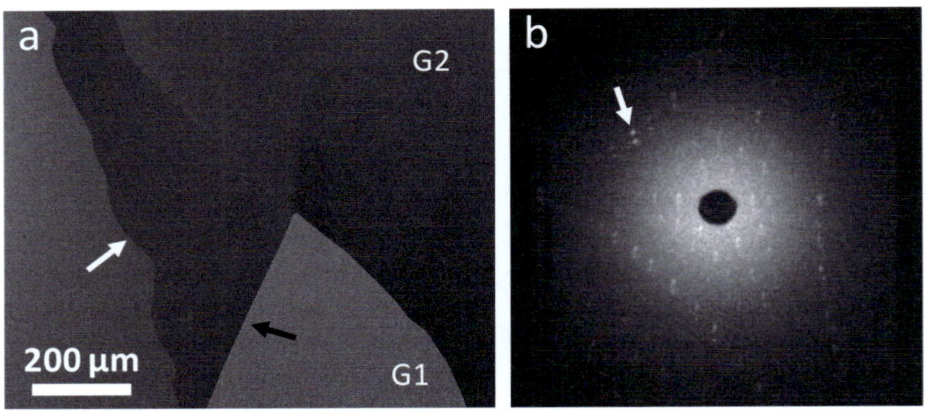

Figure 5: (a) Backscattered-electron micrograph of a FeCoCrMnNi crystal. Two grains, G1 and G2 are present, which are separated by a grain boundary (black arrow). In grain G2, additional subgrain boundaries (white arrow) are present. (b) Back reflection X-ray Laue pattern taken in a surface area containing a subgrain boundary displaying double spots (arrow).

Outstandingly large single crystals of the Cantor alloy were produced by investment Bridgman casting [52]. Investment casting is a well-established manufacturing process based on lost-wax casting, routinely used on an industrial scale, e. g., for the production of single-crystalline intermetallic turbine blades. The alloy is inductively melted in vacuum in a rotatable crucible and then poured into a pre-heated mold, which is then, corresponding to the Bridgman process, withdrawn from a temperature gradient to induce solidification. For the growth of the Cantor alloy, a zirconia shell mold with a spiral grain selector was used. The latter separates the primary solidification volume of the mold, the starter block, from the region in which the actual crystal grows. Of the several grains nucleating in the starter block, only a single one survives the competitive growth through the spiral [53].

A crystal was grown under the following conditions: an equiatomic melt was composed of high-purity elements and pre-alloyed by arc melting under 500-mbar argon atmosphere. Bridgman investment casting was carried out in vacuum (10^{-4} mbar) with a temperature gradient in the casting mold of 6 K mm^{-1} and a withdrawal rate of 3 mm min^{-1}. Figure 6 displays the resulting crystal after removal from the crucible. The part on the left-hand side solidified in the starter block, the middle part in the spiral grain selector, and the actual crystal extends to the right-hand side. The crystal has a length of about 110 mm and a diameter of 15 mm. It was found to be entirely single crystalline with a growth direction parallel to the <0 0 1> direction of the fcc lattice.

Figure 6: Equiatomic FeCoCrMnNi single crystal produced by investment Bridgman casting. Image courtesy by U. Glatzel, University of Bayreuth [52].

The final crystal had an average composition of $Fe_{21.8}Co_{24.8}Cr_{20.4}Mn_{9.1}Ni_{23.9}$ measured by micro X-ray fluorescence analysis. The Mn content varies during growth due to evaporation: at the beginning of the growth, i. e., at the bottom of the crystal, 11.1 at.% and at the end of the growth 7.8 at.% Mn were found.

Figure 7(a)–(b) show a back-reflection Laue X-ray pattern and an electron backscatter diffraction (EBSD) pole figure taken in the SEM, respectively. Both were taken perpendicular to the growth direction and display the fourfold pattern of the (0 0 1) plane. The reflections in the Laue pattern are sharp, and the poles in the EBSD figure are narrow, which reflects the high structural quality of the sample. The mosaicity is small with a deviation from the <0 0 1> growth direction below 5°.

Further analysis shows that one of the slices taken from the crystal contains a small-angle grain boundary. This can be seen as a change in the etch pattern in Figure 7(c), where the small-angle grain boundary is located between the two arrows. X-ray Laue investigation of the two subgrains demonstrates that their relative angle is 4°. The etch pattern itself (Figure 7(d)) reflects the fourfold symmetry of the <0 0 1> direction in the fcc structure.

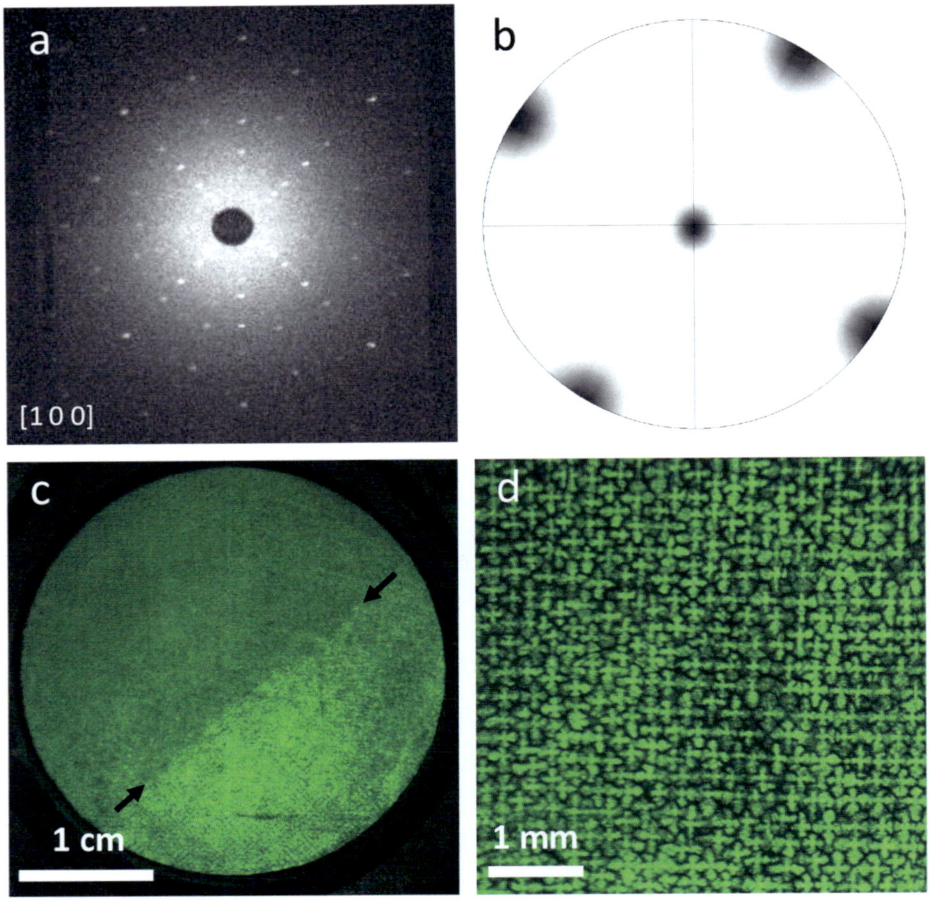

Figure 7: Characterization of the FeCoCrMnNi single crystal shown in Figure 6. (a) Back-reflection X-ray Laue diffraction perpendicular to the growth axis, showing the fourfold pattern of a (0 0 1) plane. (b) Corresponding {0 0 1} pole figure obtained by EBSD-SEM of the texture displaying the <0 0 1> growth direction and narrow mosaicity. The linear grey scale spans from 0 (white) to about 36 (dark grey). (c) Optical micrograph of the surface of an etched slice taken from the crystal. The change of the pattern indicates the presence of a small-angle grain boundary (arrows). (d) Etch pattern at higher magnification. The pattern reflects the symmetry of the (0 0 1) surface.

5.2 Al–Co–Cr–Fe–Ni

Al–Co–Cr–Fe–Ni is another well investigated HEA-forming alloy system. Near equiatomic HEAs in this system are known to solidify as a two-phase material with an ordered B2 matrix rich in Al and Ni and disordered bcc inclusions rich in Cr and Fe [25]. At Al contents below about 0.3, the alloys solidify as a single phase with fcc structure, and a transition with co-existing bcc and fcc phases extents from Al contents of about

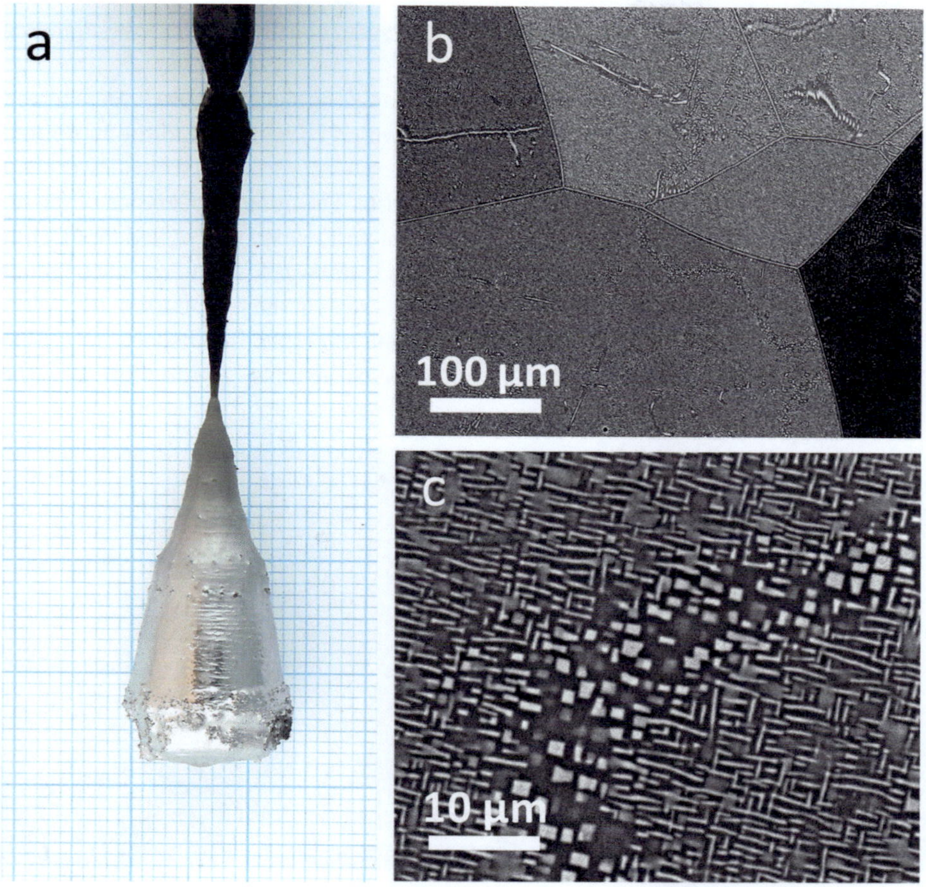

Figure 8: (a) Equiatomic AlCoCrFeNi crystal produced by Czochralski growth. (b) Backscattered-electron micrograph displaying a set of grain boundaries and the presence of a two-phase microstructure. (c) Backscattered-electron micrograph at higher magnification displaying the presence of a brighter Cr and Fe rich phase in a darker Al and Ni rich matrix.

0.3–0.88 [33]. Single-crystal growth approaches were carried out for three different compositions in this system.

Figure 8(a) shows a crystal of equiatomic composition AlCoCrFeNi grown by the Czochralski technique. The growth was carried out using pre-alloyed high-purity elements under a 600 mbar argon atmosphere at a pulling speed of 10 mm/h from an alumina crucible. As a seed, a pointed tungsten rotating at 20 rpm was used. After pulling of a thin neck, the crystal cursorily looks good, but closer inspection reveals that it is polycrystalline and contains multiple phases. Figure 8(b) is a SEM backscattered-electron micrograph showing a set of grain boundaries, the grain size in the crystal is of the order 500 µm to 1 mm. It also displays the presence of a two-phase microstructure. At higher magnification (Figure 8(c)), the presence of a brighter Cr and Fe rich phase in a darker Al and Ni rich matrix is clearly seen, consistent with investigations in the literature (e. g., [25]).

A crystal of different composition in this system is shown in Figure 9. Here a melt of composition $Al_{28}Co_{20}Cr_{11}Fe_{15}Ni_{26}$ was used, which corresponds to the measured matrix composition of the equiatomic crystal. The growth was carried out slower than that of the latter, at a speed of 1 mm/h after necking. The outer appearance of the resulting crystal is less shiny (besides the lowermost part, which was pulled rapidly out of the melt at the end of the growth process). The crystal was found to consist of about four to six large grains. A set of grain boundaries is seen in the high-current SEM electron backscatter image in Figure 9(b). On a 100-micron scale, the inspection by SEM using an electron backscatter detector shows no features (see Figure 9(c)), i. e., the

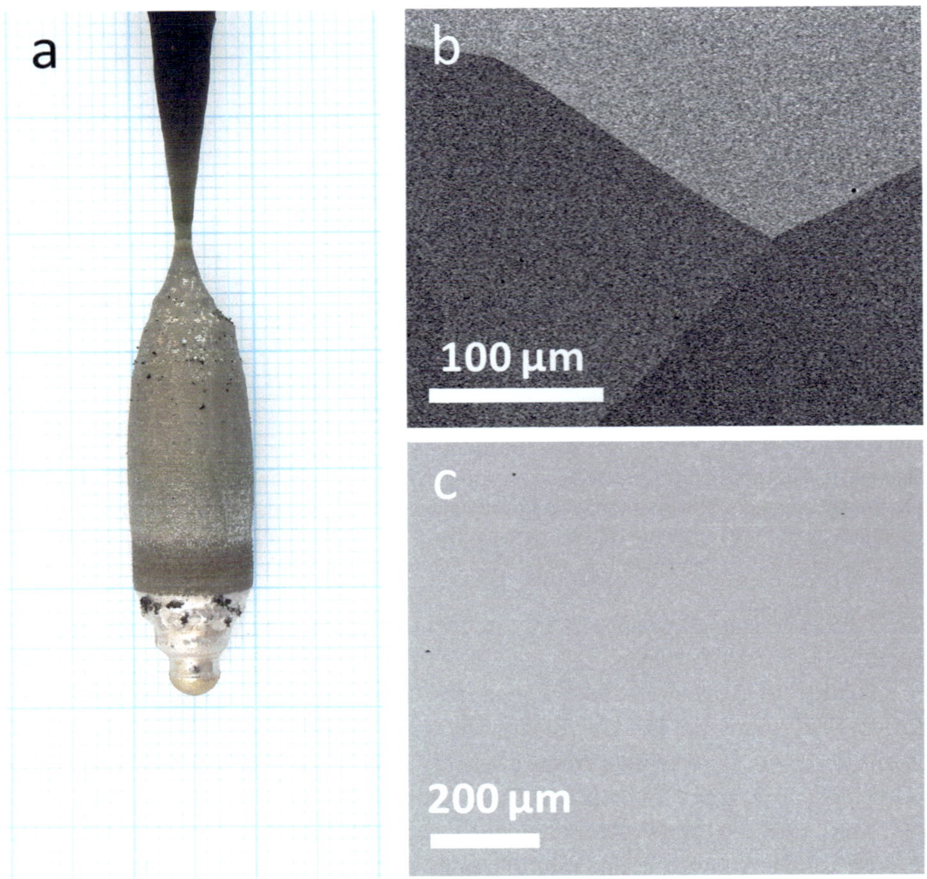

Figure 9: (a) $Al_{28}Co_{20}Cr_{11}Fe_{15}Ni_{26}$ crystal produced by Czochralski growth. (b) Backscattered-electron micrograph displaying a set of grain boundaries. (c) Featureless backscattered-electron micrograph at higher magnification.

material appears like a single phase. Closer inspection by means of high-resolution scanning transmission electron microscopy (STEM) with a high-angle annular dark-field detector, however, reveals that the material consists of an Al–Ni rich matrix with spherical inclusions of a density of about 2 vol.% (Figure 10). The matrix was identified by high-resolution STEM and electron diffraction as a B2 structure [54]. The spherical inclusions are rich in Cr and Fe and of disordered bcc structure. Their size distribution is highly monodisperse, the diameter of most particles amounting to about 17 nm. Their lattice is perfectly congruent with that of the matrix (Figure 10(b)). This was also observed for equiatomic AlCoCrFeNi HEAs [55], which is composed of the same phases but with a higher proportion of inclusions [25].

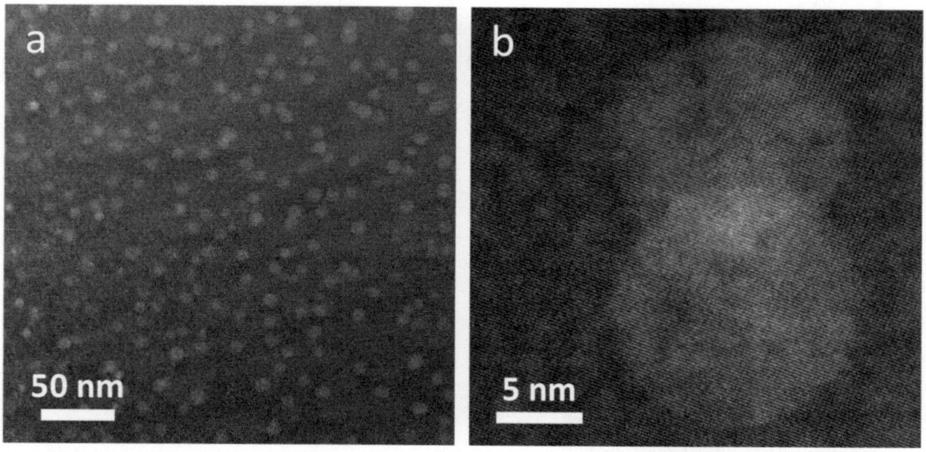

Figure 10: (a) High-resolution STEM image of $Al_{28}Co_{20}Cr_{11}Fe_{15}Ni_{26}$ displaying the presence of spherical inclusions. (b) At higher resolution the congruent interface between the inclusions and the matrix is seen.

Ma et al. [56] successfully grew single crystals of an alloy on the Al-poor side of the system, with a composition $Al_{0.3}CoCrFeNi$. In this region of the alloy system, the material solidifies as a single phase with fcc structure [33]. These authors employed a Bridgman setup, with a tube crucible 3 mm in diameter, which was inductively heated and withdrawn at a velocity of 5 µm/s through a temperature gradient of about 45 K/mm into a water-cooled Ga–In–Sn liquid. The outstanding feature of their approach is that two subsequent Bridgman runs are carried out, in between which the crucible tube is turned by 180°.

Figure 11 on the left-hand side schematically shows the resulting crystal after the first Bridgman run. At the bottom of the tube crucible, a dendritic microstructure (D) corresponding to the as-cast state of the pre-melted alloy resides. Along the direction of growth (arrow), a varying microstructural morphology developing from equiaxed grains (E) with grain sizes 50–300 µm, to extended columnar grains (C) in the mil-

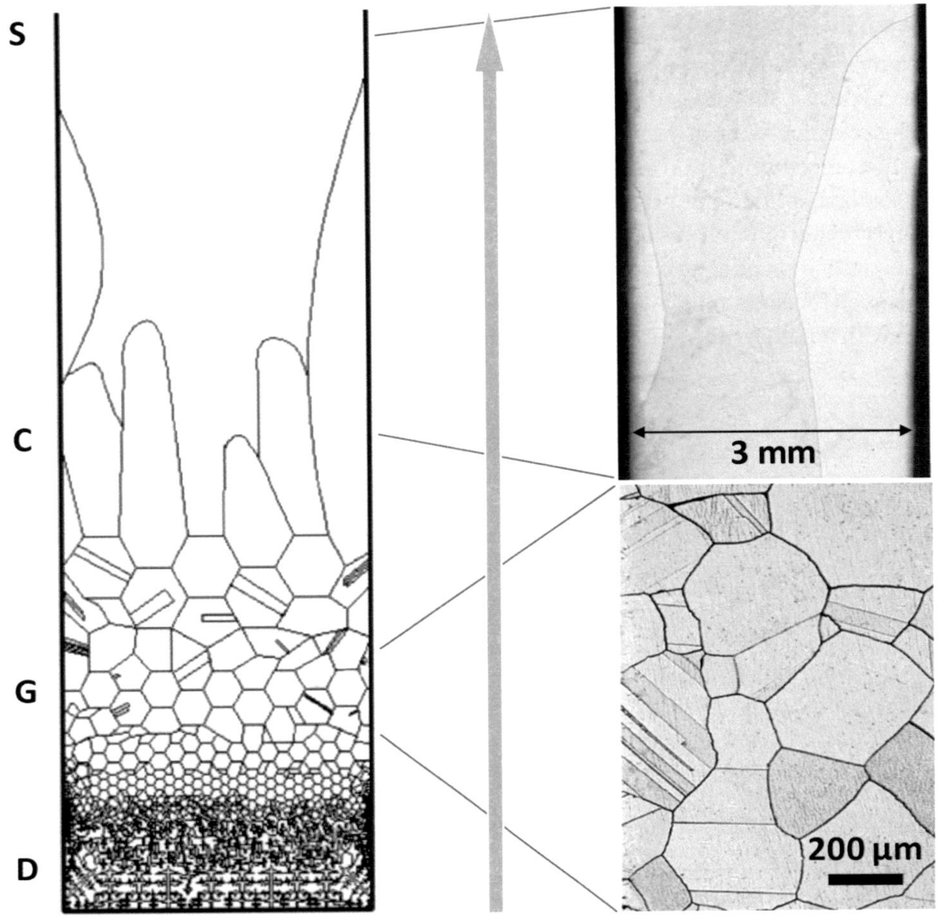

Figure 11: Growth of an $Al_{0.3}CoCrFeNi$ crystal by the Bridgman technique according to Ma et al. [56] (see text). Image courtesy by Y. Zhang, University of Science and Technology, Beijing.

limeter range, is found. Towards the upper end of the crucible, a single grain (S) extending over the full diameter of the tube prevails. On the right-hand side, two SEM micrographs of the equiaxed-grain and the columnar-grain region, respectively, are shown.

For the second Bridgman run, the tube crucible is inverted and the procedure is repeated under the same conditions. The single-crystalline region is now at the bottom and not entirely melted, so that it acts as a seed for the solidification of the melt. As a result, the entire crystal solidifies in the form of a single grain. Due to the use of a rather small crucible, the final single crystal has a diameter of 3 mm and a length of 50 mm, i. e., its volume of 0.35 cm^3 is rather small. However, this ingenious approach most likely can be scaled up and be used for the growth of larger single crystals.

5.3 Zr–Nb–Ti–Ta–Hf

Senkov [9, 22] was the first producing HEAs using refractory metals, which is a promising idea since the binary enthalpies of mixing for these elements are mostly very low and the mutual solubilities high. An issue for the growth of refractory HEAs is their melting point which is typically very high, thus ruling out diverse crucible materials. Zirconia and Alumina crucibles, e. g., can only used up to about 1,750 and 2,000 °C, respectively. Here we discuss HEAs in the system Zr–Nb–Ti–TaHf, which solidify as a single-bcc phase at an estimated melting point of 2,250 °C [22] (for the equiatomic case). Single-crystal growth experiments have, up to now, been carried out exclusively by zone-melting on HEAs of two different compositions.

Figure 12(a) shows a view into the growth chamber during the production of a Zr–Nb–Ti–Ta–Hf crystal by zone melting. The crystal is heated via the induction coil in the center of the hot zone and molten locally over a length of about 5 mm. In this ex-

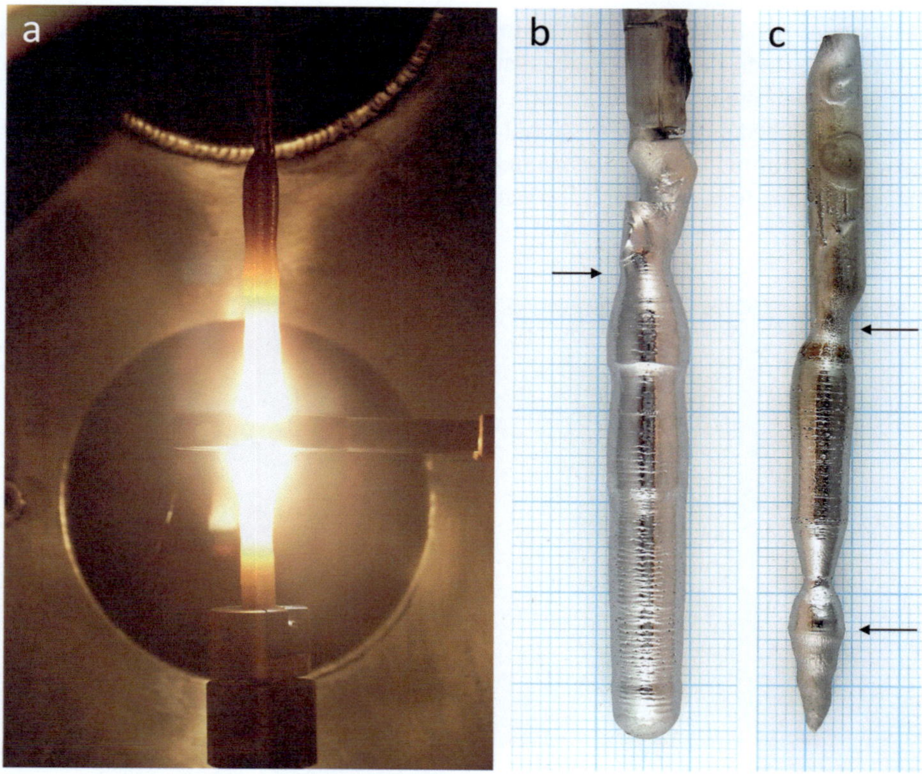

Figure 12: (a) Growth of a Zr–Nb–Ti–Ta–Hf crystal by zone melting. (b) Resulting crystal of the equiatomic composition. (c) Resulting crystal of the composition $Zr_{12.7}Nb_{30.8}Ti_{17.7}Ta_{30.8}Hf_{8.0}$. The actual crystal grown is below the arrow (in b) and between the arrows (in c).

periment, the feed rod, which was in a first step produced by arc-melting and casting, is moved upwards with respect to the fixed coil, so that the crystal grows from top to bottom.

A resulting crystal of the equiatomic composition ZrNbTiTaHf is displayed in Figure 12(b). In the upper part of the figure, the feed rod is seen. The growth was started at the position of the arrow and continued downwards with a rate of 5 mm/h with synchronized velocities of the upper and lower feed-rod attachments, resulting in an approximately constant diameter. The growth is carried out under rotation of the upper feed rod. The growth was terminated after about 50 mm by fast movement of the upper feed-rod attachment. Figure 12(c) displays a similarly grown crystal of a non-equiatomic crystal with the composition $Zr_{12.7}Nb_{30.8}Ti_{17.7}Ta_{30.8}Hf_{8.0}$. During this growth, after an initial phase of constant diameter for about 21 mm, we experimented with relative velocities of the upper and lower feed-rod attachments in order to manipulate the crystal diameter. From about 21–28 mm the upper feed-rod attachment was moved faster than the lower, in order to decrease the crystal diameter. Afterwards, the velocities were synchronized again, in order to adjust the diameter back to that of the feed rod. After five additional mm (arrow), the growth experiment was terminated.

SEM analysis shows that both crystals are single phase with a composition close to that of their melt, and both contain grain boundaries. Figure 13 are optical micrographs of etched surfaces of sectional cuts of the crystal shown in Figure 12(c) and provide insight to its three-dimensional grain structure. Figure 13(a) displays a slice taken from the upper part of the crystal perpendicular to the growth direction. The

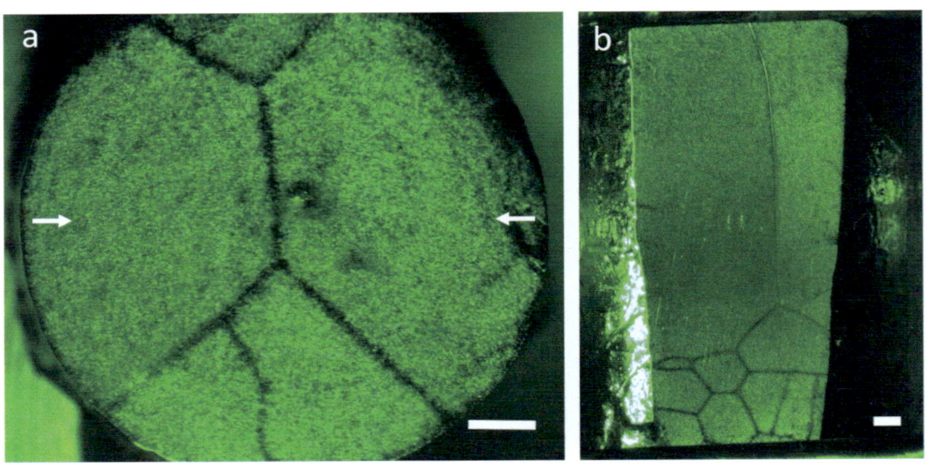

Figure 13: Optical micrographs of etched sectional cuts of the crystal shown in Figure 12(c). (a) A slice taken from the upper part of the crystal perpendicular to the growth direction displaying the presence of five grains. (b) Longitudinal cut between the arrows in (a), showing the presence of two grains in the upper, and multiple grains in the lower part. The length of the scale bars is 1 mm.

arrows indicate the position of the longitudinal cut shown in Figure 13(b). Here, the growth direction is from top to bottom and the upper edge corresponds to the location of the slice shown in Figure 13(a). The grain boundaries are seen as dark lines.

At the position of the upper slice, the crystal consists of five grains. Correspondingly, two large grains are seen in the longitudinal cut. The central grain boundary runs almost straight through the crystal until multiple branching after about 11 mm of growth occurs, which separates the crystal into several grains. This demonstrates that the crystal is not entirely single crystalline but contains large grains with volumes up to about 0.1–0.15 cm^3. Further improvement of the growth process, particularly by including a thin neck for grain selection, is required to produce larger, single-grained crystals.

5.4 FeCoCrMnAl

Marshal et al. [32] recently reported on the presence of HEAs in the system Fe–Co–Cr–Mn–Al. Single-crystal growth of the equiatomic phase in this system was carried out by the Bridgman technique [13]. An equiatomic melt was pre-alloyed using high-purity elements and cast into a cylindrical ingot. The growth was carried out at 1,530 °C under an argon atmosphere of 650 mbar and with a lowering velocity of 5 mm/h.

Figure 14 shows an equiatomic FeCoCrMnAl Bridgman crystal from several directions for an overview of its entire surface. The length of the crystal is 66 mm, the diameter is 11 mm at the bottom and 9 mm at the top. The grain boundaries located according to visual inspection are marked on the surface. The crystal consists of a large primary grain marked “P”, the volume of which is about 4.8 cm^3. Additionally, two small secondary grains marked “S” are found in the upper part and in the center.

Figure 15 depicts a series of back-reflection X-ray Laue diffraction patterns taken from the primary grain (represented using an inverted greyscale for better clarity). All Laue patterns show sharp singular spots and are consistent concerning their relative tilting angles. Figure 15(a) is the fourfold pattern of the (1 0 0) plane. Figure 15(b) is the twofold pattern of the (0 1 −1) plane, and Figure 15(c) is the threefold pattern of the (1 1 −1) plane. Figure 15(d) shows the pattern of the (2 1 −1) plane. The long axis of the crystal, i. e., the growth direction, is tilted by about 14° with respect to the [0 1 1] direction.

Figure 16(a)–(b) show bright-field TEM micrographs taken close to the [1 1 0] zone axis with the reflections (0 0 1) and (0 0 2) simultaneously excited. The images display the typical microstructure of the material on a submicron scale. The material consists of a matrix and inclusions, both existing at an approximately equal volume fraction. The inclusions can be identified as {0 0 1} platelets of about 65-nm thickness and a varying lateral extension of the order of 500 nm. Under dark-field conditions using the (0 0 1) reflection (Figure 16(c)), the inclusions are bright while the matrix

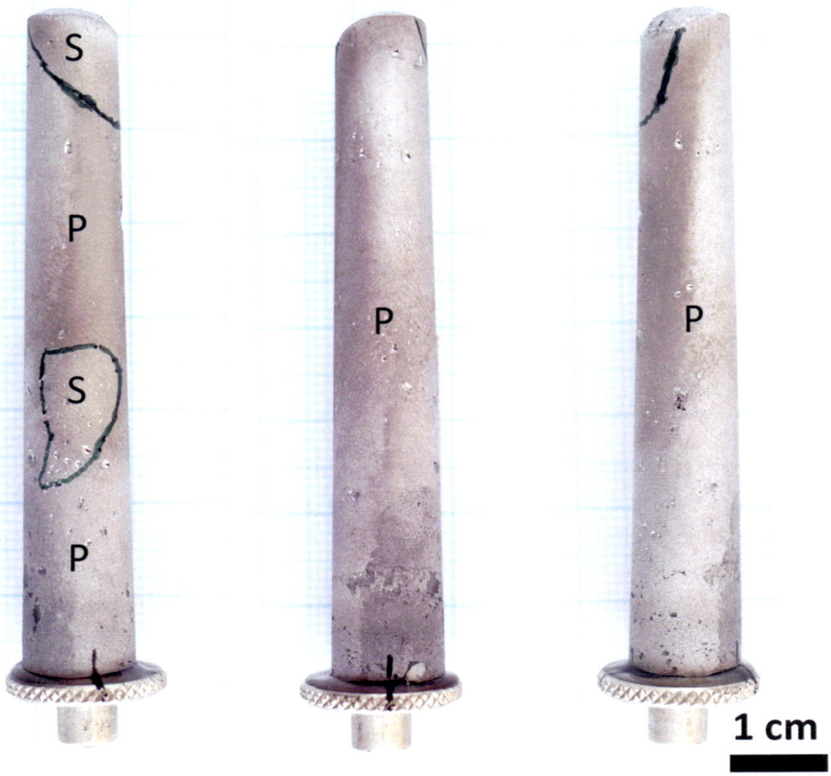

Figure 14: FeCoCrMnAl Bridgman crystal from several directions, so that its entire surface can be seen. The grain boundaries are marked on the surface. The primary and the secondary grains are labelled “P” and “S”, respectively.

is dark, which demonstrates that the inclusions possess B2 structure, and the matrix is bcc.

These conclusions are confirmed by high-resolution STEM. Figure 16(d) is a micrograph taken along the [1 1 0] direction with a Z-contrast sensitive high-angle annular dark-field detector. The upper part of the image corresponds to the matrix area, the lower part to an inclusion. The phase boundary is horizontally located in the center of the micrograph and is rather sharp. In the upper part of the image the typical pattern corresponding to the (1 1 0) projection of the bcc, structure is seen. The contrast is very homogeneous, and there is no indication of short- or medium-range order. The lower part of the image, corresponding to the inclusion, clearly shows the presence of B2 ordering: every other atomic row corresponding to a (0 0 1) plane is significantly

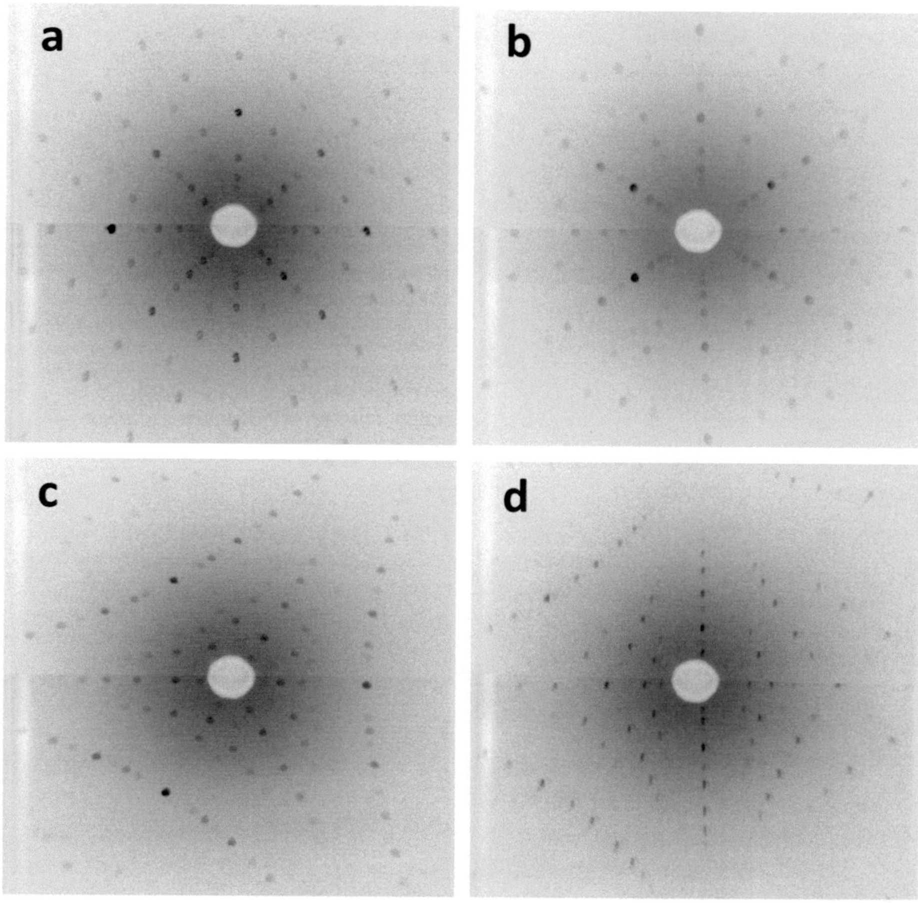

Figure 15: Back-reflection Laue X-ray patterns taken in the centre of the primary grain of the crystal shown in Figure 14. (a) (1 0 0) plane. (b) (0 1 −1) plane. (c) (1 1 −1) plane. (d) (2 1 −1) plane. The images are represented using an inverted grayscale for better clarity.

darker in contrast, which demonstrates the occupation of the corresponding atomic positions by lighter atoms. Furthermore, it can be seen that the phase boundary between the bcc and B2 phase is perfectly coherent.

The two-phase structure is similar to that of equiatomic AlCoCrFeNi HEAs, where bcc and B2 phases are also present [25]. However, the matrix/inclusion setting is inverted: while the FeCoCrMnAl HEA has a bcc matrix and B2 inclusions, in AlCoCrFeNi HEAs the matrix is B2 ordered and the inclusions are bcc (see Figure 10). Chemical analysis by means of energy-dispersive X-ray spectroscopy mapping shows that the inclusions predominantly contain Al and Co, while the matrix is rich in Fe and Cr [13].

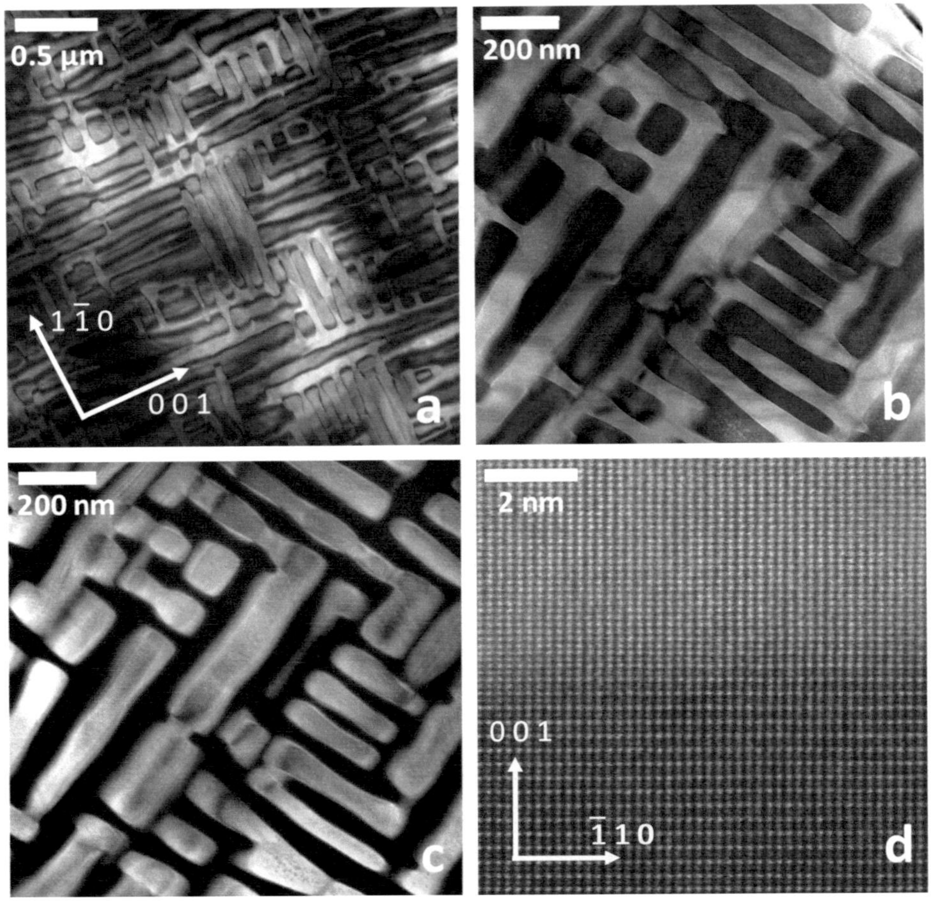

Figure 16: (a, b) Bright-field TEM micrographs taken close to the [1 1 0] zone axis displaying the typical microstructure with the reflections (0 0 1) and (0 0 2) simultaneously excited. (c) Dark-field image of the region shown in (b) using the (0 0 1) reflection. (b) High-resolution STEM micrograph along the [1 1 0] direction. Upper part bcc: matrix; lower part: B2 inclusion.

6 Conclusions

In this chapter, we demonstrate that single-crystal growth of HEAs, employing common techniques such as Bridgman and Czochralski growth, is generally feasible. To date, only a handful of HEA-forming systems have been addressed, and in the literature only few reports of successful single-crystal growth approaches are available. Single crystals of the fcc materials $Al_{0.3}CoCrFeNi$ and FeCoCrMnNi, and of the bcc/B2 two-phase material FeCoCrMnAl can be manufactured today, partly in the form of large cm^3-sized crystals.

HEAs represent a novel class of intermetallic phases, which is located between ordered intermetallics and metallic glasses—HEAs uniquely combine the presence of the chemical disorder of an ideal solid solution with the topological order of a crystalline material in a multicomponent system. The basic physical properties and mechanisms resulting from these salient features are not yet understood. Therefore, the availability of single-crystalline sample materials is of high importance and will make possible the determination of intrinsic materials properties without stray influences by grain boundaries or other impurities. Only then, an understanding of structure property relations for this class of materials can be achieved.

Large single-crystalline samples are also demanded by several characterization techniques such as neutron scattering, requiring sample volumes of the order of 1 cm^3, surface investigation techniques such as scanning tunneling microscopy, corrosion and tribology studies, requiring sample surface areas of several tens of mm^2 and lateral sample surface extensions of up to 10 mm.

The growth attempts described here demonstrate that further efforts are required to enlarge the spectrum of HEA materials available in single-crystalline form. In a medium-term perspective, all elemental groups in which HEA-forming systems exist and all structure classes should be covered. Today, for example, large HEA single crystals of hcp and bcc phases, and of rare-earth or refractory systems are not yet available. Growth experiments in the system ZrNbTiTaHf, i. e., addressing a bcc refractory phase, are ongoing but require further development.

Bibliography

[1] Yeh JW, Chen SK, Lin SJ, Gan JY, Chin TS, Shun TT, Tsau CH, Chang SY. Nanostructured high-entropy alloys with multiple principal elements: novel alloy design concepts and outcomes. Adv Eng Mater 2004, 6, 299–303.

[2] Cantor B, Chang ITH, Knight P, Vincent AJB. Microstructural development in equiatomic multicomponent alloys. Mater Sci Eng A 2004, 375–377, 213–8.

[3] Gao MC, Yeh JW, Liaw PK, Zhang Y. High-Entropy Alloys. Fundamentals and Applications. Switzerland, Springer International Publishing, 2016.

[4] Zhang Y, Zuo TT, Tang Z, Gao MC, Dahmen KA, Liaw PK, Lu ZP. Microstructures and properties of high-entropy alloys. Prog Mater Sci 2014, 61, 1–93.

[5] Youssef KM, Zaddach AJ, Niu C, Irving DL, Koch CC. A novel low-density, high-hardness, high-entropy alloy with close-packed single-phase nanocrystalline structure. Math Res Lett 2015, 2, 95–9.

[6] Zhu JM, Fu HM, Zhang HF, Wang AM, Li H, Hu ZQ. Microstructures and compressive properties of multicomponent $AlCoCrFeNiMo_x$ alloys. Mater Sci Eng A 2010, 527, 6975–9.

[7] Zhou YJ, Zhang Y, Wang YL, Chen GL. Solid solution alloys of $AlCoCrFeNiTi_x$ with excellent room-temperature mechanical properties. Appl Phys Lett 2007, 90, 181904.

[8] Zhang Y, Yang X, Liaw PK. Alloy design and properties optimization of high-entropy alloys. JOM 2012, 64, 830–8.

[9] Senkov ON, Wilks GB, Scott JM, Miracle DB. Mechanical properties of $Nb_{25}Mo_{25}Ta_{25}W_{25}$ and $V_{20}Nb_{20}Mo_{20}Ta_{20}W_{20}$ refractory high entropy alloys. Intermetallics 2011, 19, 698–706.
[10] Gludovatz B, Hohenwarter A, Catoor D, Chang EH, George EP, Ritchie RO. A fracture-resistant high-entropy alloy for cryogenic applications. Science 2014, 345, 1153–8.
[11] Yeh JW, Yeh AC, Chang SY. Potential applications and prospects. In: Gao MC, Yeh JW, Liaw PK, Zhang Y, eds. High-Entropy Alloys. Fundamentals and Applications. Switzerland, Springer International Publishing, 2016, 493–512.
[12] Ma SG, Zhang SF, Gao MC, Liaw PK, Zhang Y. A successful synthesis of the $CoCrFeNiAl_{0.3}$ single-crystal, high-entropy alloy by Bridgman solidification. JOM 2013, 65, 1751–8.
[13] Feuerbacher M, Würtz E, Kovacs A, Thomas C. Single-crystal growth of a FeCoCrMnAl high-entropy alloy. Math Res Lett 2017, 5, 128–34.
[14] Raghavan R, Kirchlechner C, Jaya BN, Feuerbacher M, Dehm G. Size effects in an equiatomic FeCrCoMnNi high entropy alloy. Scr Mater 2017, 129, 52–5.
[15] Zhang Y, Zhou JY, Lin JP, Guo LC, Liaw PK. Solid-solution phase formation rules for multi-component alloys. Adv Eng Mater 2008, 10, 534–8.
[16] Senkov ON, Miller JD, Miracle DB, Woodward C. Accelerated exploration of multi-principal element alloys with solid solution phases. Nat Commun 2014, 6, 6529.
[17] Tian F, Varga LK, Chen N, Shen J, Vitos L. Empirical design of single phase high-entropy alloys with high hardness. Intermetallics 2015, 58, 1–6.
[18] Zhang C, Zhang F, Diao H, Gao MC, Tang Z, Poplawsky JD, Liaw PK. Understanding phase stability of Al–Co–Cr–Fe–Ni high entropy alloys. Mater Des 2016, 109, 425–33.
[19] Zhang C, Zhang F, Chen S, Cao W. Computational thermodynamics aided high-entropy alloy design. JOM 2012, 64, 839–45.
[20] Hsu CY, Juan CC, Chen ST, Sheu TS, Ye JW, Chen SK. Phase diagrams of high-entropy alloy system Al-Co-Cr-Fe-Mo-Ni. JOM 2013. https://doi.org/10.1007/s11837-013-0773-2.
[21] Yeh JW. Recent progress in high entropy alloys. Ann Chim Sci Mater 2006, 31, 633–48.
[22] Senkov ON, Scott JM, Senkova SV, Miracle DB, Woodward CF. Microstructure and room temperature properties of a high-entropy TaNbHfZrTi alloy. J Alloys Compd 2011, 509, 6043–8.
[23] Feuerbacher M, Heidelmann M, Thomas C. Hexagonal high-entropy alloys. Math Res Lett 2015, 3, 1–6.
[24] Cantor B. Multicomponent and high-entropy alloys. Entropy 2014, 16, 4749–68.
[25] Manzoni A, Daoud H, Völkl R, Glatzel U, Wanderka N. Phase separation in equiatomic AlCoCrFeNi high-entropy alloy. Ultramicroscopy 2013, 132, 212–5.
[26] Jensen JK, Welk BA, Williams REA, Sosa JM, Huber DE, Senkov ON, Viswanathan GB, Fraser HL. Characterization of the microstructure of the compositionally complex alloy $Al_1Mo_{0.5}Nb_1Ta_{0.5}Ti_1Zr_1$. Scr Mater 2016, 121, 1–4.
[27] Freudenberger J, Rafaja D, Geissler D, Giebeler L, Ullrich C, Kauffmann A, Heilmaier M, Nielsch K. Face centred cubic multi-component equiatomic solid solutions in the Au-Cu-Ni-Pd-Pt system. Metals 2017, 7, 135.
[28] Senkov ON, Wilks GB, Scott JM, Miracle DB. Mechanical properties of $Nb_{25}Mo_{25}Ta_{25}W_{25}$ and $V_{20}Nb_{20}Mo_{20}Ta_{20}W_{20}$ refractory high entropy alloys. Intermetallics 2011, 19, 698–706.
[29] Takeuchi A, Amiya K, Wada T, Yubuta K, Zhang W. High-entropy alloys with a hexagonal close-packed structure designed by equi-atomic alloy strategy and binary phase diagrams. JOM 2014, 66, 1984–92.
[30] Gao MC, Zhang B, Guo SM, Qiao JW, Hawk JA. High-entropy alloys in hexagonal close-packed structure. Metall Mater Trans A. https://doi.org/10.1007/s11661-015-3091-1.
[31] Rogal L, Czerwinski F, Jochym PT, Litynska-Dobrzynska L. Microstructure and mechanical properties of the novel Hf25Sc25Ti25Zr25 equiatomic alloy with hexagonal solid solutions. Mater Des 2016, 92, 8–17.

[32] Marshal A, Pradeep KG, Music D, Zaefferer S, De PS, Schneider JM. Combinatorial synthesis of high entropy alloys: introduction of a novel, single phase, body-centered-cubic FeMnCoCrAl solid solution. J Alloys Compd 2016, 691, 683–9.
[33] Kao YF, Chen TJ, Chen SK, Yeh JW. Microstructure and mechanical property of as-cast, -homogenized, and –deformed AlxCoCrFeNi ($0 \leq x \leq 2$) high-entropy alloys. J Alloys Compd 2009, 488, 57–64.
[34] Kozak R, Sologubenko A, Steurer W. Single-phase high-entropy alloys – an overview. Z Kristallogr 2015, 230, 55–68.
[35] Ye YF, Wang Q, Lu J, Liu CT, Yang Y. High-entropy alloy: challenges and prospects. Mater Today 2016;19:349–62.
[36] Miedema AR, de Chatel PF, de Boer FR. Cohesion in alloys – fundamentals of a semi-empirical model. Physica B 1980, 100, 1–28.
[37] Takeuchi A, Inoue A. Classification of bulk metallic glasses by atomic size difference, heat of mixing and period of constituent elements and its application to characterization of the main alloying element. Mater Trans A 2005, 46, 2817–29.
[38] Zhang Y, Zhou Y. Solid solution formation criteria for high entropy alloys. Mater Sci Forum 2007, 561–565, 1337–9.
[39] Porter DA, Easterling KE. Phase Transformations in Metals and Alloys. Cheltenham, United Kingdom, Nelson Thornes Ltd, 2001.
[40] Yang X, Zhang Y. Prediction of high-entropy stabilized solid-solution in multi-component alloys. Mater Chem Phys 2012, 132, 233–8.
[41] Ye YF, Wang Q, Lu J, Liu CT, Yang Y. The generalized thermodynamic rule for phase selection in multicomponent alloys. Intermetallics 2015, 59, 75–80.
[42] Senkov ON, Miracle DB. A new thermodynamic parameter to predict formation of solid solution or intermetallic phases in high entropy alloys. J Alloys Compd 2016, 658, 603–7.
[43] Mizutani U. Hume-Rothery Rules for Structurally Complex Alloy Phases. Boca Raton, USA, CRC Press, 2016.
[44] Wang Z, Guo S, Liu CT. Phase selection in high-entropy alloys: from nonequilibrium to equilibrium. JOM 2014, 66, 1966–72.
[45] Zhang Y, Zhou YJ, Lin JP, Chen GL, Liaw PK. Solid-solution phase formation rules for multi-component alloys. Adv Eng Mater 2008, 10, 534–8.
[46] Guo S, Liu CT. Phase selection rules for complex multi-component alloys with equiatomic or close-to-equiatomic compositions. Chin J Nat 2013, 35, 85–96.
[47] Miracle DB, Miller JD, Senkov ON, Woodward C, Uchic MD, Tiley J. Exploration and development of high entropy alloys for structural applications. Entropy 2014, 16, 494–525.
[48] Landolt-Börnstein – Group IV Physical Chemistry 5J (Pu-Re – Zn-Zr) (Accessed June 29, 2017, at http://materials.springer.com/lb/docs/sm_lbs_978-3-540-70705-9_2846.10.1007/10551312_2846.
[49] Zhang Y, Koch CC, Ma SG, Zhang H, Pan Y. Fabrication routes. In: Gao MC, Yeh JW, Liaw PK, Zhang Y, eds. High-Entropy Alloys. Fundamentals and Applications. Switzerland, Springer, 2016, 151–79.
[50] Vaidya M, Trubel S, Murty BS, Wilde G, Divinski SV. Ni tracer diffusion in CoCrFeNi and CoCrFeMnNi high entropy alloys. J Alloys Compd 2016, 688, 994–1001.
[51] Wilke KT. Kristallzüchtung. Berlin, Germany, VEB Deutscher Verlag der Wissenschaften, 1988.
[52] Glatzel U. Private communication, 2017.
[53] Dai HJ, D'Souza N, Dong HB. Metall Mater Trans A 2011, 42, 3430.
[54] Li C, Zhao M, Li JC, Jiang Q. B2 structure of high-entropy alloys with addition of Al. J Appl Phys 2008, 104, 113504.

[55] Welk BA, Williams REA, Viswanathan GB, Gibson MA, Liaw PK, Fraser HL. Nature of the interfaces between the constituent phases in the high entropy alloy CoCrCuFeNiAl. Ultramicroscopy 2013, 134, 193–9.
[56] Ma SG, Zhang SF, Gao MC, Liaw PK, Zhang Y. A successful synthesis of the $CoCrFeNiAl_{0.3}$ single-crystal, high-entropy alloy by Bridgman solidification. JOM 2013, 65, 1751–8.

Zygmunt Henkie

Crystal growth of the filled skutterudite arsenides by the flux method under enhanced vapor pressure

1 Introduction

The flux method is becoming more and more widely used to grow high-quality single crystals. The flux is a material that is liquid at a moderately high temperature and can dissolve noncongruent or high-melting-temperature compounds, or its constituent elements. Crystals grow when the solution is oversaturated at a low rate. A comprehensive review on the principles and applied growth techniques is given by Fisk and Remeika [1] as well as Canfield and Fisk [2].

The simplest technique of the flux method consists of slow cooling a high-temperature liquid solution of the compound to be grown in solvent that is, at the same time, one of the constituent elements of the grown crystal, taken with excess in relation to stoichiometric content. This is named the self-flux method. The compounds' solubility vs. temperature dependence determines the composition of the liquid solution and temperature range for slow cooling. The self-flux method has been used for growing crystals of numerous binary rare-earth compounds RM_3 where R = f-element (La, Ce, Pr, Nd, and Lu), M = main group metal (Sn, Pb, In, and Ga) [3–6]. A common feature of these compounds is that they are the richest in M binary phases in R–M systems. After cooling down the solution, one obtain crystals of the binary compounds in M solid matrix. Crystals of USb_2 and UBi_2 compounds are other examples grown in this way [7, 8].

The molten-metal-solution evaporation method was used to grow U_3Bi_4 single crystal [9]. This compound is in equilibrium with the U_3Bi_4–Bi liquid solution at temperature range between 1,150 °C and 1,010 °C. Keeping this system at the fixed temperature of 1,080 °C and setting the evaporation rate of Bi at 0.14 g/h by using crucible cover and He pressure (~0.173 bar), we create a free crystal in the crucible and Bi at the cold bottom of the ampoule. The U_3Bi_4 crystals burn explosively in the air. About a 20-mm^3 piece was separated from the bigger crystal, encapsulated in an aluminum container with a helium atmosphere and examined by neutron diffraction to determine its quality and magnetic structure. The examination confirmed its good crystal stoichiometry and its low mosaicity.

Acknowledgement: The author thanks Professor M. Brian Maple and his team from the University of California, San Diego, for fruitful discussion and collaborations with members of the Institute of Low Temperature and Structure Research, PAS, Wrocław working with skutterudite materials.

https://doi.org/10.1515/9783110496789-008

This paper present another flux growth technique, i. e., growing crystals under enhanced vapor pressure, and demonstrates its utility for growing of the filled skutterudite arsenides. As an extended introduction will be presented general information in a concise manner about the skutterudites and other techniques for growing the skutterudite crystals.

2 The filled skutterudite arsenides

The filled skutterudite compounds are the ternary transition metal (T) pnictides with the chemical formula $(A, R)T_4X_{12}$ (A = alkali metal, alkaline earth; R = rare earth, actinide element; T = Fe, Ru, Os; X = P, As, Sb), which crystallize in cubic space group Im–3 and exhibit a wide variety of strongly correlated electron phenomena [10–13] that attract scientists' interest. Many of these phenomena depend on hybridization between the rare earth or actinide f-electron states and the conduction electron states which, in some filled skutterudite systems, leads to the emergence of semiconducting behavior.

These compounds were derived from the binary pnictides TX_3, the prototype of which is $CoAs_3$, a diamagnetic zero gap semiconductor [14]. The unit cell of $CoAs_3$ is a cube formed with eight octants, the corners of which are occupied by Co^{3+} cation atoms. The centers of six of the octants are filled by square radicals $[As_4]^{4-}$, while the two remaining octants are empty. Therefore, the valence ratio Co:$[As_4]$ is equal to 4:3 and the total structure remains electrically neutral, favoring a semiconducting electronic structure.

The TX_3 compounds belong to the family of Zintl phases, i. e., solid compounds of metals with nonmetals or metalloids forming polyanions [15] According to the Zintl concept, 24 valence electrons per TX_3 formula, or 96 valence electrons per $(A, R)T_4X_{12}$, stabilize such skutterudite compounds, affording a tendency for semiconducting behavior. Following this concept, Luo et al. [16] discovered a large family of filled skutterudites, based on the group 9 transition metals Co, Rh, and Ir, the alkali, alkaline-earth, and rare-earth elements, and $[Sb_4]^{4-}$ polyanions. Moreover, electronic structure calculations—for some representatives of this group—showed they are semiconductors. Replacing Fe (as well as by Ru or Os) with Co, $[As_4]^{4-}$ with $[Sb_4]^{4-}$ polyanions, and filling the two empty octants by Th or tetravalent Ce leaves the structure electrically neutral with the 4:3 valence ratio and 96 valence electrons. Therefore, the semiconducting or semi-metallic properties are expected for the modified skutterudites. This applies to the properties of the phosphide and antimonide skutterudites, as well. Indeed, $CeOs_4As_{12}$ [17], $CeRu_4As_{12}$ [18, 19] and $CeFe_4As_{12}$ [20] become semiconductors at low temperatures and $ThFe_4As_{12}$ [21] becomes a semimetal. In the aforementioned compounds, presumably the tetravalent states of the filler atoms and the semiconducting state are a result of the hybridization of one f-electron of Ce [22].

Their high potential for application originates in the possibilities for the modification of their electronic properties. This can be demonstrated by their thermoelectric properties characterized with the dimensionless thermoelectric figure of merit, ZT (for definition see, for instance, [23], also in this book; Chap. 10, p. 217). The highest ZT observed before 1997, barely exceed one, was for $Ce_{0.9}Fe_3CoSb_{12}$ [24], while decade later ZT exceed 1.8, observed for In-doped multifilled n-type skutterudites [25]. Based on a recent review on thermoelectricity of skutterudites, G. Rogl and P. Rogl [26] concluded that the filled skutterudites are "among the most promising materials for a widespread thermoelectric application in a waste heat gradient from room temperature to about 900 K". This is the reason to increase of our knowledge of the electronic and other physical properties of the skutterudites and it can be done in the best way by examining their single crystals.

3 Growing skutterudites

The flux method under ambient pressure has already been used by Jeitschko and Braun [27–29], for growing light rare-earth filled skutterudite phosphides and antimonides. This method was then slightly modified and used by numerous Japanese scientists to grow crystals for intensive examination of skutterudites. Shirotani [30] found that high pressure is useful for synthesis of the skutterudites filled with heavy rare earths. A comprehensive review on the results of growing filled skutterudite pnictides is compiled by Sato et al. [31]. Basic information on these growing methods is given in the following section as an introduction to a comprehensive presentation of the methods of crystal growth under enhanced vapor pressure.

Owing to the low vapor pressure of antimony (much lower than one bar below 1,050 °C), crystals of skutterudite antimonides can be grown by the so-called self-flux method under ambient pressure. So the RT_4Sb_{12} crystals for T = Ru and Os can be successfully grown in the molar ratio R:T:Sb = 1:4:20 in a process similar to that described earlier [28], i. e., annealing for 3 h at 850 °C, followed by one week at 730 °C. The Sb-flux was dissolved in aqua regia. This process requires special care, since aqua regia also attacks the sample. Typically, a few hours are enough for removing Sb-flux. In the case of T = Fe, single crystal growth has to be promoted by a solid-state reaction since RFe_4Sb_{12} is not stable above ~700 °C [32]. Single crystals might be mechanically isolated from the Sb-flux.

The skutterudite phosphides were grown in the Sn-flux. The components of the skutterudite and flux were encapsulated in a quartz tube with a molar ratio R:T:P:Sn = 1:4:20:40. The ampoule is heated up to 1,050 °C in a furnace, kept at this temperature for 100 hours, cooled down to 650 °C at the rate of 1 °C/hour and then the furnace is cooled down to room temperature [31]. A sufficient amount of Sn dissolves all components and reduces the very high vapor pressure of phosphorus at a high temperature. Reduction of Sn content to the ratio R:T:P:Sn = 1:4:20:20 caused an explosion of the ampoule at around 850 °C [31]. The Sn-flux was removed by dissolving bathing in concentrated hydrochloric acid that does not dissolve the grown crystals.

The above-mentioned methods do not work for both P- and Sb-based skutterudites filled with trivalent heavy rare earths and for growing crystals of all rare-earth filled skutterudite arsenides. Shirotani et al. [33, 34] showed that it is possible to obtain polycrystalline samples of heavy rare earth-filled skutterudites by applying the high-pressure technique. They obtained RFe_4P_{12} compounds for all the rare earth elements by wedge-type cubic anvil pressure cell up to ~4 GPa.

Next, the single crystals of RFe_4As_{12} (R = La, Ce, Pr) were grown. Such skutterudite compounds have been grown under pressure 4–5 GPa [31] by Tatsuoka et al., [35] getting initially good but polycrystalline samples of $LaFe_4As_{12}$. The constituent elements R, T and As in the ratio of 1:4:(20–30) are put into a cylindrical boron nitride crucible, which was surrounded by a cylindrical graphite tube heater. These, along with a pair of thermocouples, are set into a cylindrical hole in a cube made of pyrophyllite and subjected to high pressures up to 3–5 GPa at room temperature. Next, the crucible was heated to 1,100 °C, kept for one hour and quickly decreased to 950 °C, then cooled down slowly to 80 °C at a rate of 6 °C/hour, and finally the furnace was cooled down to room temperature.

4 Crystal growth under enhanced vapor pressure technique

Vapor pressure of arsenic is too high to grow As-based skutterudite crystals by the self-flux method in a quartz ampoule under ambient pressure and at high temperature. To overcome this problem, the first As-based skutterudite crystals, i. e., $PrOs_4As_{12}$, have been grown in As-rich Cd:As flux [36, 37] by utilizing the enhanced pressure technique. The ampoule, sealed with the skutterudite components and the flux, was heated in a pressure cell with the pressure controlled to compensate an estimated inner ampoule pressure. Such a technique may also be applied for synthesis and crystals' growing in other enhanced vapor-pressure systems. Therefore it is presented in detail.

Reproducible results of growing crystals were observed if crystals of $PrOs_4As_{12}$ were formed in Cd–As flux with Cd:As ratio $< 1/2$. The flux could be removed from the grown crystals by means of sublimation in an evacuated 15-cm-long quartz ampoule placed into a two-zone furnace. The sublimation is done from zone with temperature 600 °C to zone with about 300 to 400 °C. The flux condenses in the cold zone in two spatially separated sections, which consisted of As, and $CdAs_2$, respectively. The $CdAs_2$ condensed just beyond the hot zone, while the As condensed at the coldest end of the ampoule. Then $CdAs_2$ and As, mixed in a proper ratio, could be repeatedly used for growing other crystals. This procedure reduces possible environment pollution, whereas the flux behavior suggests it can be treated as the liquid solution of As in $CdAs_2$ solvent.

The vapor pressure of $CdAs_2$ at 850 °C, $p(CdAs_2)$, was shown [38] to be negligibly low as compared to that of As, and for simplicity it is neglected in further estimations of the flux vapor pressures. It is also assumed that the Raoult law can be used to determine the vapor pressure of As over the As–$CdAs_2$ liquid solution, $p(\text{As})$,

$$p(\text{As}) = p^*(\text{As})[n(\text{As})/(n(\text{As}) + n(\text{CdAs}_2))] \tag{1}$$

The temperature dependence of the arsenic vapor pressure can be approximated using the well-known Clausius Clapeyron law,

$$\lg(p/\text{bar}) = \text{A}/T + \text{B} \tag{2}$$

where A = −6892.4 and B = 7.7704 are constants determined for arsenic vapor pressure $p^*(\text{As})$ equal to one bar and 28 bar at temperatures of 887 and 1090 K, respectively [39].

Equations (1) and (2) predict the vapor pressures for As, $CdAs_3$ and $CdAs_{2.5}$ fluxes at 850 °C to be equal to 43.1, 21.5 and 14.3 bar, respectively. The composition $CdAs_3$ was used for major experiments, which were done in ampoules made of quartz tube of inside/outside diameters equal to 15/19 mm. The chemical formulas are used only to give the fluxes' composition. The ampoule, with flux and skutterudite components inside, was heated in a custom-built pressure cell.

Two pressure cells, made of creep-resisting steel tubes, were used. The smaller and simpler cell, of inside/outside diameters equal to 27.5/38 mm and 1,500 mm long, is welded from the top and a has closing unit at the bottom. The unit is equipped with a pressure gauge, a straight-run valve and metallic capillary for Ar supply. The decrease of pressure strength, p_S, of the cells with increasing walls' temperature, θ_w, is given by eq. (3):

$$\lg(p_S/\text{bar}) = 4.4374 - 3.5357 \times 10^{-3} \times \theta_w \tag{3}$$

Where p_S is in bars and θ_w in °C. It delimits Ar pressure, $p(\text{Ar})$, in the cell controlled to compensate for inner vapor pressure, $p(\text{As})$, in the ampoule heated in the cell. The Ar pressure should follow the relationship (4):

$$p(\text{Ar})(\theta) = (p(\text{As}) + \text{E}) \le p_S \tag{4}$$

where E is an excess of Ar pressure allowed by relation (4) for $0 \le \text{E} \le \Delta$ and Δ denotes a strength of the empty quartz ampoule against outside pressure. It was found that Δ depends on temperature and is equal at least 100 bar at room temperature and 18 bar at 900 °C and was approximated with dependence, $\lg(\Delta/\text{bar}) = 2.3724 - 1.2412 \times 10^{-3} \times \theta$,

(θ in °C). The high quartz-ampoule strength against outside pressure facilitates control of the safe value for $p(\mathrm{Ar})$. Additionally, an ampoule strength for inside pressure, which has not been taken into consideration, increases the ease of using this method.

A rough estimation of the upper limit of pressure/temperature (bar/°C) for $p(\mathrm{Ar})$ of the small cell, i. e., when the ampoule's strength is neglected, equals 21.0/881, 24.7/861, and 33.2/825 for $CdAs_{2.5}$, $CdAs_3$ and As fluxes, respectively. The sealed ampoule is loaded into the cell from the bottom to the top with a ceramic support. The top cell is placed into a three-zone furnace, providing a low temperature gradient and temperature control satisfying the relationship (4).

The crystal growth process, referred to particular compound, will be given later.

The larger cell, of inside/outside diameters equal to 40/48.5 mm and 1,500 mm long, is welded from the bottom and has a closing unit at the top end. This unit has the same devices as that of the smaller cell and is additionally equipped with a pass for the Ni/Cr–Ni thermocouple and electric wires to power an inner resistance furnace. The larger cell has a water-cooled system, just below the top end, to protect the closing unit from overheating. The ampoule is loaded into the inner furnace, and both are loaded to the bottom end of the pressure cell. The wall temperature and low gradient are controlled by a two-zone outside furnace. The wall temperature is stabilized at 600 °C, setting the upper limit for the pressure cell at $p(\mathrm{Ar}) = 207$ bar.

The cell with the inner furnace enabled us to test the growth of As-based filled skutterudite crystals by the self-flux method. The quartz ampoule (4–5 cm^3 inner volume) used here, sealed with skutterudite-composing elements and As-flux inside, resisted when heated up to 920 °C under the compensating pressure $p(\mathrm{Ar}) = 105$ bar. The flux's vapor is formed with As_4 molecules, at the expense of liquid As. Its mass equals $m(\mathrm{As_4}, \mathrm{g}) = M(\mathrm{As_4}) \times (V_\mathrm{A}/V_0) \times (T_0/T) \times (p(\mathrm{As})/p_0)$, in an ideal gas approximation. There are: $M(\mathrm{As_4}) = 300$ g/mol, standard molar volume of the ideal gas $V_0 = 22\,414$ cm^3 at $T_0 = 273$ K and $p_0 = 1$ bar. Then $m(\mathrm{As_4}, \mathrm{g})$ is equal to 1.2 g in the free volume of the ampoule $V_\mathrm{A} = 4$ cm^3, at temperature $T = 1{,}193$ K and a corresponding vapor pressure of 98.5 bar. Estimated quantity of As leaving flux for vapor, typically, ~20 to 40 % of the liquid As flux, should be completed.

It is important to inform readers of an attribution of the previous approximation (eq. (1) in Ref. [38]) for overestimation of the vapor pressure of As. The previous approximation predicts 137 bar of vapor pressure over As at 920 °C. But present experiments have shown that such compensating Ar pressure flattens out the ampoule with grown skutterudite crystals in arsenic flux, though the flattened ampoule remains hermetic. Unfortunately, an increased transverse dimension of the flattened ampoule broke the inner furnace. It was also found that the ampoule resisted either under 70 atm and under 95 atm of the compensating Ar pressures at 890 °C, i. e., that is consistent with this and the previous approximation, respectively. Furthermore, careful examination of the strengths of the ampoule against outside and inside pressure may considerably extend the possibility for easy control of the compensating pressure for higher temperature ranges.

4.1 Details on growing of selected skutterudite

Typically, the growth of the crystals was performed in the quartz ampoules described previously. The inside walls of the bottom end of the ampoule were covered with a layer of pyrolitic carbon. The elemental components of the $R_{1.1}T_4As_{12}$ skutterudite and the flux were incorporated in the atomic ratio R:T:Cd:As = 1.1:4:20:(72 + $m(As_4, g)/g$), i. e., ($R_{1.1}T_4As_{12}$+20$CdAs_3$+$m(As_4, g)/g$), in the case of Cd:As flux, or R:T:As = 1.1:4:(37+ $m(As_4, g/g)$), i. e., ($R_{1.1}T_4As_{12}$ + (25As + $m(As_4, g/g)$)), in the case of self-flux, and were loaded into the quartz ampoule and sealed in a vacuum.

Generally, the ampoule was heated in one of two pressure cells, which was filled with a controlled compensating Ar pressure. The latter exceeded the inner ampoule pressure, by 0–18 bar at 900 °C. The heating rate was 20 °C/h from 350–650 °C, and next the rate was increased to ~50 °C/h for heating up to the selected temperature θ_{low} > 730 °C. The ampoule was then heated and cooled down repeatedly between θ_{low} and θ_{low} + ~15 °C, for a period lasting between a few days and weeks, at the rate 3 °C/h. Finally, we obtained an ingot of solidified flux with filled skutterudite crystals after the ampoule was cooled to the ambient temperature at the rate 20 °C/h.

The flux was removed from the ingot by means of sublimation in the two-zone furnace, as was described before. The released skutterudite crystals were washed in a mixture of several acids which does not attack the skutterudite crystals but may easily dissolve rare-earth arsenides and remove them. Other impurities that appear in the grown material are crystals of diarsenides of transition metals—Fe, Ru and Os. These crystals can be sometimes hardly distinguished from the skutterudite crystals by shape and color. Unfortunately, the washing mixture does not dissolve them. However, diarsenides of the transition metals are semiconductors that, at room temperature, show thermoelectric power by over one order of magnitude higher than that shown by the grown skutterudites and can be easily identified by using thermoelectric sound.

A difficult problem produces overgrowing of the iron diarsenide crystals by crystals of the skutterudite phase. Such a case may be revealed by cracks on the skutterudite crystal's surface. The cracks were observed on $PrFe_4As_{12}$ crystal overgrowing $FeAs_2$ phase (see Figure 8 in Ref. [38]) or $LaFe_4As_{12}$ overgrowing $FeAs_2$ crystal [40]. Figure 1 shows the latter case. Careful inspection of the crystal surfaces makes possible noticing that the surface of $LaFe_4As_{12}$ crystal, full of etching holes, contrast with fine polished-like surface of the $FeAs_2$ crystal. It means that $LaFe_4As_{12}$ skutterudite was dissolved in acid easier than $FeAs_2$ crystal, and the latter cannot be removed by dissolving it in the acid mixture.

This problem can be partially reduced synthesizing the filled skutterudite with about 5 to 10 % excess of rare earth component and carrying the synthesis in low temperature gradients. This reduces contamination of the grown crystals with the diarsenides of the transition metals. On the other hand, a small excess of the rare earth increases the completeness of filling the skutterudite cages. The excess of the rare earths forms arsenides that can be easily dissolved in acids. The latter is espe-

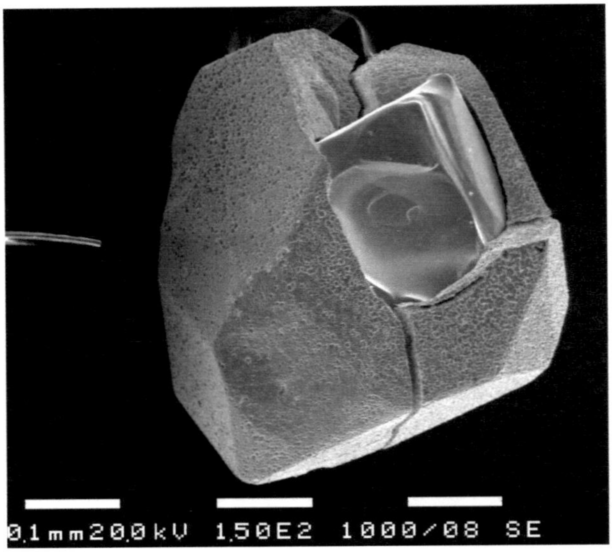

Figure 1: Small $FeAs_2$ crystal overgrown by $LaFe_4As_{12}$ one, after etching in the washing mixture of acids.

cially true in the case when the crystallization is carried out in the Cd:As flux. In this case, the rare-earth arsenides co-crystallize with flux components in the form of parallelepipeds. The latter ones, following sublimation of the flux, are composed of irregular rare-earth-diarsenide powder with a large amount of interstitial space (see Figure 2 in Ref. [38] for $PrAs_2$) responsible for the rapid dissolving.

Crystals of $PrOs_4As_{12}$ compound were the first ones from the filled skutterudite arsenides grown by Cd:As flux method under the flux's vapor pressure [36, 37]. Naturally, a primary question is what ratio of Cd:As optimizes the filled-skutterudite crystal growth. It was found that when the Cd:As ratio $< 4/10$, the results of $PrOs_4As_{12}$ crystal growth becomes reproducible. A decrease of the ratio increases the growth rate and reproducibility of the growing results but simultaneously increases also the vapor pressure, hence causing technical difficulties. Taking into account these two opposed effects, the $CdAs_3$ flux composition seemed to be optional.

In order to explore the effect of growth temperature on the synthesis of filled skutterudite arsenides, $PrOs_4As_{12}$ crystals were grown at 750, 775, 800, 825, 835 and 850 °C where each growth lasted two weeks. In each case, the elemental composition was $Pr_{1.1}Os_4As_{12} + 12CdAs_3$. The resulting material was analyzed, after the flux was removed by sublimation, followed by washing in the acid mixture and splitting the material into two fractions: $PrOs_4As_{12}$ crystals and sand-like formations.

The growth experiments done at 750 °C yielded about 10 % fraction of crystals with sizes up to ~0.6 mm. As the growth temperature was increased from 750 °C to 825 °C, the amount and size of the $PrOs_4As_{12}$ crystals increased gradually, at the expense of the sand-like formation. Finally, at 850 °C, no skutterudite growth was observed. Next, two experiments on the recrystallization of small $PrOs_4As_{12}$ crystals, at

835 and at 860 °C, were performed. Although the recrystallization at 835 °C caused an increase in crystal size, the crystals also showed clear signs of etching, such as cavities at faces and rounding the crystal edges. This was presumably due to insufficient stability of the temperature, as dissolving and growing processes had taken place at different times. A similar result was obtained at this temperature when the $PrOs_4As_{12}$ to $CdAs_3$ ratio was increased or decreased by a factor of two. A recrystallization of $PrOs_4As_{12}$ at 860 °C resulted in the transformation of the skutterudite crystals back into osmium and praseodymium diarsenides.

To examine the growth mechanism of the filled skutterudite arsenides, 0.5 g (2.8×10^{-4} moles) of $PrOs_4As_{12}$ single crystals with sizes 0.06–0.25 mm were sealed in an ampoule with 33×10^{-4} moles of $CdAs_3$. The ampoule was heated to 820 °C, annealed for five hours, and then cooled at a rate 1.5 °C/h down to 620 °C. In the theoretical case when the, crystals have sufficient solubility in the flux, the smaller crystals are expected to dissolve upon the increasing temperature, and larger crystals should form upon cooling. However, an increase in crystal size was not clearly observed, presumably due to low solubility of $PrOs_4As_{12}$ in the flux. Therefore, it was concluded that the optimal growth of crystals take place when a synthesis process is completed, with a mineralization process, i. e., when long term heating is accompanied by the temperature oscillations around the optimal growth temperature.

Another attempt to overcome the problem of the low solubility of the skutterudites in a Cd:As flux was an experiment done for growing $LaRu_4As_{12}$ crystals [41]. A fraction of fine-crystalline $LaRu_4As_{12}$ was used as a substrate for growing bigger crystals in the molten flux placed in a temperature gradient. The substrate, dissolved in a zone with a temperature of about 820 °C, diffused to the lower zone with somewhat lower temperature where $LaRu_4As_{12}$ crystals with dimensions up to ~0.7 mm grew for four weeks. However, at the bottom of the ampoule, where the lowest temperature, i. e., about 805 °C, existed, grew $RuAs_2$ crystals. Presumably due to this effect, there were found, in this batch, the skutterudite crystals showing a deficit of Ru up to 2.8 at.%. This Ru-deficit crystal had both volume lower by 0.8 % and reaches superconducting transition from a normal state of positive thermoelectricity. The stoichiometric crystals reaches the superconducting transition from a normal state of negative thermoelectricity [41].

Neodymium is the heaviest rare-earth filler with which the filled-skutterudite arsenide crystals were grown by the described method. The $NdOs_4As_{12}$ crystals were grown from Cd:As flux, as well as from the self flux. For the first case, the skutterudite components and the flux were taken in the atomic ratio Nd:Os:Cd:As = 1.2:4:14:54, sealed in the ampoule and heated up to 875 °C, with the rate of 20 °C/h next, after three h cooled to 730 °C, at the rate of 1.2 °C/h, then to room temperature, at the rate of 20 °C/h. A characteristic form of the largest grown crystals is a regular bi-pyramid with summits cut out along [100] planes and dimensions up to ~0.5 mm, as shown in Figure 2a. Another form, representing smaller crystals (~0.3 mm) from the same batch, is shown in Figure 2d.

Figure 2: Here are shown different forms of skutterudite arsenide crystals filled with Nd filler; a) The characteristic form of the biggest $NdOs_4As_{12}$ crystals, grown in the molten Cd:As flux, is a regular bi-pyramid with summits cut out along the [100] planes. b) Representative of the biggest $NdOs_4As_{12}$ crystals grown in the As flux. c) A dendrite like inner structure of $NdRu_4As_{12}$ single crystal, resembling a partially destructed pyramid from outside. d) Another form of $NdOs_4As_{12}$ crystals grown in the Cd:As flux. This form resembles the crystal grown in As flux (Figure 2b).

In the case of the self-flux method, the components were taken in the atomic ratio Nd:Os:As = 1.2:4:50 and heated, in the ampoule, with the rate 20 °C/h, up to 875 °C, kept at this temperature for three h, then cooled down to 820 °C at the rate of 1.2 °C/h, abruptly cooled to about 700 °C, followed by cooling to room temperature at the rate of 20 °C/h. The typical form of the largest crystals (of dimension ~0.9 mm) are shown in Figure 2b. Its form slightly resembles the form of smaller crystal, grown in the molten Cd:As flux, presented in Figure 2d.

Several experiments on growing of $NdRu_4As_{12}$ single crystals have been performed. One of them resulted in crystals only of size up to ~0.6 mm, like that shown in Figure 2c. It resembles a partially destructed pyramid. Its X-ray diffraction pattern is like one expected for a single crystal. This happens, although crystal slices reveal several branches suggesting a sort of dendrite like growth. (A more complete pyramid of a single crystal of $NdRu_4As_{12}$ is shown in Figure 1 of Ref. [42].) Such dendrite-like growth is presumably due to instability in the connection with an excessively fast crystal-growth rate.

The dendrite-like crystals were grown by the self-flux method. Elemental components in the atomic ratio Nd:Ru:As = 1.2:4:60, sealed in a quartz ampoule, were heated up to 850 °C at the rate 20 °C/h. Next, the ampoule was heated up to 890 °C, at the rate

6 °C/h then, after three h cooled back to 850 °C, at the rate of 2 °C/h, abruptly cooled to about 700 °C, followed by cooling to room temperature at the rate of 20 °C/h. Low-temperature specific heat and magnetic properties examinations of these $NdRu_4As_{12}$ single crystals yielded precise data. They proved that $NdRu_4As_{12}$ becomes ferromagnetic below 2.3 K with a quartet ground state of the Nd^{3+} ions. Deep in the ferromagnetic state, the heat capacity data revealed a Schottky anomaly, which was ascribed to Zeeman splitting in a presence of a molecular field [42].

It can be summarized that skutterudite arsenides of Fe, Ru, and Os, filled with the light rare earth, i. e., La, Ce, Pr, Nd, and Th, could be grown by the enhanced vapor-pressure technique with the Cd/As or As fluxes at selected temperature regimes. These fillers, used in excess, form a diarsenide phase (in As flux) or a mixed parallelepiped form with $CdAs_2$ (in Cd/As flux), in addition to the skutterudite crystals grown.

On the other hand, nine attempts at growing $SmOs_4As_{12}$, $YbOs_4As_{12}$, $YbFe_4As_{12}$ and $LuFe_4As_{12}$ skutterudite crystals, either in As and Cd:As fluxes, under different high temperature conditions, were unsuccessful. Instead of getting the expected skutterudite crystals filled with these trivalent rare earths, we obtained rare-earth monoarsenide crystals accompanied by corresponding crystals of transition-metal di-arsenides. It is interesting that, in contrast to the light rare earth-fillers, the diarsenide phases are unknown for the Sm, Yb and Lu. The origin of this correlation is not clear yet. It might also be interesting that LuAs crystals grew in the form of rectangular parallelepipeds, while SmAs and YbAs grew in the form of bi-pyramids. However, in this case, the summits of the bi-pyramid are not cut in contrast to those at skutterudite bi-pyramids. Figure 3 shows a YbAs crystal.

Figure 3: The characteristic bi-pyramid form of YbAs crystal with well-formed summits. It unintentionally grew in a batch prepared for growing $YbFe_4As_{12}$ in the Cd:As flux.

4.2 About morphology and quality the grown crystals

The skutterudites filled with trivalent rare earth exhibit metallic electron transport properties, and therefore their quality may be described by the residual resistivity ratio (RRR) parameter. It varies from RRRs equal to about 80 for $PrOs_4As_{12}$ [36] and $PrFe_4As_{12}$ [43], through 198 for $LaRu_4As_{12}$ [44] and up to 306 observed for $NdOs_4As_{12}$ [45] grown with Cd/As flux. The high quality of these crystals made possible de Haas–van Alphen's study of $PrOs_4As_{12}$ and $LaOs_4As_{12}$ [46] and observation, for the first time, of the multiple quadrupolar spin echoes of ^{139}La nuclei in $LaOs_4As_{12}$ [40]. These two effects are very sensitive to sample quality.

Tetravalent thorium-filling $ThFe_4As_{12}$ skutterudite causes its semimetallic properties with a charge carrier density of 2×10^{20} cm^{-3}, in terms of a two-band model. The charge imbalance, originating from ionized impurities (the difference between electron and hole carriers density), was of the order 0.4×10^{20} e^-/cm^{-3} in the best examined crystals [21]. The imbalance value may be compared to the density of the formula units, f.u., (3.5×10^{21} f.u./cm^3), which means 0.011 or 0.00067 single ionized impurity per f.u. or a single atom composing $ThFe_4As_{12}$, respectively. The $CeOs_4As_{12}$ [17] and $CeFe_4As_{12}$ [20] are, in turn, hole-carrier semiconductors at the lowest temperatures. A residual-carriers density (RCD) may be used as a measure of their purity. The RCD equals 1.5×10^{17} e^+/cm^{-3} (below 0.5 K) for $CeOs_4As_{12}$ crystals [47], i. e., corresponding to $\sim 5 \times 10^{-5}$ ionized impurity per f.u. Their electrical resistivity, $\rho(T)$, shows semiconducting-type behavior and is weakly sample dependent. In contrast, the RCD for $CeFe_4As_{12}$ crystals is significantly higher and depends on procedure of growing. They differ one from the other in crystal morphology and the type of their electron-transport behavior.

The $CeFe_4As_{12}$ skutterudite crystals, shown in (Figure 4), were grown from Cd:As flux in the same temperature conditions for two batches in which skutterudite com-

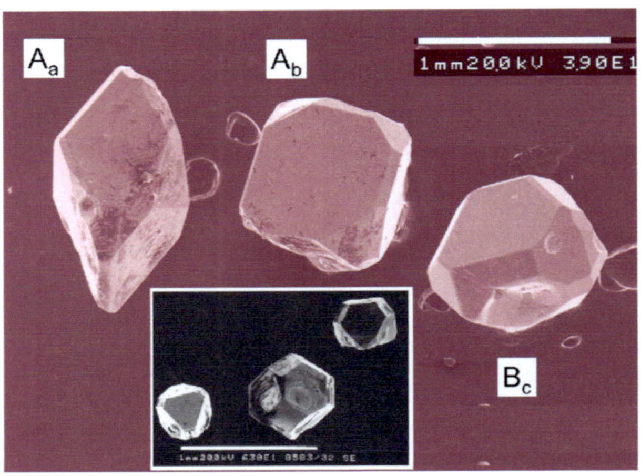

Figure 4: Three types of characteristic $CeFe_4As_{12}$ crystals that were grown; A_a – elongated diamond-shaped faces; A_b – rectangular shape of the largest faces; B_c – isometric form with many faces. *Inset*: Characteristic forms of $PrOs_4As_{12}$ crystals (Ref. [38] – Figure 3).

ponents differ in Ce content: A) – 0.85Ce:4Fe:12As and B) – 1.15Ce:4Fe:12As [20]. Two types of characteristic shapes and morphologies of the crystals grew in the batch A. They are labelled A_a and A_b. Crystals with the third type of shape grew in the batch B and are labelled B_c. Figure 4 shows crystals representing all three types of shape. The RCD for A_b crystals equals about $2.9 \times 10^{20}\ e^+/cm^{-3}$ ($0.094\ e^+$/f.u.) and show "metallic"-like behavior of electrical resistivity, i. e., their resistivity increases with temperature from the lowest temperatures. On the other hand, the RCD for A_a crystals, grown in the same batch, decreases to $\sim 1.5 \times 10^{20}\ e^+/cm^3$ ($0.042\ e^+$/f.u.) and reveals the semiconducting $\rho(T)$ behavior of these crystals. The increase of Ce content in batch B by 15 %, with respect to stoichiometric content, affects the decrease of RCD for B_c sample of $CeFe_4As_{12}$ to $\sim 1.0 \times 10^{20}\ e^+/cm^3$ ($0.031\ e^+$/f.u.). It improves semiconducting properties, as the data of resistivity and carrier density in Figure 5 show.

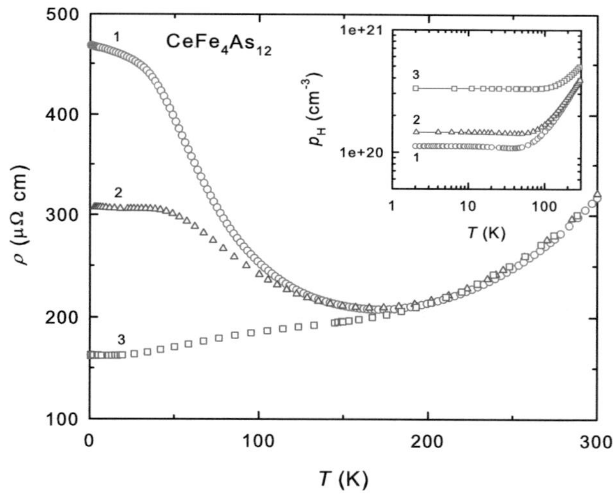

Figure 5: Electrical resistivity ρ vs temperature T for $CeFe_4As_{12}$ crystal A_b (curve 3), A_a (curve 2) and B_c (curve 1). *Inset:* Density of hole carrier p_H vs temperature T for the same crystals as in the main figure, respectively.

It is worth adding that no change of crystal structure was revealed by X-ray diffraction (XRD) examination of the A_a and A_b crystals down to 12 K. The determined occupation factors point to 0.22, 0.47 and 0.16 vacancies per f.u. for A_a, A_b and B_c samples, respectively. EDS measurements showed that the greatest differences in the elemental content between Ce, Fe, and As are equal to 0.5 %, 1.5 % and 0.5 %, respectively [20]. Thus, growing the $CeFe_4As_{12}$ crystals, with the Cd:As flux and the skutterudite components with off-stoichiometric Ce content, cause more residual hole carriers and less clearly exhibit the semiconducting $\rho(T)$ behavior in the case of a deficit of Ce than in the case of its excess. There is also a visible relationship between the form of the grown crystals and their residual carrier density.

5 Summary

A new technique for the flux method, adequate for crystals growing in a system composed of moderately high vapor-pressure components, is presented. The technique consists of using the pressure cell for heating the system encapsulated in quartz ampoule. The argon pressure in the cell is controlled to compensate for the estimated inner-ampoule vapor pressure. Construction of the equipment and techniques for pressure control are described. Using quartz tubes with 2-mm-thick walls for the system encapsulation and cells with 5-mm-thick walls made of creep-resisting steel tubes, it was possible to carry out all growth processes up to ~100 bar of vapor pressure at a temperature of 920 °C.

Using this technique and As or its solutions in $CdAs_2$ compound as the flux, it was possible to grow of Fe-, Ru- and Os-based skutterudite arsenide crystals, filled with light rare earths and actinide, i. e., La, Ce, Pr, Nd, and Th. These fillers can form diarsenide compounds. On the other hand, there were unsuccessful attempts to grow the skutterudite arsenides filled with trivalent heavy rare earths, like Sm, Yb, and Lu. Instead, diarsenide phases are formed for these fillers.

Bibliography

[1] Fisk Z, Remeika JP. Growth of single crystals from molten metal fluxes. In: Gschneidner KA, Jr, Eyring L, eds. Handbook on the Physics and Chemistry of Rare Earths. vol. 12. Amsterdam, North Holland, 1989, 53, 18 pages.

[2] Canfield PC, Fisk Z. Growth of single crystals from metallic fluxes. Philos Mag B 1992, 65, 1117, 7 pages.

[3] Staliński B, Kletowski Z, Henkie Z. Electrical resistivity of $RESn_3$ single crystals (RE = La, Ce, Pr, and Nd). Phys Status Solidi A 1973, 19, K165, 4 pages.

[4] Staliński B, Kletowski Z, Henkie Z. Electrical resistivity of antiferromagnetic $RESn_3$ single crystals (RE = Sm, Eu, Gd). Bull Acad Polon Sci, Sér Sci Chim 1975, 23, 827, 6 pages.

[5] Kletowski Z, Iliew N, Henkie Z, Staliński B. Single crystal growth of (Rare Earth) Me_3 compounds were Me = Sn, In, and Pb. J Less-Common Met 1985, 110, 235, 4 pages.

[6] Kletowski Z, Fabrowski R, Slawiński P, Henkie Z. Resistance of some $REMe_3$ compounds, RE = La and Lu, Me = Sn, Pb, In, and Ga. J Magn Magn Mater 1997, 166, 361–4.

[7] Henkie Z, Misiuk A. Growth of USb_2 single crystals and their structural perfection. Krist Tech 1979, 35, 539, 5 pages.

[8] Henkie Z. Single crystal growth and some electron transport properties of uranium pnictides. In: Drescher A, ed. High-Purity Materials in Sciences and Technology – Supplement. Dresden, 1985, 105–19.

[9] Henkie Z, Wiśniewski P, Gukasow A. U_3Bi_4 single crystal growth by the molten metal solution evaporation method. J Cryst Growth 1997, 172, 459, 7 pages.

[10] Maple MB, Bauer ED, Frederick NA, Ho P-C, Yuhasz WM, Zapf VS. Strongly correlated electron phenomena in filled skutterudite compounds. Physica B, Condens Matter 2003, 328, 29, 5 pages.

[11] Aoki Y, Sugawara H, Harima H, Sato H. Novel Kondo behaviours in the filled skutterudite structure. J Phys Soc Jpn 2005, 74, 209–13.

[12] Maple MB, Henkie Z, Baumbach RE, et al. Correlated electron phenomena in Ce- and Pr-based filled skutterudite arsenides and antimonides. J Phys Soc Jpn 2008, 77(Suppl A), 7, 7 pages.

[13] Sato H, Kikuchi D, Tanaka K, et al. Novel features in the filled skutterudite structure. J Phys Soc Jpn 2008, 77(Suppl A), 1–6.

[14] Anno H, Matsubara K, Caillat T, Fleurial J-P. Valence-band structure of the skutterudite compounds $CoAs_3$, $CoSb_3$, and $RhSb_3$. Phys Rev B 2000, 62, 10737, 7 pages.

[15] Sevov SC. Zintl phases. In: Westbrook JH, Freisher RL, eds. Intermetallic Compounds – Principles and Practice. Progress, 3. John Wiley & Sons, LTD, 2002, 113–20.

[16] Luo H, Krizan JW, Muechler L, et al. A large family of filled skutterudites stabilized by electron count. Nat Commun 2015, 6, 6489, 10 pages.

[17] Baumbach RE, Ho PC, Sayles TA, et al. The filled skutterudite $CeOs4As_{12}$. A hybridization gap semiconductor. Proc Natl Acad Sci USA 2008, 105, 17307, 5 pages.

[18] Sekine C, Hoshi N, Takeda K, et al. Electrical and magnetic properties of Kondo semiconductor $CeRu_4As_{12}$. J Magn Magn Mater 2007, 310, 260–3.

[19] Baumbach RE, Ho P-C, Sayles TA, et al. Non-Fermi liquid behavior in a fluctuating valence system, the filled skutterudite compound $CeRu_4As_{12}$. J Phys Condens Matter 2008, 20, 075110, 7 pages.

[20] Wawryk R, Henkie Z, Pietraszko A, et al. Filled skutterudite $CeFe_4As_{12}$: disclosure of a semiconducting state. Phys Rev B 2011, 84, 005100, 11 pages.

[21] Wawryk R, Żogał O, Rudenko A, Cichorek T, Henkie Z, Maple MB. Electron transport and magnetic properties of the filled skutterudite compound ThF_4As_{12}. J Alloys Compd 2016, 688, 478, 11 pages.

[22] Dordevic SV, Dilley NR, Bauer ED, Basov DN, Maple MB, Degiorgi L. Optical properties of MFe_4P_{12} filled skutterudites. Phys Rev B 1999, 60, 11321, 8 pages.

[23] Mahan G, Sales B, Sharp J. Thermoelectric materials: new approaches to an old problem. Phys Today 1997, 50, 42, 6 pages.

[24] Sales BC, Mandrus D, Chakoumakos BC, Keppens V, Thompson JR. Electron crystals phonon glasses: filled skutterudite antimonides. Phys Rev B 1997, 56, 15081–9.

[25] Rogl G, Grytsiv A, Yubuta S K, et al. In-doped multifilled n-type skutterudites with ZT = 1.8. Acta Mater 2015, 95, 201–11.

[26] Rogl G, Rogl P. Skutterudites a most promising group of thermoelectric materials. Curr Opin Green Sustain Chem 2017, 4, 50–8.

[27] Braun DJ, Jeitschko W. Ternary arsenides with $LaFe_4As_{12}$-type structure. J Solid State Chem 1980, 32, 357.

[28] Braun DJ, Jeitschko W. Thorium-containing pnictides with the $LaFe_4P_{12}$ structure. J Less-Common Met 1980, 76, 33–8.

[29] Jeitschko W, Braun D. $LaFe_4P_{12}$ with filled $CoAs_3$-type structure and isotypic lanthanoid-transition metal poliphosphides. Acta Crystallogr B 1977, 33, 3401, 6 pages.

[30] Shirotani I, Adachi T, Tachi K, et al. Electrical conductivity and superconductivity of metal phosphides with skutterudite type structure prepared at high pressure. J Phys Chem Solids 1996, 57, 211–6.

[31] Sato H, Sugawara H, Aoki Y, Harima H. Magnetic properties of filled skutterudites. In: Buschow KHJ, ed. Handbook of Magnetic Materials, vol. 18. 1st ed. Amsterdam, The Netherland, North-Holland is an imprint of Elsevier, 2009, 1–110.

[32] Morelli DT, Meisner GP. Low temperature properties of the filled skutterudite $CeFe_4Sb_{12}$. J Appl Phys 1995, 77, 3777, 11 pages.

[33] Shirotani I, Shimaya Y, Kihou K, Sekine C, Yagi T. Systematic high-pressure synthesis of new filled skutterudites with heavy lanthanide, $LnFe_4P_{12}$ (Ln = heavy lanthanide, including Y). J Solid State Chem 2003, 17, 32–3.
[34] Shirotani I, Araseki N, Shimaya Y, et al. Superconductivity of the new skutterudite compound $La_xRh_4P_{12}$ prepared at high pressure. J Phys Condens Matter 2005, 17, 4383–5.
[35] Tatsuoka S, Sato H, Kikuchi D, et al. Itinerant-electron weak ferromagnetism in La-based skutterudite $LaFe_4As_{12}$. J Phys Soc Jpn. 2008, 77, 033701, 4 pages.
[36] Yuhasz WM, Butch NP, Sayles TA, et al. Multiple ordered phase in the filled skutterudite compound $PrOs_4As_{12}$. Phys Rev B 2006, 73, 144409, 5 pages.
[37] Maple MB, Butch NP, Frederick NA, et al. Field-dependent ordered phases and Kondo phenomena in the filled skutterudite compound $PrOs_4As_{12}$. Proc Natl Acad Sci USA 2006, 103, 6783, 7 pages.
[38] Henkie Z, Maple MB, Pietraszko A, et al. Crystal growth and properties of filled skutterudite arsenides. J Phys Soc Jpn 2008, 77(Suppl A), 128–6 pages.
[39] Jefferson Lab, Science Education, It's Elemental, The Periodic Table of Elements, The Element Arsenic, https//education.jlab.org/itselemental/ele033.
[40] Nowak B, Żogał O, Henkie Z, Maple MB. Multiple quadrupolar spin echos in La-filled skutterudite $LaOs_4As_{12}$. Solid State Nucl Magn Reson 2009, 36, 209, 3 pages.
[41] Juraszek J, Wawryk R, Bochenek Ł, et al. Superconductivity of $LaRu_4As_{12}$ emerging from normal states of alternatively different thermoelectricity. To be published.
[42] Rudenko A, Henkie Z, Cichorek T. Low-temperature specific heat and magnetic properties of the filled skutterudite ferromagnet $NdRu_4As_{12}$. Solid State Commun 2016, 242, 21, 4 pages.
[43] Sayles TA, Yuhasz WM, Paglione, et al. Thermodynamic and transport studies of the ferromagnetic filled skutterudite compound $PrFe_4As_{12}$. Phys Rev B 2008, 77, 144432, 12 pages.
[44] Sayles TA, Baumbach RE, Yuhasz WM, et al. Superconductivity and crystalline electric field effects in the filled skutterudite $PrRu_4As_{12}$. Phys Rev B 2010, 82, 104513, 7 pages.
[45] Cichorek T, Rudenko A, Wiśniewki P, et al. Schottki-like anomaly on the border of localized ferromagnetism in the filled skutterudite $NdOs_4As_{12}$. Phys Rev B 2014, 90, 195123, 8 pages.
[46] Ho P-C, Singleton J, Maple MB, et al. A de Haas-van Alphen study of the filled skutterudite compounds $PrOs_4As_{12}$ and $LaOs_4As_{12}$. New J Phys 2007, 9, 269, 15 pages.
[47] Bochenek Ł. Ph.D thesis, Institute of Low Temperature and Structure Research, 213, available at http://bibliteka.int.pan.wroc.pl/lib/exe.fetch.php?media=doktoraty-b:l.bochenek.pdf, p. 151.

Yoshichika Ōnuki, Masashi Kakihana, Masato Hedo, and
Takao Nakama

High-quality single crystal growth in heavy fermion compounds

1 Introduction

Various techniques are applied to grow single crystals of transition metal, lanthanide and actinide compounds, depending on the melting point and the degree of the vapor pressure of the melt. Ce, Yb and U compounds are typical in exhibiting a variety of characteristic properties including spin and charge orderings, spin and valence fluctuations, heavy fermions and anisotropic superconductivity [1]. This variety is due to the result of competition between the Ruderman–Kittel–Kasuya–Yosida (RKKY) interaction and the Kondo effect. The RKKY interaction enhances the long-range magnetic order, where the f electrons with magnetic moments are treated as localized electrons, and the indirect f-f interaction is mediated by the spin polarization of the conduction electrons. On the other hand, the Kondo effect quenches the magnetic moments of the localized f-electrons by the spin polarization of the conduction electrons, producing a spin singlet state with a binding energy $k_B T_K$. T_K is the Kondo temperature. Localized f electrons are thus changed into f-itinerant electrons at low temperatures, forming heavy fermions.

A meaning of "heavy fermions" can be simply understood from the lattice constant a-value dependence of the electronic specific-heat coefficient γ in UX_3 (X: Si, Ge, Sn and Pb) with the well-known $AuCu_3$-type cubic structure, as shown in Figure 1. The a-value corresponds to the lattice spacing between the U atoms or the 5f electrons. USi_3 and UGe_3 are Pauli paramagnets, USn_3 is a spin-fluctuating compound and UPb_3 is an antiferromagnet with a Néel temperature $T_N = 30$ K. This trend is closely related to the lattice constant (USi_3: 4.035 Å, UGe_3: 4.206 Å, USn_3: 4.626 Å, and UPb_3: 4.792 Å). The corresponding electronic specific-heat coefficient γ increases with increasing the lattice constant, but decreases in magnitude in UPb_3 because the antiferromagnetic ordering occurs in UPb_3. This means that the largest γ value is formed when the electronic state is very close to the magnetic ordering. The electronic state with the large γ value is named "heavy fermions". An electronic state of $T_N \to 0$ (T_N: Néel temperature) corresponds to the quantum critical point. The heavy fermions are formed in the quantum critical region.

Acknowledgement: We would like to thank T. Komatsubara, H. Sato, H. Harima, H. Yamagami, A. Sumiyama, R. Settai, K. Sugiyama, Y. Inada, T. Takeuchi, D. Aoki, F. Honda, Y. Haga, E. Yamamoto, T. D. Matsuda, A. Nakamura and A. Teruya for fruitful collaborations and discussions.

https://doi.org/10.1515/9783110496789-009

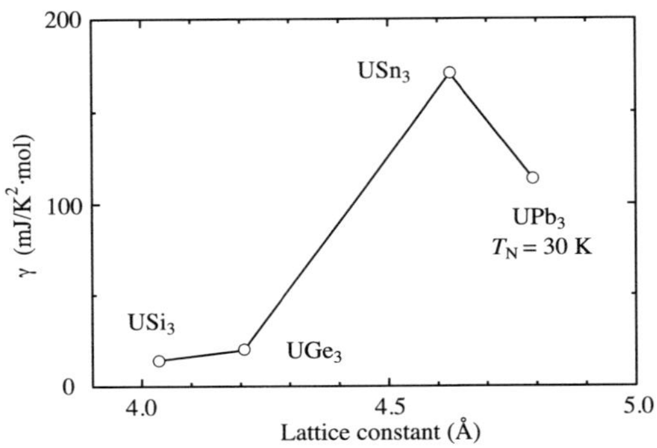

Figure 1: Lattice constant a dependence of the electronic specific heat coefficient γ in UX_3 (X: Si, Ge, Sn and Pb) with the $AuCu_3$-type cubic structure.

We note that the 5f electrons in U compounds are slightly different from the 4f electrons in Ce or Yb compounds in nature. At low temperatures, the 5f electrons are rather itinerant as in the 3d electrons in the transition-metal compounds even if the corresponding U compounds are in the magnetically ordered state. On the other hand, the 4f electrons of Ce and Yb compounds are mainly localized in the magnetically ordered state, but are changed into the "heavy fermions" under pressure. When the heavy fermions are formed, the energy scale in the electronic state is changed from the Fermi energy $\varepsilon_F = 10^4$ K to the Kondo temperature T_K, which is less than 100 K. Since the energy scale is small in magnitude, the electronic state can be remarkably changed by temperature, magnetic field and pressure.

Here, the 3d-electrons in transition-metal compounds are itinerant in the whole temperature range and become conduction electrons. The Fermi surfaces are thus mainly composed from the 3d electrons. Correlations between these conduction electrons are reflected in a relatively large γ value, together with the magnetically ordered moment. This is derived from a large partial density of states in the 3d electrons. The ordered moment is thus produced from a difference between the spin-up (majority) and spin-down (minority) 3d-partial densities of states.

We emphasize that high-quality single crystals are needed to clarify the characteristic properties in heavy-fermion compounds, especially the superconducting property. Cooper-pairs are broken by impurities and crystalline defects, and the existence of a residual density of states in superconductors is crucial in determining the superconducting property from the temperature dependence of nuclear spin-lattice relaxation rate $1/T_1$ below a superconducting temperature T_{sc}. Moreover, the amplitude of the de Haas–van Alphen (dHvA) oscillations, which is proportional to $\exp[-\alpha(m_c^*/H)(T + T_D)]$, is closely related to the sample quality, where $\alpha = 2\pi^2 ck_B/e\hbar$. If the cyclotron effective mass m_c^* is a rest mass of an electron m_0, the temperature of 1 K is usual in the dHvA experiment. A much lower temperature of 0.01 K is needed for $m_c^* = 100m_0$. Even if the low temperature is realized using a dilution refrigerator, the reduction of $\exp[-\alpha(m_c^*/H)T_D]$ is inevitable in the dHvA amplitude. Here, T_D is a so-called Dingle temperature, which is inversely proportional to the scattering lifetime of the carrier τ. Therefore, sample quality with a large τ value is essentially important to observe the dHvA oscillations for the heavy-fermion compounds.

Followings are our typical examples in growing high-quality heavy-fermion compounds. These are classified into seven methods, as shown in Figure 2, on the basis of the degree of the vapor pressure of the melt.

- Single crystal growth
  - Open type (low vapor pressure of the melt)
    - 1) Czochralski method
    - 2) zone-melting method
    - 3) floating-zone method
    - 4) Solid state electrotransport method
  - Closed type (high vapor pressure of the melt)
    - 5) flux method
    - 6) Bridgman method
    - 7) chemical transport method

Figure 2: Seven methods of single crystal growth.

2 Single-crystal growth

2.1 Czochralski method

Various techniques for single-crystal growth have been applied to heavy-fermion compounds, depending on the melting point and the degree of the vapor pressure of the melt. If the vapor pressure is low, the Czochralski method is a powerful method to obtain a single-crystal ingot of large size. The rf-furnace is used for the Czochralski method when the melting point T_m is less than 1,500 °C, and the tungsten crucible is available for $CeCu_6$ (melting point T_m = 938 °C), for example, as shown in Figures 3(a) and (c). In this case, the tungsten crucible, which is surrounded by an rf-working coil, becomes a heater for the $CeCu_6$ melt. A large ingot 10-mm in diameter is available in this temperature-stable rf-furnace. In the furnace shown in Figure 4, the tungsten crucible is surrounded by the heat-shielding materials such as a Tamman tube or a quartz tube. Since $CeCu_6$ has a low melting point, the quartz tube is the best material because gasses are not emitted from the quartz tube during heating. Thus, the temperature around the crucible becomes stable by setting the heat-shielding material. The pulling speed is 10 mm/hr. Usually, the furnace gas is filled with argon gas, but helium

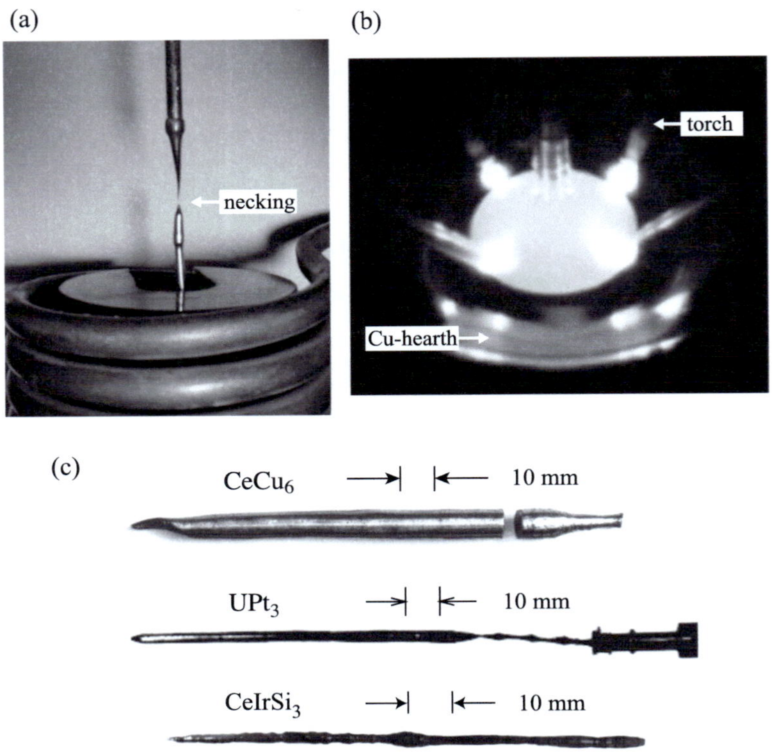

Figure 3: (**a**) $CeCu_6$ ingot in the rf-furnace, (**b**) pulling process of a UPt_3 ingot in the tetra-arc furnace, (**c**) $CeCu_6$, UPt_3, and $CeIrSi_3$ ingots grown by the Czochralski method.

gas is much better because a pipe line through which helium gas flows is easily cooled by liquid nitrogen. This becomes a getter for impurity gasses contained in the helium gas. Here, choosing the crucible material is a challenge because the transition metals and rare-earth metals often react with the crucible. So, tungsten or boron nitride is usually used as the crucible.

Now, we present a change of electronic states in $Ce_xLa_{1-x}Cu_6$ as a function of the constitution x. $Ce_xLa_{1-x}Cu_6$ crystallizes in the monoclinic structure at low temperature, as shown in Figure 5. Note that $CeCu_6$ crystallizes in the orthorhombic structure (*Pnma*: $a = 8.105$ Å, $b = 5.105$ Å, $c = 10.159$ Å) at room temperature, but it changes into the monoclinic one ($P2_1/c$: $a = 5.080$ Å, $b = 10.121$ Å, $c = 8.067$ Å, $\beta = 91.36°$) below about 200 K [2], similarly below about 500 K in $LaCu_6$ ($a = 5.143$ Å, $b = 10.204$ Å, $c = 8.144$ Å, $\beta = 91.40°$). In the following, we use the notation of the monoclinic structure. $Ce_xLa_{1-x}Cu_6$ ($x = 0.094$) corresponds to an impure Kondo system, while $CeCu_6$ belongs to the coherent Kondo lattice or a typical heavy fermion, with $\gamma = 1600$ mJ/(K^2·mol). Figures 6(a) and (b) show the logarithmic scale of temperature dependences of elec-

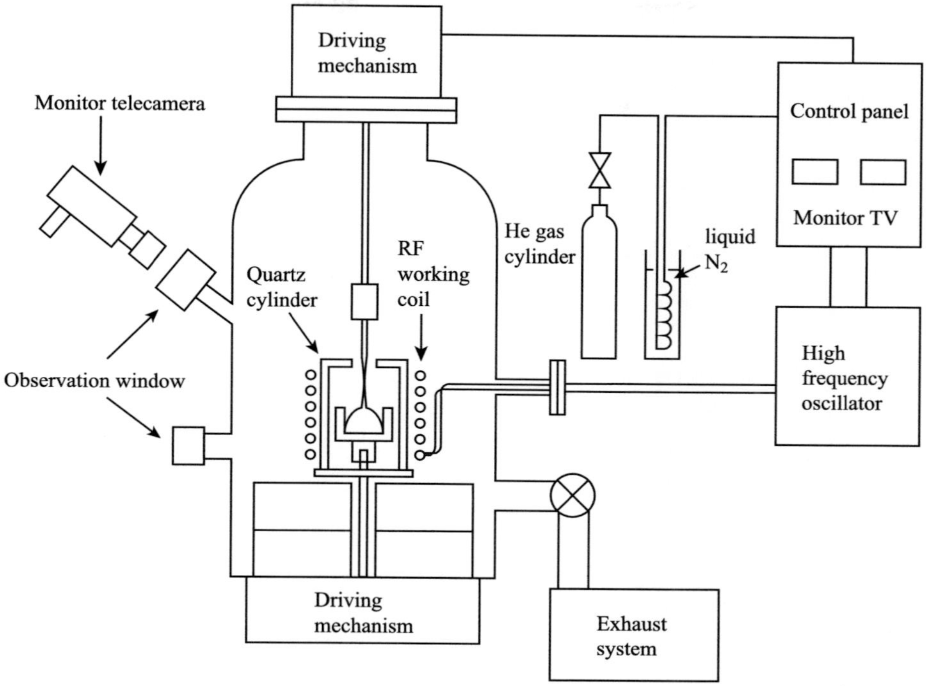

Figure 4: Schematic view of the rf-furnace.

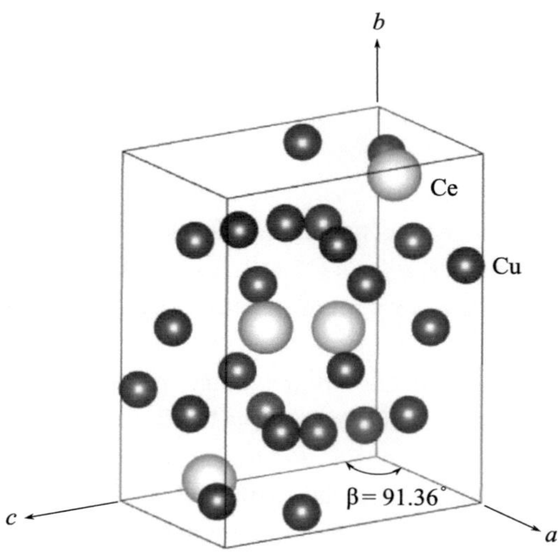

Figure 5: Monoclinic structure of $CeCu_6$.

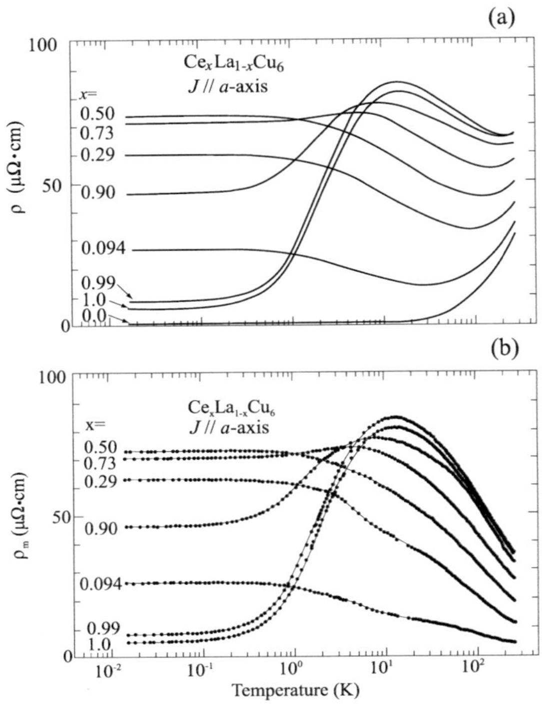

Figure 6: Temperature dependences of (**a**) the electrical resistivities and (**b**) magnetic resistivities in $Ce_xLa_{1-x}Cu_6$, where $\rho_m = \rho_{Ce_xLa_{1-x}Cu_6} - \rho_{LaCu_6}$, cited from ref. [3].

trical resistivities ρ and ρ_m, respectively, in $Ce_xLa_{1-x}Cu_6$ [3]. We define here the magnetic resistivity $\rho_m = \rho_{Ce_xLa_{1-x}Cu_6} - \rho_{LaCu_6}$. The electron scattering with phonon and crystalline defects are thus subtracted in ρ_m. As shown in Figure 6(a), the electrical resistivity for $x = 0.094$ shows a resistivity minimum around 30 K and a $-\log T$ dependence at lower temperatures, which is characteristic of the Kondo effect, and finally reaches a constant resistivity, or the so-called unitarity limit. The Kondo temperature for the $x = 0.094$ sample is estimated as $T_K = 3.7$ K from the resistivity below 20 K. On the other hand, $CeCu_6$ also indicates the $-\log T$ dependence below 300 K and a peak around 10–20 K, decreases steeply with decreasing temperature and finally follows a Fermi-liquid relation of the resistivity $\rho = \rho_0 + AT^2$, with an extremely large value of $A = 42\,\mu\Omega\cdot\text{cm/K}^2$. Note that a large $\sqrt{A}$ value correlates with a large γ value of 1600 mJ/(K^2·mol).

Figures 7(a) and (b) show the cerium concentration dependence of the residual resistivities ρ_{m_0} in $Ce_xLa_{1-x}Cu_6$. The residual resistivity in $CeCu_6$ is larger than that of $LaCu_6$, but is the same order as $LaCu_6$. Here, we emphasize that the residual resistivity of $Ce_xLa_{1-x}Cu_6$ can be simply explained by an alloy model, with each Ce atom contributing a resistivity equal to the unitarity limit 320 μΩ·cm. As shown in Figure 7(b), the residual resistivity per mole cerium decreases almost linearly with concentration x and tends to become zero in $CeCu_6$: 5 μΩ·cm in the present sample. The residual re-

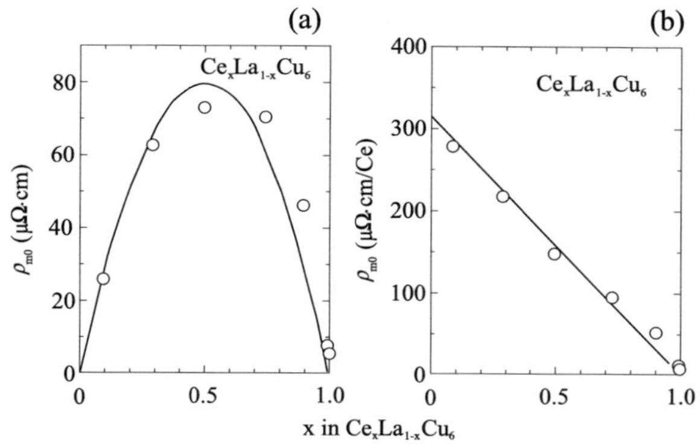

Figure 7: Cerium concentration dependences of (**a**) the residual resistivities and (**b**) the molar residual resistivities. Solid lines in (a) and (b) are $320x(1-x)$ μΩ·cm and $320(1-x)$ μΩ·cm, respectively, cited from ref. [3].

sistivity thus follows the so-called Nordheim law, $320x(1-x)$ or $320(1-x)$ as expected for an alloy model.

The Czochralski method is also applied to the compounds with high melting points, for example, about 2,500 °C in the tetra-arc furnace, as shown in Figure 3(b) and Figure 8. In this case, as shown in Figure 3(b), the crucible is unnecessary. The starting materials are set on the water-cooled Cu-hearth plate, and the single crystals are grown by the Czochralski method in the tetra-arc furnace. The heating temperature is continuously changed because the amount of the melt is reduced by pulling an ingot from the melt. It is necessary to adjust the heat current and the distance between the tip of the torch and the melt. The pulling speed is 10–15 mm/hr. It is better to use a single crystal as a seed.

UGe_2 with the orthorhombic crystal structure is a congruent compound, as shown in Figure 9(a) from a U-Ge alloy phase diagram [4]. A pulled ingot of UGe_2 is shown in Figure 9(b). The single crystal was not good in quality, revealing the residual resistivity ratio (RRR = ρ_{RT}/ρ_0, ρ_{RT}: resistivity at room temperature, ρ_0: residual resistivity) RRR = 10–50 for the first trial but reaching ρ_0 = 0.19 μΩ·cm and RRR = 910 in the final trial, as shown Figure 9(c) [5]. UGe_2 is a ferromagnet with a Curie temperature T_C = 52 K, but becomes a spin-triplet superconductor under pressure [6].

Here, the typical compounds with high melting points are UPt_3 (T_m = 1700 °C) [7] shown in Figure 3(c), $CeRu_2Si_2$ [8], $CeRh_3B_2$ [9], UB_2 (T_m = 2385 °C) [10] and UC (T_m = 2530 °C) [11]. The melting points of $CeRu_2Si_2$ and $CeRh_3B_2$ are unknown, but are estimated to be 2,000–2,500 °C. These compounds are most likely to be congruent compounds as in UGe_2.

The Czochralski method is also applied to the incongruent compounds such as $CeRu_2$ [12], USi_3 [13] and $CeIrSi_3$ [14]. For these compounds, the off-stoichiometric starting materials of $CeRu_{1.8}$, $USi_{4.6}$ and $CeIrSi_{3.5}$ were used in the tetra-arc furnace. Large single-crystal ingots are shown in Figure 3(c) for $CeIrSi_3$. In general, ternary alloy phase diagrams are unknown. $CeIrSi_3$ single crystals are not obtained by the Czochralski method using the starting materials with the stoichiometric com-

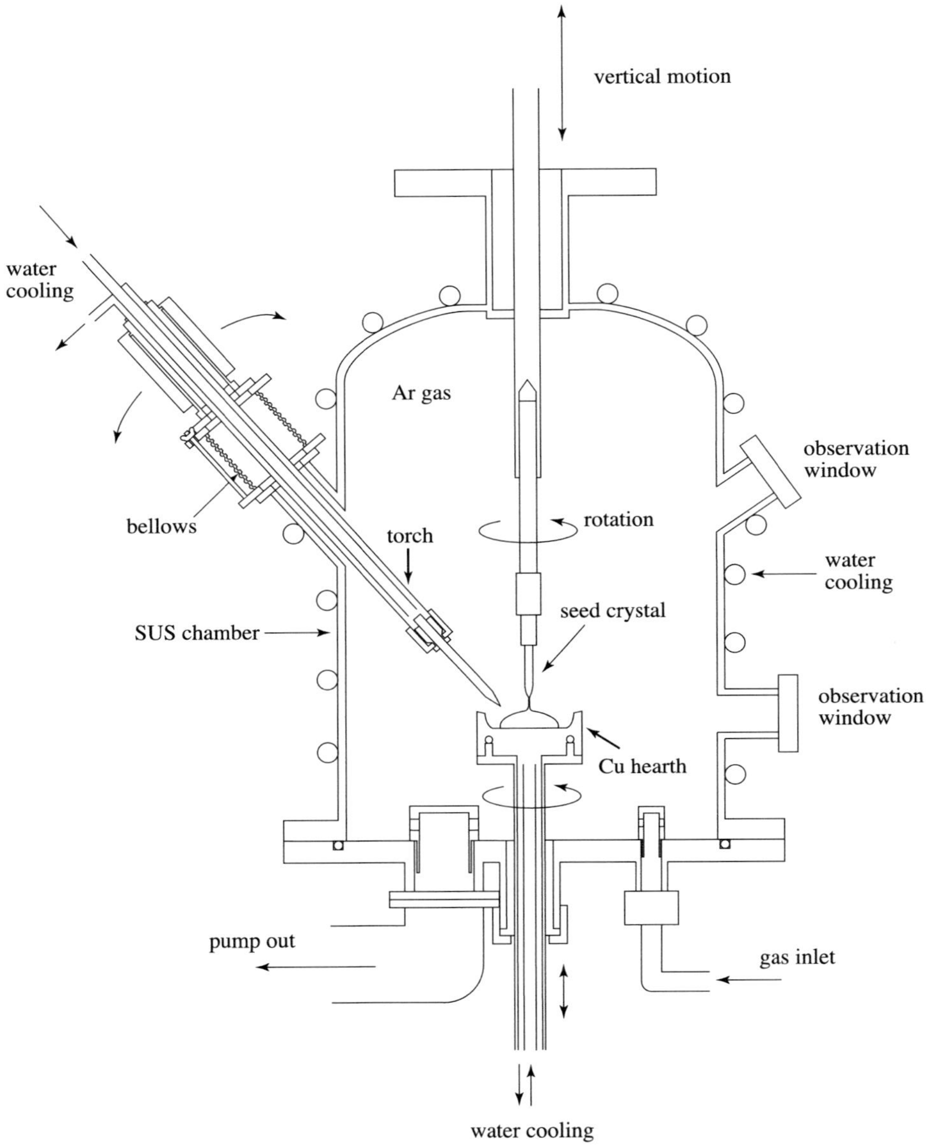

Figure 8: Schematic view of the tetra-arc furnace.

position of Ce:Ir:Si=1:1:3. We simply supposed the incongruent phase diagram for $CeIrSi_3$, as schematically shown in Figure 10(a). We selected the starting materials of Ce:Ir:Si=1:1:3.5. This is because there exist no intermetallic compounds in the Si-rich side in the R-T-Si-alloy phase diagram, as shown in Figure 10(b).

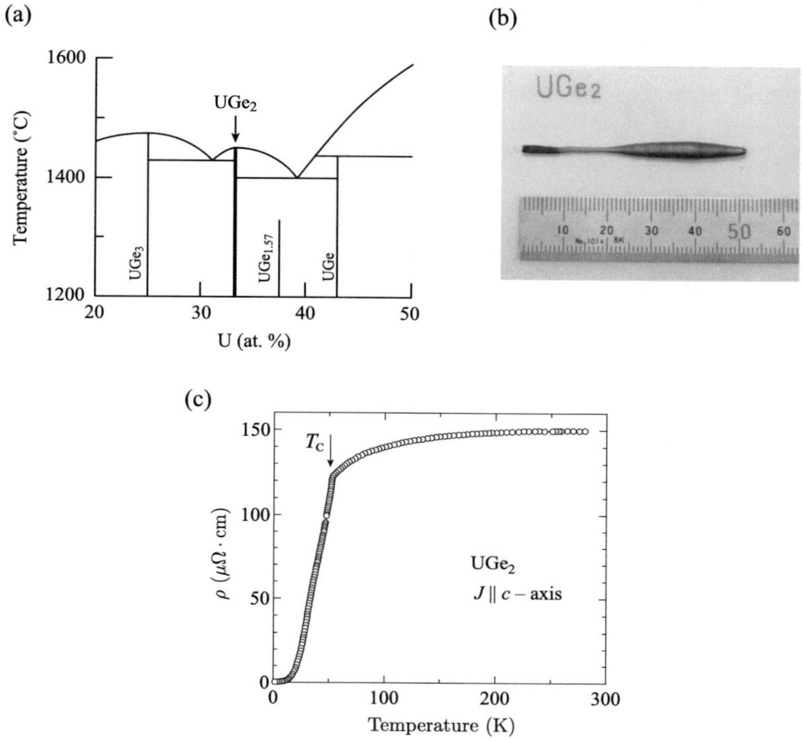

Figure 9: (**a**) U-Ge-alloy phase diagram, (**b**) a pulled single-crystal ingot of UGe_2, and (**c**) the corresponding temperature dependence of the electrical resistivity, cited from ref. [5].

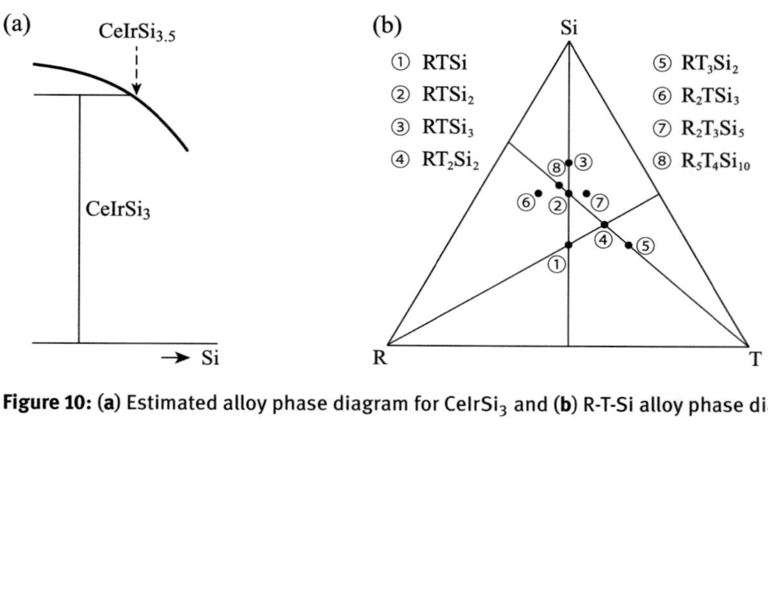

Figure 10: (**a**) Estimated alloy phase diagram for $CeIrSi_3$ and (**b**) R-T-Si alloy phase diagram.

2.2 Zone-melting method

The zone-melting method is simple but is useful to grow single crystals. Figure 11 depicts one of the examples. First, we prepare a long polycrystal ingot, which is placed on a water-cooled Cu boat. An rf-working coil moves from the left-hand side to the right-hand side, with a speed of 5–10 mm/hr. For example, we grew single crystal of $CeRu_2$ by this method, with a stoichiometric constitution of Ce:Ru=1:2, whereas we could not succeed in growing single crystals by the Czochralski method, with the stoichiometric constitution, as just mentioned.

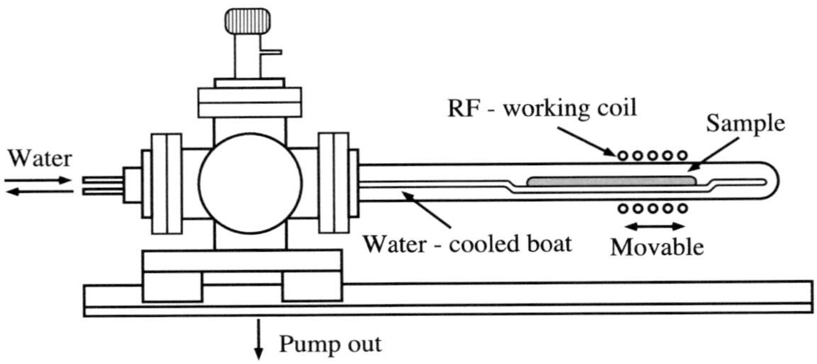

Figure 11: Schematic view of the zone-melting apparatus.

A simple technique based on the zone-melting method is as follows. We use a horizontal electric furnace. Polycrystals with various sizes are placed in a carbon boat, which is encapsulated in a quartz tube and is set into the electric furnace with a temperature gradient. The temperature is heated up to a temperature higher than a melting point of the compound, e. g., $AuAl_2$, and is slowly cooled down, taking four days in total. This is a simple and effective method in the case of $AuAl_2$ because the volume of $AuAl_2$ expands in the process of crystallization.

2.3 Floating-zone method

First, we explain the single-crystal growth of rare earth hexaborides RB_6 (R: La, Ce, Pr, Nd) with a cubic structure using the rf-furnace under a high Ar gas pressure of 5–8 kg/cm^2. The rf-furnace is the same one used in the Czochralski method, as shown in Figure 4. A melting point $T_m = 2,550$ °C in CeB_6 is extremely high. To heat up and melt a sintered rod of CeB_6, an rf-working coil is small in size compared with the coil in the Czochralski method, as shown in Figure 12(a). Here, the sintered rod is about 10 mm in diameter. The working coil is fixed and two sintered rods are shifted down during the growing process. First, we melt the lower rod partially, connect an upper

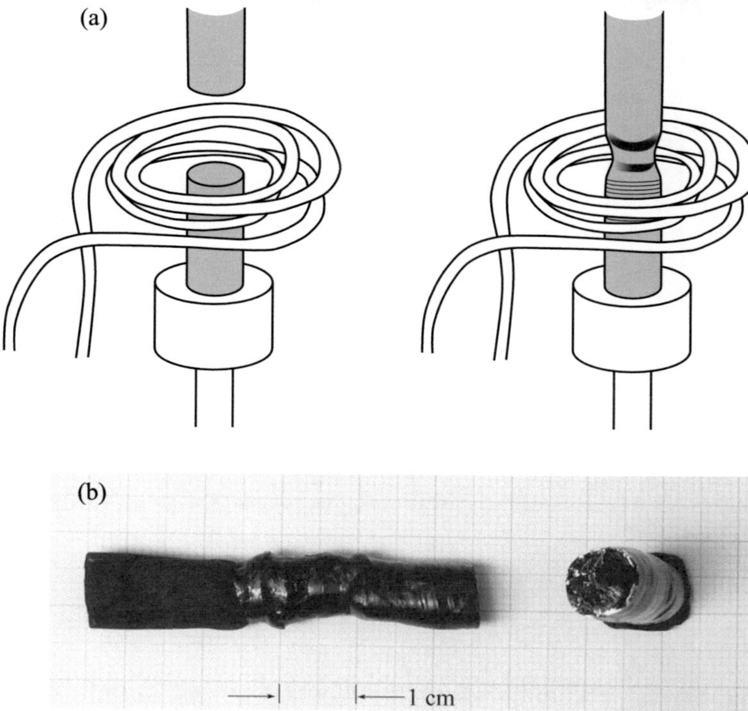

Figure 12: (**a**) Floating-zone for CeB_6 and (**b**) an ingot of CeB_6.

rod with the lower rod and shift down them with a speed of 10–20 mm/hr. Note that the speed of the upper rod is slightly faster than that of the lower rod, and the upper rod becomes single crystals. Later, this single-crystal ingot is used as the lower rod which corresponds to a seed. A single-crystal ingot of CeB_6 is shown in Figure 12(b).

In this process of single-crystal growth by means of the floating-zone method, the sintered rods are essentially important, especially in shape, to grow a high-quality single crystal. The sintered rods are produced as follows: (1) synthesis: $CeO_2 + 8B \rightarrow CeB_6 + B_2O_2 \uparrow$; (2) making fine powders of CeB_6 by a process of grinding; (3) a rod is produced by pressing a rubber tube containing fine powders of CeB_6 in water; and (4) a sintered rod is formed by heating.

Single crystals of a spin-triplet superconductor Sr_2RuO_4 are also grown by a floating-zone technique using an infrared-radiation image furnace [15], as shown in Figure 13(a). The growing process of single crystals is the same as just described. Synthesis of $2SrCO_3 + RuO_2 \rightarrow Sr_2RuO_4 + 2CO_2 \uparrow$ is carried out at 950 °C for 24 hrs. The powder is reground and pressed into rods of about 5 mm in diameter and 70 mm in length. The rods are sintered at 1,350 °C for 3 hrs under a flow of oxygen and/or air. The speed of growing in the infrared-radiation image furnace is rather slow, at 3–5 mm/hr.

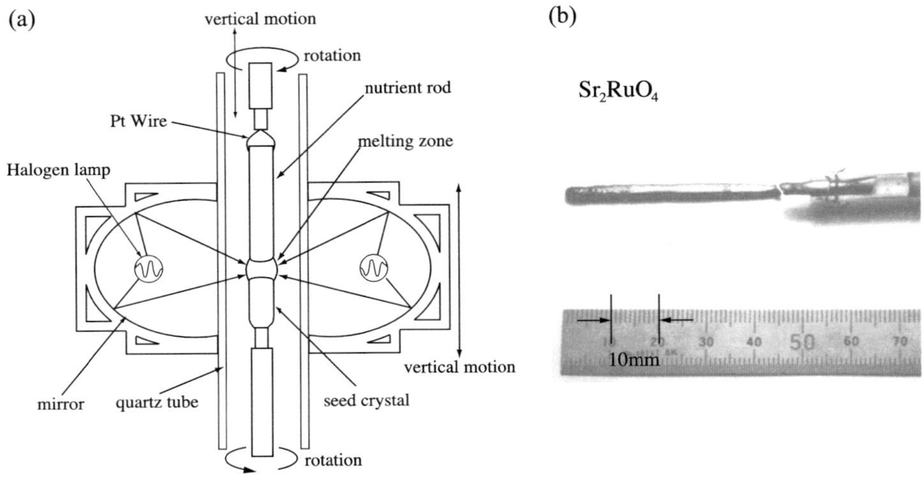

Figure 13: (**a**) Schematic view of the infrared-radiation image furnace and (**b**) a single-crystal ingot of Sr_2RuO_4.

2.4 Solid-state electrotransport method

Here, the starting material of a uranium ingot is not of high quality, roughly 99.9 % (3N), and must be annealed under a high vacuum of 10^{-7} Pa via the solid-state electrotransport method. A starting uranium metal was a rectangular rod of $4{\times}4{\times}150$ mm^3; the melting temperature of uranium is 1,135 °C. Figure 14(a) shows a schematic view for the solid-state electrotransport apparatus. The ingot was heated up to 1,120 °C by floating the DC current for the ingot under high vacuum for a long period of two weeks. Note that the temperature of 1,120 °C was the highest at around a midpoint of the ingot because the ends of the ingot were fixed by water-cooled electrodes.

We analyzed impurities of the ingot by means of an inductively coupled plasma-atomic emission spectrometer [16]. Figure 14(b) shows the concentration of impurities of Mg, Al, Ti, V, Cr, Mn, Fe, Co, Ni, Cu and Zn for five parts of the ingot. By this annealing, an Fe impurity of 40 ppm in the starting uranium ingot is reduced to less than 2 ppm in the center part of the rod, and the Cu impurity is completely removed. Impurities with high vapor pressure such as Mn and Zn were evaporated. The impurities are reduced qualitatively by three effects: sweeping out by the current flow, diffusion and evaporation.

Subsequent annealing of the uranium compounds under high vacuum on the basis of the solid-state electrotransport method drastically improves the quality of the single crystal and were applied to UPt_3 and URu_2Si_2, together with $CeRu_2$, as shown in Figure 14(c) for UPt_3. An annealing temperature is found to be important in UPt_3: RRR = 600–650 (ρ_0 = 0.19 μΩ·cm) in the temperature range of 900–1,000 °C is re-

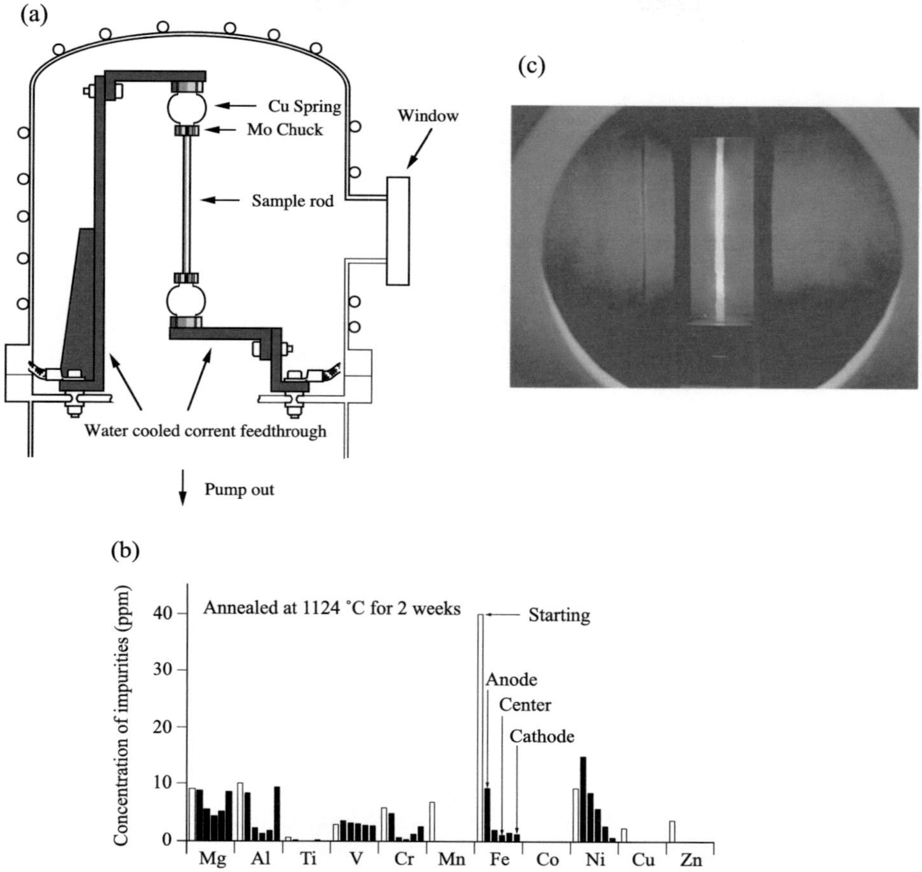

Figure 14: (**a**) Schematic view of the solid state electrotransport apparatus, (**b**) concentration of impurities in the U ingot annealed at 1124 °C for two weeks, cited from ref. [16] and (**c**) subsequent annealing of UPt_3.

duced to RRR = 500 at 1,200 °C. RRR reaches more than 500 in URu_2Si_2 [17]. In the case of $CeRu_2$, the ρ_0 and RRR values are 0.6–0.8 μΩ·cm and 240–270, respectively.

2.5 Flux method

Figure 15(a) indicates single crystals of $CeRhIn_5$ grown by the In-flux method, where the starting materials and flux were inserted into an alumina crucible, sealed in a quartz tube, heated up to about 1,000 °C in an electric furnace and cooled down to lower temperatures, taking about 20 days.

In flux was removed at 300 °C by spinning the ampoule in a centrifuge, as shown in Figure 15(b). Usually, metals with low-temperature melting points such as Zn, Cd,

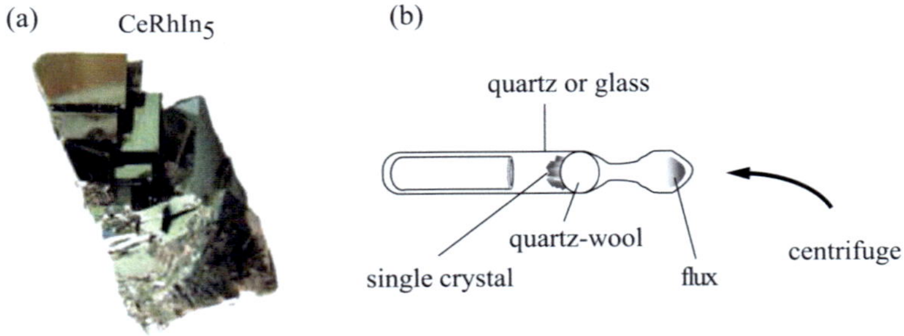

Figure 15: (**a**) $CeRhIn_5$ grown by the In-flux method and (**b**) separation of flux and single crystals by spinning the ample in the centrifuge.

Al, Ga, In, Sn, Pb, Sb, and Bi are available. In the case of $CeRhIn_5$, excess In itself becomes flux, where the ratio of Ce:Rh:In is 1:1:15. In the lanthanide compounds, it is possible to grow a series of single crystals from Y to Lu or the corresponding actinide compounds. The following is one of the example in RCu_2Si_2 grown by using the Sn flux, with an off-stoichiometric constitution of R:Cu:Si:Sn=1:15:3:50, which is the same ratio for a single-crystal growth of UCu_2Si_2 [18]. Figure 16 shows single-crystal ingots of RCu_2Si_2, except for $EuCu_2Si_2$ [19]. Note that single crystals of $EuCu_2Si_2$ were not grown by the Sn-flux method. If we use In as the flux for $EuCu_2Si_2$, single crystals are grown,

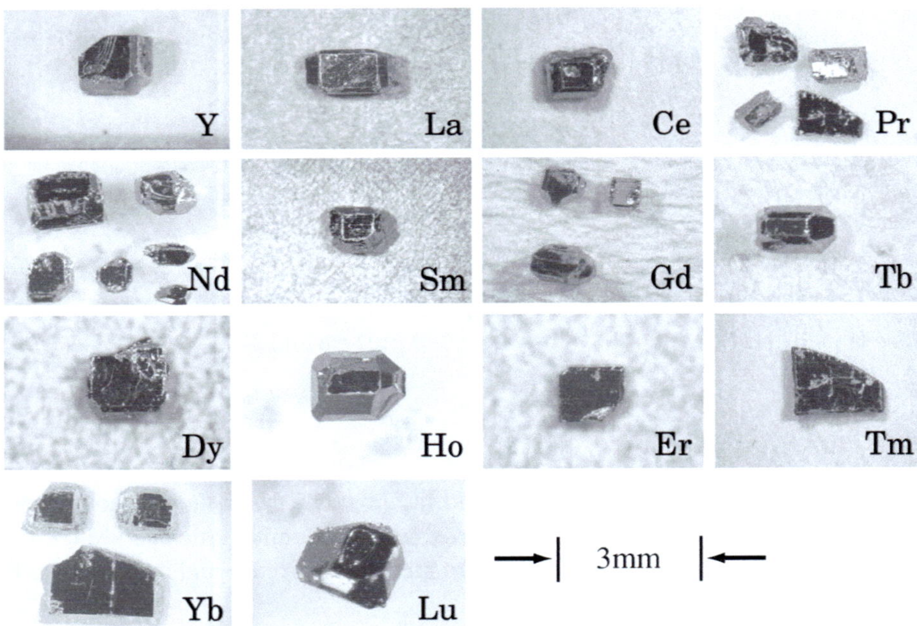

Figure 16: Single crystal ingots of RCu_2Si_2 grown by the Sn-flux method, cited from ref. [19].

but these single crystals are found to be off-stoichiometric, revealing vacancies with a Cu-filling rate of 91.4 %, with the corresponding spin-glass magnetic susceptibility at low temperatures. A single crystal, shown next, of stoichiometric $EuCu_2Si_2$ is grown by the Bridgman method. This compound is close to the Eu-trivalent compound.

We present here the experimental results of rare-earth compounds RCu_2Si_2 with the $ThCr_2Si_2$-type tetragonal structure [19], as shown in Figure 17, where the 4*f* electrons are localized. Note that $CeCu_2Si_2$, $EuCu_2Si_2$ and $YbCu_2Si_2$ are excluded in the present discussions. It is found that the Néel temperatures T_N roughly follow the de Gennes factor, as shown in Figure 18(a), namely

$$T_N \sim (g_J - 1)^2 J(J+1) \tag{1}$$

where g_J is the Landé *g* factor and *J* is the total angular momentum. The Néel temperature of $PrCu_2Si_2$ highly deviates from this scaling. This is due to the quadrupole interaction. The magnetic resistivities ρ_{mag} are obtained by subtracting the resistivity of a non-4*f* reference compound $LuCu_2Si_2$, $[\rho(LuCu_2Si_2) - \rho_0(LuCu_2Si_2)]$ from $[\rho(RCu_2Si_2) - \rho_0(RCu_2Si_2)]$, namely, $\rho_{mag} = [\rho(RCu_2Si_2) - \rho_0(RCu_2Si_2)] - [\rho(LuCu_2Si_2) - \rho_0(LuCu_2Si_2)]$. The temperature dependence of $\rho_{mag}(T)$ is shown in Figure 18(b). Note that $\rho_{mag}(T)$ above 200 K is almost temperature-independent because the temperature above 200 K is larger than Δ. Namely, the overall splitting energy in the 4*f* crystal electric field (CEF) scheme Δ in RCu_2Si_2 is smaller than 200 K. ρ_{mag}(200 K) is roughly explained by the de

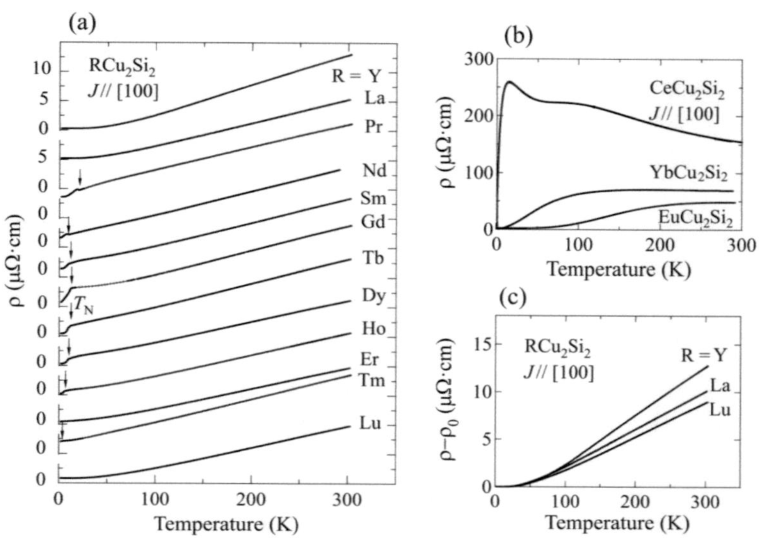

Figure 17: Temperature dependences of the electrical resistivities in (**a**) RCu_2Si_2 (R: Y, La, Pr, Nd, Sm, Gd, Tb, Dy, Ho, Er, Tm and Lu), (**b**) $CeCu_2Si_2$, $EuCu_2Si_2$ and $YbCu_2Si_2$ and (**c**) YCu_2Si_2, $LaCu_2Si_2$ and $LuCu_2Si_2$, where the residual resistivities ρ_0 are subtracted in these compounds, cited from ref. [19].

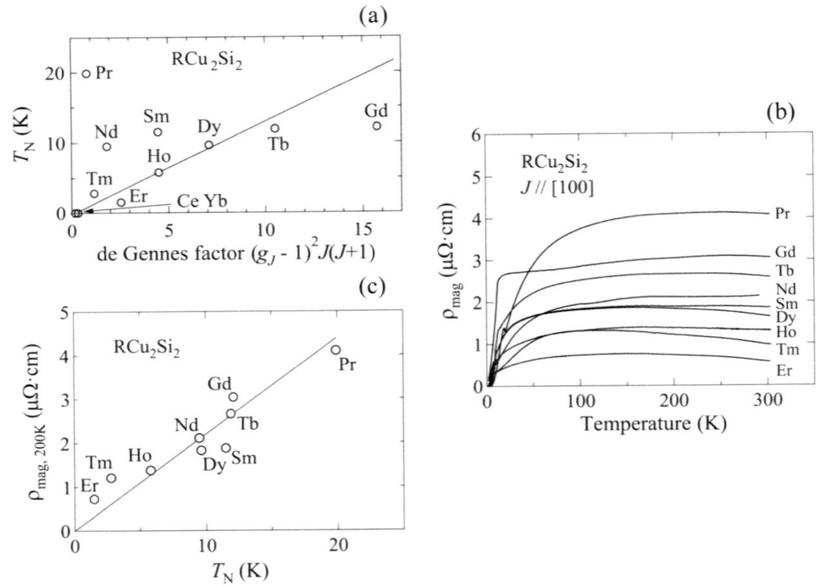

Figure 18: (**a**) Néel temperature T_N vs de Gennes factor, (**b**) temperature dependence of the magnetic resistivity ρ_{mag} and (**c**) ρ_{mag} at 200 K vs T_N in RCu_2Si_2, cited from ref. [19].

Gennes factor, but is well-scaled by T_N, as shown in Figure 18(c), meaning that both ρ_{mag}(200 K) and T_N are the same in magnetic origin.

In the heavy fermion state, the magnetic specific heat C_{mag} is changed into the electronic specific heat C_e. This means that the localized 4*f* electrons move in the crystal. The Fermi liquid state with an extremely large effective mass is realized at low temperatures. The magnetic resistivity is correspondingly changed into the resistivity due to the electron-electron scattering

$$\rho_{e-e} = AT^2 \tag{2}$$

where A is extremely large, $A = 1.0\,\mu\Omega\cdot cm/K^2$ in a heavy fermion compound $CeCu_2Si_2$ with $\gamma = 1.1\,J/(K^2\cdot mol)$ and $A = 0.040\,\mu\Omega\cdot cm/K^2$ in $YbCu_2Si_2$ with $\gamma = 150\,mJ/(K^2\cdot mol)$. Note that the A values in the usual *s* and *p* electrons systems are extremely small in magnitude, e. g., $A = 3\times10^{-7}\,\mu\Omega\cdot cm/K^2$ in Al, and it is not easy to measure the A value correctly. A relationship between the A and γ values is called a Kadowaki–Woods plot, revealing a relation of $\sqrt{A} \sim \gamma$ [20].

A series of RIn_3 (R: rare earth) and $AcIn_3$ (Ac: actinide, Th, U, Np, Pu) were grown mainly by the In-flux method [21]. YIn_3, $LaIn_3$, $CeIn_3$, $PrIn_3$ and $NdIn_3$ were, however, grown by the Czochralski method in the rf furnace. A small amount of In impurity is found to be included in these compounds grown by the flux method, but is not included in the samples grown by the Czochralski method.

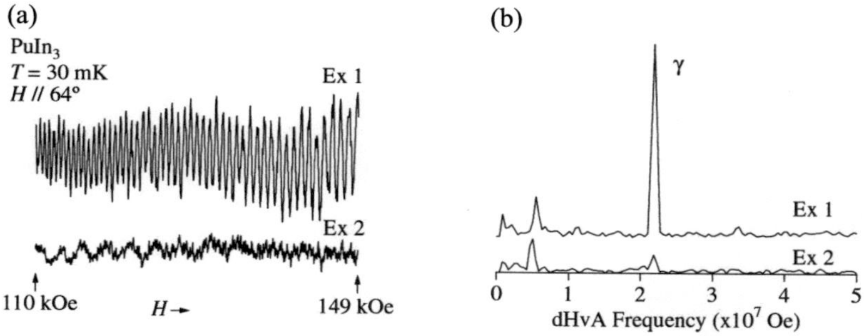

Figure 19: de Haas-van Alphen oscillations and the corresponding fast Fourier transformation (FFT) spectra in $PuIn_3$, cited from ref. [23].

Following is an example in the case of $PuIn_3$ [22]. Figure 19 shows the de Haas-van Alphen (dHvA) oscillations and the corresponding fast Fourier transformation (FFT) spectra. The angular dependence of the observed dHvA frequency named branch γ agrees well with the result of the $5f$-itinerant band calculations, namely, a nearly spherical band 27-electron Fermi surface at Γ. Note the two kinds of dHvA experiments Ex1 and Ex2. Ex2 was carried out for nine days after Ex1. The dHvA amplitude of branch γ is extremely reduced as a function of time, namely at Ex2, which is due to point defects produced by α-decay of ^{239}Pu. On the other hand, the dHvA amplitude of $F = 5.5 \times 10^6$ Oe is the same as that between Ex1 and Ex2, being unaffected by the radiation damage. This dHvA signal is not due to $PuIn_3$ but due to In impurities that might be included in sub-grain boundaries of the $PuIn_3$ single crystal sample. Finally, we note UIn_3 [23], which was not grown by the usual self-In flux method, but was grown in the growing process of $UPtIn_5$.

A change of Fermi surfaces based on the dimensionality of the electronic states could be realized when the c/a ratio of the tetragonal structure is continuously changed. Figure 20 shows several different tetragonal structures in T-Ga (T: transition metal) binary compounds [24]. The following is a main point in this study. The smaller the tetragonal c-value is, the more enhanced the one-dimensionality of the electronic state is. The larger the c-value is, the more enhanced two-dimensionality is. We expect that nearly one-dimensional characteristic Fermi surfaces can be realized in V_2Ga_5 (space group No. 127, a = 8.936 Å, c = 2.683 Å, formula units per cell Z = 2), and two-dimensional Fermi surfaces in $ZrGa_3$ (No. 139, a = 3.965 Å, c = 17.461 Å, Z = 4) and $ZrAl_3$. Fermi surfaces in $CoGa_3$ (No. 136, a = 6.230 Å, c = 6.431 Å, Z = 4) are most likely three-dimensional because the c-value is almost the same as the a-value. In contrast to these compounds, $TiGa_3$ (No. 139, a = 3.789 Å, c = 8.734 Å, Z = 2) crystallizes in a typical tetragonal structure as in the $ThCr_2Si_2$-type tetragonal structure.

Single crystals of these compounds were grown by the Ga (Al) self-flux method, and de Haas-van Alphen (dHvA) experiments were carried out, together with the full-potential linearized augmented plane-wave (FLAPW) energy-band calculations. Inter-

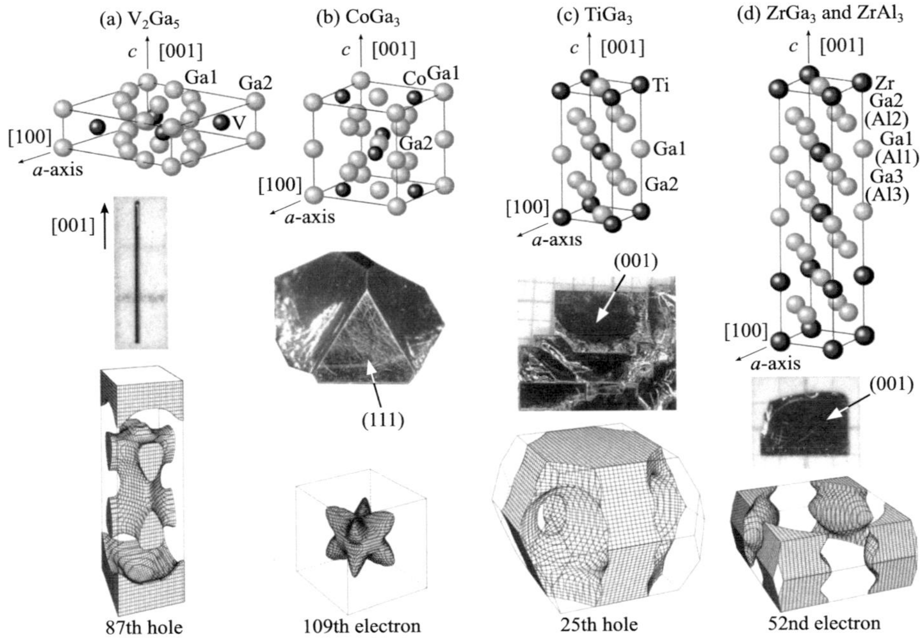

Figure 20: Tetragonal crystal structure, a single crystal, and the corresponding Fermi surface in (**a**) V_2Ga_5, (**b**) $CoGa_3$, (**c**) $TiGa_3$ and (**d**) $ZrGa_3$ and $ZrAl_3$, cited from ref. [24].

esting are the characteristic shapes of single crystals in these compounds, as shown in Figure 20. The characteristic features of the crystal structures and single crystals are summarized as follows:

1) In V_2Ga_5, the crystal structure is flat along the tetragonal [001] direction (c-axis) and the corresponding single crystal is of needle shape along the [001] direction.
2) $CoGa_3$ is tetragonal in the crystal structure, but is approximately cubic, as shown in Figure 20(b). Correspondingly, $CoGa_3$ is of pyramidal shape, which is characteristic of the crystals with the fcc structure, and also with the diamond-type structure. The flat plane of the pyramid corresponds to the (111) plane.
3) The shape of a single crystal in $TiGa_3$ is typically tetragonal. The flat plane of a rectangular single crystal corresponds to the tetragonal (001) plane (c-plane).
4) In $ZrGa_3$ and $ZrAl_3$, the crystal structure is elongated along the tetragonal [001] direction, and the corresponding single crystal is of thin-plate shape of which the flat plane corresponds to the tetragonal (001) plane.

Reflecting the crystal structures or the corresponding Brillouin zones, the Fermi surfaces, which are clarified by the dHvA experiment and band calculations, are characteristic, are summarized as follows:

1) A nearly one-dimensional plate-like Fermi surface is obtained in the band calculations, as shown in Figure 20(a). The plate is, however, wavy in shape, and the electronic state is not one-dimensional but nearly three-dimensional. Therefore, Peierls instability is not realized in this compound.
2) The Fermi surfaces in $CoGa_3$ are very similar to those of Ni_3Ga with the $AuCu_3$-type cubic structure. The band 109th electron Fermi surface is typical between these two compounds, as shown in Figure 20(b).
3) The Fermi surfaces in $TiGa_3$ are also very similar to those of YCu_2Si_2 with the $ThCr_2Si_2$-type tetragonal structure. The band 25th hole Fermi surface is typical between these two compounds, as shown in Figure 20(c).
4) The flat Brillouin zone often produces the cylindrical Fermi surface, which is realized in $ZrGa_3$ and $ZrAl_3$. The band-52nd electron Fermi surface is typical, possessing concave and convex shapes in the cylinder, as shown in Figure 20(d).

2.6 Bridgman method

A single crystal of $CePt_3Si$ with the non-centrosymmetric tetragonal structure (*P*4*mm*) is grown by the Bridgman method in a Mo crucible [25], as shown in Figure 21. First, polycrystals of $CePt_3Si$ are prepared by arc-melting the starting materials with a ratio of Ce:Pt:Si=1:3:1 and the polycrystals are inserted into a Mo-crucible. A top-plate of the Mo-crucible is sealed using the tetra-arc furnace. Note that the usual Cu-hearth plate is changed into another water-cooled Cu-holder, as shown in Figure 21. The crucible is heated up to 1,350 °C in Ar-gas and cooled down slowly in a furnace, taking 7 days in total. We used a siliconit furnace reaching a maximum temperature of 1,500 °C, as shown in Figure 22. Usually, the crucible is shifted down slowly, and the melt is enhanced to crystallize from the bottom to the top of the melt. In the present case, the crucible is set in the temperature gradient and a bottom of crucible is narrow in size, enhancing that the melt crystallizes as a single crystal. Note that we could not succeed in growing single crystals of $CePt_3Si$ by the Czochralski method in a tetra-arc furnace, suggesting an incongruent compound of $CePt_3Si$.

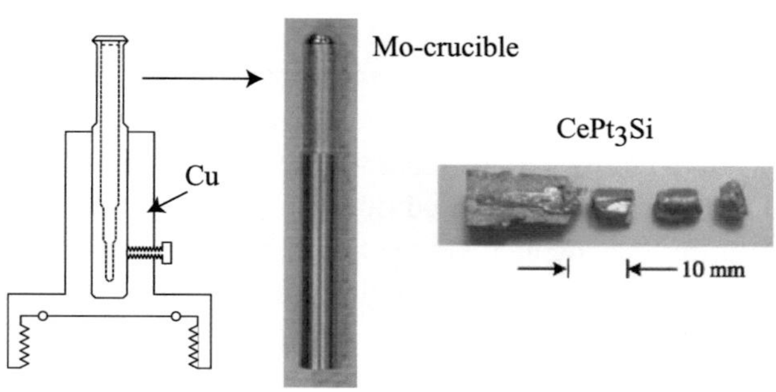

Figure 21: Sealing set of a top plate of the Mo-crucible using a tetra-arc furnace and a $CePt_3Si$ ingot.

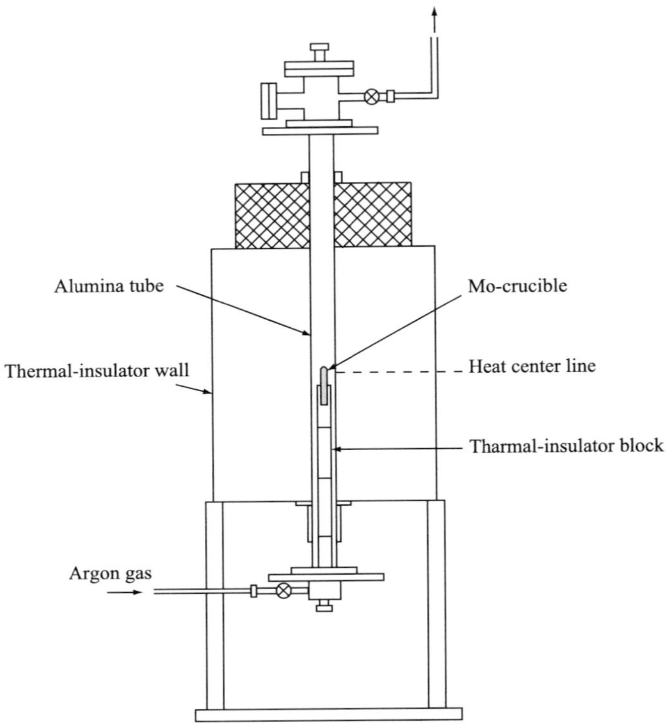

Figure 22: Schematic view of the siliconit furnace.

In the lanthanide compounds, Eu, Sm and Yb are high in vapor pressure. It is impossible to grow these single crystals by the open-type techniques such as the Czochralski method. The flux method is most likely the best technique if we know the flux and the starting materials' constitution. It is, however, difficult to grow single crystals with the stoichiometric constitution of the starting materials. Eu compounds are one example. Most of the Eu compounds are different from the other rare-earth compounds including the cerium compounds in the electronic state. Namely, the rare-earth compounds are trivalent, but a Eu-divalent (Eu^{2+}) electronic state is more stable than a Eu-trivalent (Eu^{3+}) state. Large lattice parameters of the Eu compounds, compared with those of the other rare earth compounds with the trivalent electronic state, are a good indication of the divalence of Eu. The energy difference between Eu^{2+} and Eu^{3+} states is, however, not significantly large. Therefore, the Eu^{2+}-electronic state can be changed by applying pressure because high pressure is applicable in the technique just treated.

Single crystals of EuT_2Si_2 and EuT_2Ge_2 were mainly grown by the Bridgman method using the Mo-crucible at temperatures heated up about 1,500 °C, as shown in Figure 23. In the case of $EuRh_2Si_2$, we arc-melted the starting materials of Rh and Si with a stoichiometric constitution, Rh:Si=1:1 [26]. Note that a melting point of Rh is extremely high, T_m = 1963 °C, while T_m = 1452 °C in RhSi. Then, Eu and RhSi materials with 1.8:2 were inserted into a Mo-crucible and sealed by arc-melting a cap made

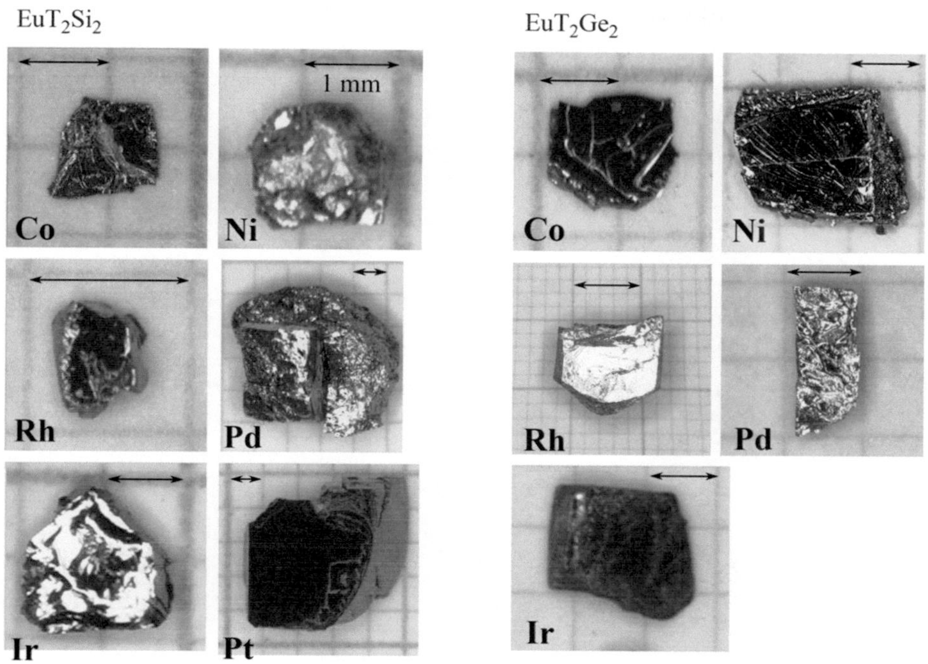

Figure 23: Single crystals of EuT_2Si_2 and EuT_2Ge_2 grown by the Bridgman method using the Mo-crucible, except $EuIr_2Ge_2$, $EuNi_2Ge_2$ and $EuPt_2Si_2$ grown by the In-flux method.

of Mo in Ar atmosphere. An excess of Eu is necessary to reduce the melting point of $EuRh_2Si_2$, which is supposed to be larger than 1,500 °C. The crucible was heated to about 1,500 °C and then cooled down to room temperature, taking 8–9 days in total. As-grown single crystals are small in size, 1 mm or less than 1 mm, but we obtained a large number of such crystals.

Figure 24(a) shows the temperature dependences of electrical resistivities under pressure P in $EuRh_2Si_2$. Valence instability occurs from a nearly divalent electronic state, $Eu^{2+\delta}$ ($\delta < 1$) ($4f^7$ of Eu^{2+}: $S = 7/2$, $L = 0$ and $J = 7/2$) at high temperatures, to a nearly trivalent electronic state, $Eu^{3-\delta'}$ ($\delta' < 1$) ($4f^6$ of Eu^{3+}: $S = L = 3$ and $J = 0$) at low temperatures at 0.96 and 1.27 GPa. Figure 24(b) shows the corresponding P-T phase diagram, where T_v is called the valence-transition temperature. The critical end point (CEP) in the valence transition, where the first-order transition terminates, is established as $P_{CEP} = 2.1$ GPa and $T_{CEP} = 176$ K. The resistivity of $EuRh_2Si_2$ in the CEP electronic state, or the resistivity at 2.09 GPa, is similar to the resistivity in $EuPd_2Si_2$, as shown in Figure 24(c). With further increasing pressure for $EuRh_2Si_2$, the resistivity is changed into the resistivity in the nearly trivalent or trivalent electronic state in $EuNi_2Si_2$, $EuCo_2Si_2$ and $EuPd_3$ in Figure 24(d). It is characteristic that an electronic state between Eu^{2+} and Eu^{3+}, namely, an intermediate valence state, or the moderate

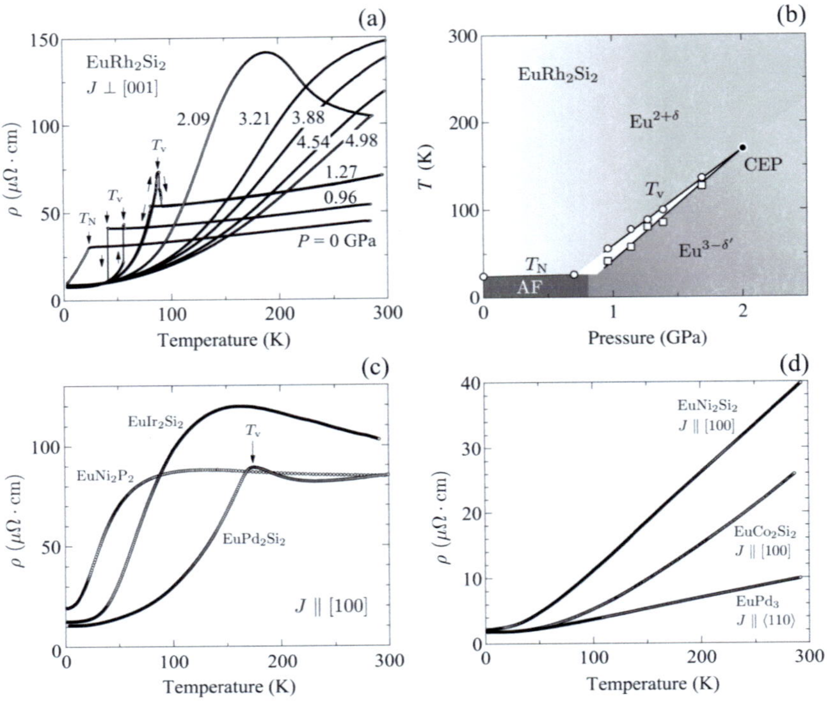

Figure 24: (**a**) Temperature dependences of electrical resistivities under pressure P in an antiferromagnet of $EuRh_2Si_2$, indicating the first-order valence transition, (**b**) P-T phase diagram in $EuRh_2Si_2$, cited from ref. [26], (**c**) temperature dependences of electrical resistivities in a temperature-induced valence transition compound $EuPd_2Si_2$ and moderate heavy fermion compounds of $EuNi_2P_2$ and $EuIr_2Si_2$ and (**d**) nearly trivalent or trivalent compounds of $EuNi_2Si_2$, $EuCo_2Si_2$ and $EuPd_3$.

heavy-fermion state is observed in $EuNi_2P_2$ and $EuIr_2Si_2$, as shown in Figure 24(c). The heavy-fermion state in $EuNi_2P_2$ is formed on the basis of the Kondo effect [27].

2.7 Chemical transport method

Single crystals of $U_3As_4(U_3P_4)$ and $UAs_2(UP_2)$ are grown by the chemical transport method using iodine as a transport agent in an electric furnace with a distinct temperature gradient [28, 29], as shown in Figure 25(a). Starting materials of powders $U_3As_4(UAs_2)$ are placed in the low-temperature side (830 °C), and single crystals are grown in the high-temperature side (900 °C), as shown in Figure 25(b). Note that, in the transition-metal compounds such as CoS_2, powders of CoS_2 are placed in at a high-temperature side (700 °C), and single crystals of CoS_2 are grown in the low-temperature side (650 °C), as shown in Figure 25(c). The transport agent is powders of $CoBr_2$, which change into Br_2 gas at high temperatures [30]. In this case, CoS_2 pow-

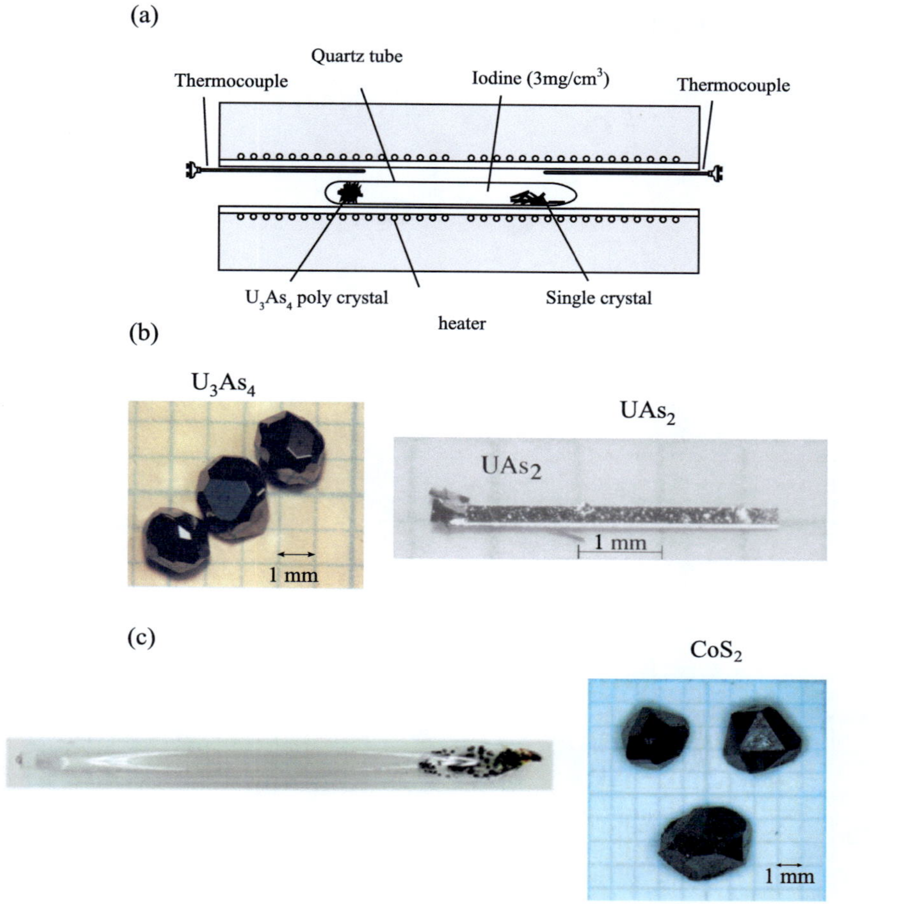

Figure 25: (**a**) Schematic view of the chemical transport furnace, (**b**) single crystals of U_3As_4 and UAs_2 and (**c**) a quartz tube in which many single crystals of CoS_2 are grown.

ders, which are placed in the left-hand side of a quartz tube, are completely changed into single crystals at the right-hand side. Pyramid-like single crystals are shown in enlarged scale. A high-quality sample with $\rho_0 = 0.42\,\mu\Omega{\cdot}\text{cm}$ and RRR = 410 is obtained.

The pyrite-type compounds TX_2, including CoS_2 and $CoSe_2$ already mentioned, crystallize in the cubic crystal structure. The T atom is occupied in the face-centered positions, where each T atom is surrounded by six nearest-neighbor X atoms which form a trigonally distorted octahedron, as shown in Figure 26(a). The $3d$ levels in TX_2 split into e_g and t_{2g} orbitals by the octahedral CEF. The e_g level is higher than the t_{2g} level in energy. The T-e_g ($d\gamma$: $x^2 - y^2$, $3z^2 - r^2$) orbitals in TS_2 expand along the S atoms in the octahedron, hybridize with S-$3p$ orbitals and form the antibonding and bonding orbitals, separated by a large gap. On the other hand, the T-t_{2g} ($d\varepsilon$: xy, yz, zx)

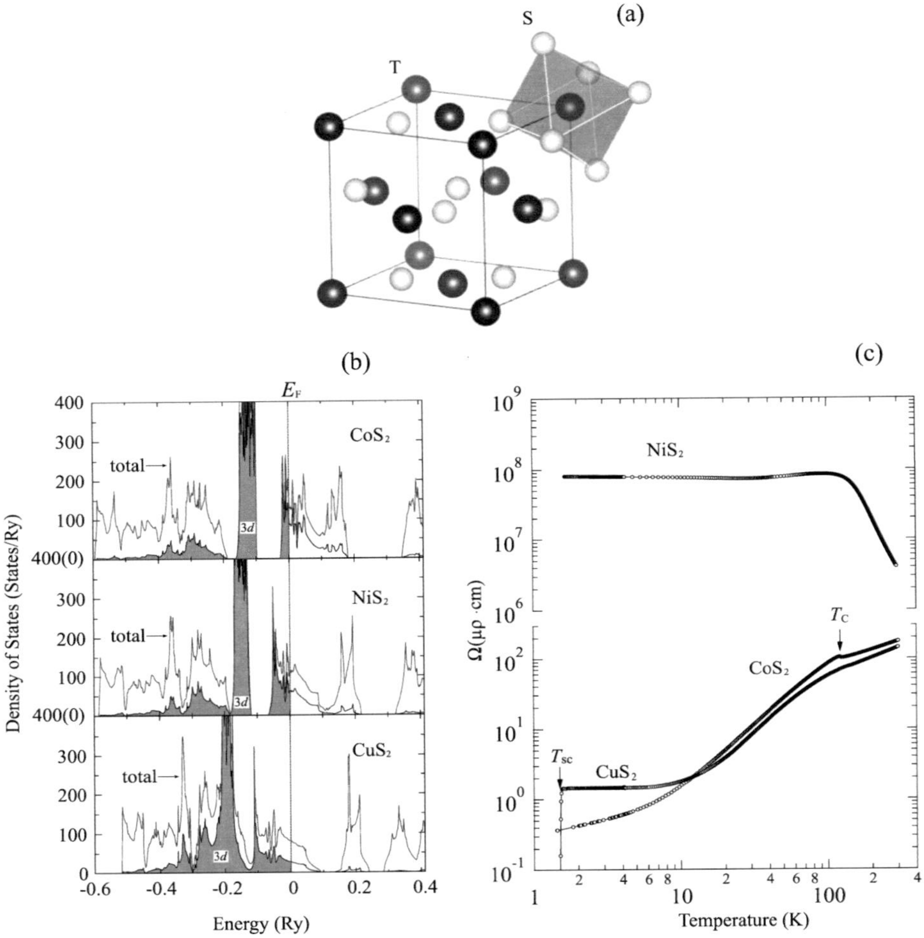

Figure 26: (**a**) Pyrite-type cubic crystal structure in TX_2 (T: transition metal, X: S, Se, Te), (**b**) theoretical total and partial (3*d* electrons) densities of states in the paramagnetic CoS_2, NiS_2 and CuS_2, after courtesy of H. Harima and (**c**) the corresponding temperature dependences of electrical resistivities in the current along the [100] direction.

orbitals scarcely possess a hybridized partner, forming a localized narrow-band. The electronic properties of the pyrite-type compounds are thus understood from a progressive filling of the antibonding $3d-e_g$ band. Namely, the filling of the e_g band due to 3*d* electrons is empty in a band-gap insulator (nonmagnetic Fe) FeS_2 ($t_{2g}^6 e_g^0$), a quarter in a metallic ferromagnet CoS_2 ($t_{2g}^6 e_g^1$), a half in a Mott-Hubbard insulator NiS_2 ($t_{2g}^6 e_g^2$) (complicated antiferromagnet) based on the electron correlations, three-quarters in a superconductor CuS_2 ($t_{2g}^6 e_g^3$) and completely full in a diamagnetic wide-band-gap insulator ZnS_2 ($t_{2g}^6 e_g^4$). Figure 26(b) shows the theoretical total and partial (3*d* electrons) densities of states in CoS_2, NiS_2, and CuS_2 in the paramagnetic state. The corresponding electrical resistivities are shown in Figure 26(c), revealing a ferromagnetic ordering at a Curie temperature $T_C = 122\,K$ in CoS_2, the Mott-Hubbard insulating property in NiS_2 and a superconducting transition at $T_{sc} = 1.5\,K$ in CuS_2. Note that CuS_2 is crystallized under pressure of 5 GPa and 900 °C.

3 Concluding remarks

We have presented seven techniques for the single-crystal growth of heavy-fermion compounds containing transition metals, lanthanides and actinides. It is not easy for us to grow high-quality single crystals because crystal growth is mainly based on experimental (personal) techniques and knowledge. Solutions or methods to grow high-quality single crystal are not unique, but rather there exist many variations.

A ferromagnet MnP ($T_C = 291\,K$) was once grown by the Bridgman method [31]. Namely, powders of Mn and P, which are encapsulated in a quartz tube, are gradually heated up to 800 °C, taking one week, and then powders of MnP are obtained. These MnP powders are pulverized into fine powders and pressed into pellets. The pellets are inserted into an alumina and/or yttria crucible, encapsulated in a quartz tube, heated up to 1,170 °C and slowly cooled down in a temperature-gradient vertical- electric furnace. This corresponds to the Bridgman method. A single-crystal ingot with a large size of $10^{\phi} \times 50\,mm^3$ was obtained, with an RRR value less than 100. Later, single crystals were also grown by the chemical transport method without a transport agent [32]. Namely, MnP powders are placed in a hotter part of the quartz tube at 850 °C, and single crystals of one mm^3 are grown in a lower-temperature part at 800 °C. The RRR value was extremely improved, reaching 1,300, and then the dHvA signals were observed using this sample. Very recently, a usual Sn-flux method was applied to MnP [33], where powders of Mn and P, together with Sn, with a constitution of Mn:P:Sn=1:1:10, were inserted into an alumina crucible, encapsulated in a quartz tube, heated up to 1,150 °C and cooled down slowly, taking ten days in total. The RRR value was 800–1,100. High-quality samples are grown by this simple method, with a typical size of $0.2\times0.2\times(3–5)\,mm^3$. Single crystals of MnP are also grown by the In-flux method, revealing RRR = 100, with a typical size of $1\times1\times(3–5)\,mm^3$. MnP has recently proven to become a superconductor under pressure [33], which is a useful technique to change the electronic state in heavy-fermion compounds.

$CeIrSi_3$ in Figure 3(c) crystallizes in the tetragonal non-centrosymmetric structure (*I4mm*) as in $CePt_3Si$. The Ce atoms occupy the four corners and the body center of tetragonal structure, similar to the well-known $ThCr_2Si_2$-type tetragonal structure, but the Ir and Si atoms lack inversion symmetry along the [001] direction, which structure is called “Rashba-type”. An antiferromagnet $CeIrSi_3$ with a Néel temperature $T_N = 5.0\,K$ also becomes a superconductor under pressure [34]. The huge upper critical field H_{c2} in superconductivity is realized in $CeIrSi_3$ under pressure of 2.6 GPa: $H_{c2}(0) = 450$ kOe for $H \parallel [001]$ and $H_{c2}(0) = 95$ kOe for $H \perp [001]$, with a superconducting transition temperature $T_{sc} = 1.6$ K. Superconductivity of $CeIrSi_3$ is of the strong-coupling nature and is realized in the antiferromagnetic quantum critical region. Most important is that in the present non-centrosymmetric (Rashba-type) spin-orbit coupling, all the spins of conduction electrons are perpendicular to the magnetic-field direction of $H \parallel [001]$. This is because a degenerate Fermi surface, which consists of conduction electrons with the up- and down-spin states, is split into two Fermi surfaces. The corresponding spins of conduction electrons are rotated clockwise in the tetragonal (001) plane for one Fermi surface, but counterclockwise for the other Fermi surface. The spin degeneracy is thus lifted in a zero magnetic field. Therefore, no paramagnetic suppression in superconductivity is realized when the magnetic field is applied along the [001] direction. This is a reason why the huge H_{c2} value is realized for $H \parallel [001]$ in $CeIrSi_3$.

Superconductivity and magnetism in the non-centrosymmetric compounds have been developed into a new field in solid-state and condensed-matter physics. The corresponding Fermi surface splits into two Fermi surfaces due to the antisymmetric spin-orbit coupling. Moreover, the magnetic interaction is based on the so-called Dzyaloshinsky-Moriya interaction. Unique magnetism of "skyrmion" is observed in MnSi [35]. There exist many non-centrosymmetric compounds because the compounds belonging to 138 space groups in total 230 space groups possess no inversion symmetry. New physics awaits us in our future study.

Bibliography

[1] Ōnuki Y, Settai R, Takeuchi T, Sugiyama K, Honda F, Haga Y, Yamamoto E, Matsuda TD, Tateiwa N, Aoki D, Sheikin I, Harima H, Yamagami H. J Phys Soc Jpn 2012, 81, SB001.

[2] Asano H, Umino M, Ōnuki Y, Komatsubara T, Izumi F, Watanabe N. J Phys Soc Jpn 1986, 55, 454.

[3] Sumiyama A, Oda Y, Nagano H, Ōnuki Y, Shibutani K, Komatsubara T. J Phys Soc Jpn 1986, 55, 1294.

[4] Okamoto H. In: Massalski TB, Okamoto H, Subramanian P, Kacprzak L, eds. Binary Alloy Phase Diagramsc. ASM International, 1990, 2015–7.

[5] Yun S, Satoh K, Fujimaki Y, Umehara I, Ōnuki Y, Takayanagi S, Aoki H, Uji S, Shimizu T. Physica B 1993, 186, 129.

[6] Ōnuki Y, Settai R, Haga Y, Machida Y, Izawa K, Honda F, Aoki D. C R Phys 2014, 15, 616.

[7] Kimura N, Settai R, Ōnuki Y, Toshima H, Yamamoto E, Maezawa K, Aoki H, Harima H. J Phys Soc Jpn 1995, 64, 3881.

[8] Aoki H, Uji S, Albessard AK, Ōnuki Y. J Phys Soc Jpn 1993, 62, 3157.

[9] Okubo T, Yamada M, Thamizhavel A, Kirita S, Inada Y, Settai R, Harima H, Takegahara K, Galatanu A, Yamamoto E, Ōnuki Y. J Phys Condens Matter 2003, 15, L721.

[10] Yamamoto E, Haga Y, Honma T, Inada Y, Aoki D, Hedo M, Yoshida Y, Yamagami H, Ōnuki Y. J Phys Soc Jpn 1998, 67, 3171.

[11] Yamamoto E, Haga Y, Inada Y, Murakawa M, Ōnuki Y, Maehira T, Hasegawa A. J Phys Soc Jpn 1999, 68, 3953.
[12] Hedo M, Inada Y, Sakurai K, Yamamoto E, Haga Y, Ōnuki Y, Takahashi S, Higuchi M, Maehira T, Hasegawa A. Philos Mag B 1998, 77, 975.
[13] Tokiwa Y, Harima H, Aoki D, Nojiri S, Murakawa M, Miyake K, Watanabe N, Settai R, Inada Y, Sugawara H, Sato H, Haga Y, Yamamoto E, Ōnuki Y. J Phys Soc Jpn 2000, 69, 1105.
[14] Okuda Y, Miyauchi Y, Ida Y, Takeda Y, Tonohiro C, Oduchi Y, Yamada T, Dung ND, Matsuda TD, Haga Y, Takeuchi T, Hagiwara M, Kindo K, Harima H, Sugiyama K, Settai R, Ōnuki Y. J Phys Soc Jpn 2007, 76, 044708.
[15] Yoshida Y, Mukai A, Settai R, Ōnuki Y, Takei H. J Phys Soc Jpn 1998, 67, 2551.
[16] Haga Y, Honma T, Yamamoto E, Ohkuni H, Ōnuki Y, Ito M, Kimura N. Jpn J Appl Phys 1998, 37, 3604.
[17] Haga Y, Matsuda TD, Tateiwa N, Yamamoto E, Ōnuki Y, Fisk Z. Philos Mag 2014, 94, 3672.
[18] Matsuda TD, Haga Y, Ikeda S, Galatanu A, Yamamoto E, Shishido H, Yamada M, Yamaura J-I, Hedo M, Uwatoko Y, et al. J Phys Soc Jpn 2005, 74, 1552.
[19] Dung ND, Ota Y, Sugiyama K, Matsuda TD, Haga Y, Kindo K, Hagiwara M, Takeuchi T, Settai R, Ōnuki Y. J Phys Soc Jpn 2009, 78, 024712.
[20] Kadowaki K, Woods S. Solid State Commun 1986, 58, 507.
[21] Ōnuki Y, Settai R. Low Temp Phys 2012, 38, 89.
[22] Haga Y, Aoki D, Yamagami H, Matsuda TD, Nakajima K, Arai Y, Yamamoto E, Nakamura A, Homma Y, Shiokawa Y, Ōnuki Y. J Phys Soc Jpn 2005, 74, 2889.
[23] Tokiwa Y, Aoki D, Haga Y, Yamamoto E, Ikeda S, Settai R, Nakamura A, Onuki Y. J Phys Soc Jpn 2001, 70, 3326.
[24] Teruya A, Takeda M, Nakamura A, Harima H, Haga Y, Uchima K, Hedo M, Nakama T, Ōnuki Y. J Phys Soc Jpn 2015, 84, 054703.
[25] Yasuda T, Shishido H, Ueda T, Hashimoto S, Settai R, Takeuchi T, Matsuda TD, Haga Y, Ōnuki Y. J Phys Soc Jpn 2004, 73, 1657.
[26] Honda F, Okauchi K, Nakamura A, Li D, Aoki D, Akamine H, Ashitomi Y, Hedo M, Nakama T, Ōnuki Y. J Phys Soc Jpn 2016, 85, 063701.
[27] Hiranaka Y, Nakamura A, Hedo M, Takeuchi T, Mori A, Hirose Y, Mitamura K, Sugiyama K, Hagiwara M, Nakama T, Ōnuki Y. J Phys Soc Jpn 2013, 82, 083708.
[28] Inada Y, Wiśniewski P, Murakawa M, Aoki D, Miyake K, Watanabe N, Haga Y, Yamamoto E, Ōnuki Y. J Phys Soc Jpn 2001, 70, 558.
[29] Wiśniewski P, Aoki D, Miyake K, Watanabe N, Inada Y, Settai R, Haga Y, Yamamoto E, Ōnuki Y. Physica B 2000, 281&282, 769.
[30] Teruya A, Suzuki F, Aoki D, Honda F, Nakamura A, Nakashima M, Amako Y, Harima H, Hedo M, Nakama T, Ōnuki Y. J Phys Soc Jpn 2016, 85, 064716.
[31] Takase A, Kasuya T. J Phys Soc Jpn 1980, 48, 430.
[32] Ohbayashi M, Komatsubara T, Hirahara E. J Phys Soc Jpn 1976, 40, 1088.
[33] Cheng J-G, Matsubayashi K, Wu W, Sun J, Lin F, Luo J, Uwatoko Y. Phys Rev Lett 2015, 114, 117001.
[34] Settai R, Katayama K, Aoki D, Sheikin I, Knebel G, Flouquet J, Ōnuki Y. J Phys Soc Jpn 2011, 80, 094703.
[35] Mühlbauer S, Binz B, Jonietz F, Pfleiderer C, Rosch A, Neubauer A, Georgii R, Böni P. Science 2009, 323, 915.

Takao Mori, Jean-Baptiste Vaney, Gabin Guélou, Fainan Failamani, and Quansheng Guo

Crystal growth of intermetallic thermoelectric materials

1 Introduction

Considering that more than half of all primary energy, i. e., fossil fuels that mankind consumes is lost in the form of waste heat, the solid-state conversion of heat to electricity that thermoelectric materials represents through the Seebeck effect can have a huge benefit for society [1, 2]. However, it is not easy to achieve a high thermoelectric conversion efficiency. The conversion efficiency is a function of the figure of merit ZT which is composed of contradictory requirements for the physical properties. Namely, $ZT = \alpha^2\sigma\kappa^{-1}T$, where α is the Seebeck coefficient, σ is electrical conductivity, κ is thermal conductivity, and T is temperature. Therefore, to achieve a very high thermoelectric performance ZT, the material should have a high electrical conductivity like a metal but low thermal conductivity and large Seebeck coefficient like an insulator but high electrical conductivity. These requirements have made straightforward enhancement of the thermoelectric performance difficult. However, recently, novel principles have been proposed to try to enhance the power factor [3–10]. In regard to the paradox of conducting electricity but blocking heat, nanostructuring has been shown to be effective to achieve phonon-selective scattering and thereby enhance ZT [11, 12, 1, 13–16].

Therefore, in the thermoelectric fields, single crystals are usually not sought after so much because the grain boundaries in polycrystalline sintered bodies, especially nanostructured designed ones, can lead to selective scattering of the phonons and thereby to higher thermoelectric performance. But there are several notable examples of thermoelectric materials where the crystal growth is very important, and this will be reviewed in this chapter, which contains sections on borides: cluster compounds and layered compounds, chalcogenides like SnSe, TiS_2, and $FeSb_2$ and clathrates.

2 Borides—candidates for the promising high-temperature power-generation applications

2.1 Introduction

Borides have been of interest as thermoelectric materials since relatively good thermoelectric properties were discovered in boron carbide around a third of a century ago [17, 18]. It is also one of the few thermoelectric materials that has a history of commer-

https://doi.org/10.1515/9783110496789-010

cialization, with the Hi-Z company of the US selling modules containing boron carbide as the p-type element for some years now. Continuing research on boron carbide modules has been reported [19]. An interesting feature of boron carbide is that, despite intensive research over the years, there have still been uncertainties in the precise crystal structure, of which a recent work has elucidated some previously unknown local crystal-structure modifications [20]. Recently, borides have gained renewed interest as several promising thermoelectric applications have been proposed for the very-high-temperature region, which requires high temperature stability that few thermoelectric materials possess [21]. In particular, the topping cycles for power plants have the possibility to be implemented relatively quickly with very large gross impact for improved power generation [22]. And new phenomenon, such as thermoelectric enhancement in the samarium phase of REB_{66} (RE = rare-earth) [23, 24] and excellent p-n control in Zr-doped elemental beta-boron [25], have been discovered. Relatively novel borides for boron cluster compounds like $REB_{44}Si_2$ and layered compounds like the AlB_2-analogues have also been discovered with promising thermoelectric properties [26–31]. The topics that particularly relate to single crystals and crystal growth will be reviewed. The reason that crystals are advantageous for boron cluster compounds like $REB_{44}Si_2$ and REB_{66} is that they typically possess very large Seebeck coefficients and have relatively low thermal conductivity due to a variety of mechanisms that have been reviewed before [32–36]. Their weak point is their typically relatively low electrical conductivity from variable range hopping (VRH). Therefore, there is not much merit for nanostructuring polycrystalline material to try to engineer low thermal conductivity, but more advantageous having crystals that exhibit the highest electrical conductivity possible. VRH materials in general would not be considered as thermoelectric materials up to the mid-high temperature range, but at very high temperatures, as the hopping conduction becomes more robust, they can become competitive.

2.2 Crystal growth and thermoelectric properties of REB_{66}

For REB_{66}, extensive research was first carried out on the prototypical YB_{66} compound. As a compound with a relatively complex crystal structure with ~1, 600 atoms in the unit cell, there were initially some various reports as to the composition and detailed crystal structure of the compound, but much was clarified by Richards and Kasper, and the YB_{66} notation was adopted [37]. Small crystals sufficient for structure determination could be obtained from simple melts. The first synthesis of YB_{66} crystals was accidently obtained from ytrrium and a boron mixture melted in a vacuum induction furnace [38]. The crystal structure of YB_{66} is cubic (space group $Fm\bar{3}c$) with a large unit cell of a = 23.44 Å (Figure 1). Despite the large amount of atoms in the unit cell, the structure can be said to be made up of just three structural components: The "super-icosahedron" B12(B_{12})12, where 12B_{12} icosahedra are connected to all the apices of an

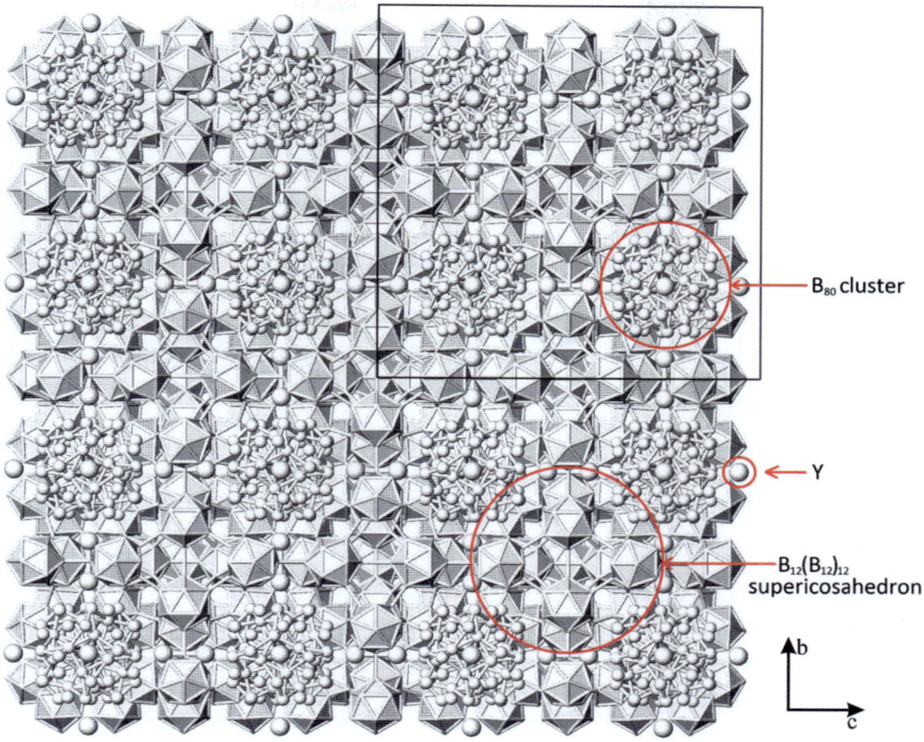

Figure 1: Crystal structure of REB_{66}.

icosahedra, B_{80} cluster which only contains ~42 B atoms due to partial occupancy, and yttrium atoms.

Growth of larger YB_{66} crystals suitable for transport properties' measurements was first successfully carried out by the pedestal growth method where crystals are pulled from cooled copper boats [39]. They report that growth attempts by a sealed crucible Bridgman technique and Czochralski technique using BN crucibles were unsatisfactory. Golikova and coworkers have grown REB_{66} crystals by zone melting [40]. Particularly high quality YB_{66} single crystals 50 mm in length and longer were grown by Tanaka and coworkers using the floating-zone (FZ) method [41, 42]. The composition of congruent melting is reported to be [B]/[Y] = 62. Details of the FZ method are described elsewhere in this book, but a typical apparatus picture and schematic used to grow such boride crystals is shown in Figure 2. Monochromator-grade extremely high-quality crystals were grown by the indirect-heating FZ method, where the heating is from the radiation of a tungsten ring heated by the inductive rf method [42]. They report high-quality growth both with a convex growth interface for congruent composition, and also growth under incongruent conditions. Application of YB_{66} crystals was envisioned in the form of using them as soft X-ray (1~2 keV) monochromators in

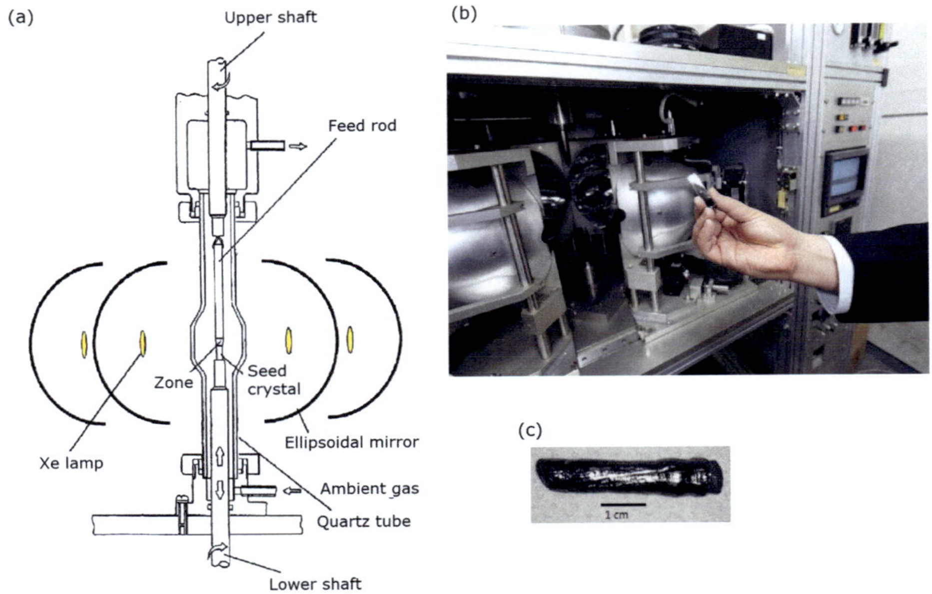

Figure 2: (***a***) Scheme and (***b***) A floating-zone crystal-growth apparatus and (***c***) SmB_{62} crystal picture.

synchrotron radiation sources by utilizing the large unit cell [43]. Usage in the case of high-intensity sources was found to be difficult because of the relatively low thermal conductivity of YB_{66} leading to crystal damage from insufficient heat dispersal.

In regard to thermoelectric properties, REB_{66} have low thermal conductivity, actually with glass-like behavior [44, 45]. The large unit cell, i. e., crystal complexity, was considered to contribute to this. A quantitative analysis showed that disorder actually may be the dominant factor, with the partial occupancies of the rare-earth possibly playing a large role [34]. REB_{66} typically exhibits large Seebeck coefficients which are advantageous but poor electrical conductivity resulting in $ZT < 0.01$ [36]. Two recent discoveries have generated renewed interest in this system. First of all, the homogeneity region of YB_{66} was for a long time considered to be $56 \leqq [B]/[Y] \leqq 68$. Accidentally from $YB_{44}Si_2$ crystal growth and afterwards purposely, a much more metal-rich YB_{48} crystal was found to be possible to be grown [46]. A new yttrium site was also observed at (1/4,1/4,1/4), replacing in part the B–B dumbbell. Assumedly from delocalization, the ZT of YB_{48} attained a value of 0.1 which is a 30 times enhancement compared to previously measured YB_{66} crystals. The second discovery is the ZT enhancement discovered in SmB_{62} leading to $ZT \approx 0.13$ which is 40 times an enhancement compared to YB_{66} (YB_{62}) [23, 24]. The mechanism is not fully clear but mixed valency of samarium has been indicated, and is considered to be playing a role. In any case, there is a strong renewed interest in this long time well-known system.

2.3 Crystal growth and thermoelectric properties of $REB_{44}Si_2$

Tanaka and coworkers discovered the novel YB_{50} compound, but crystal growth for this compound was found to be difficult because the compound decomposed into more stable boride phases before melting [47]. An addition of Si was found to stabilize the melt at lower temperatures and enable crystal growth [48]. Recently Pt doping has been reported to enable melt growth of YB_{50}-type samples without the need for Si [49]. Arc-melting of $REB_{44}Si_2$ yields polycrystalline samples with small single crystals for structural determination also able to be extracted. In regards to the nomenclature, to avoid confusion, the unified notation of $REB_{44}Si_2$ is used, but in actuality, the boron and silicon contents vary [35]. The main framework of the crystal structure of $REB_{44}Si_2$ is composed of five structurally independent B_{12} icosahedra and the $B_{12}Si_3$ polyhedron [50]. As an interesting structural feature, it has a chain of B_{12} icosahedra and adjacent bond-alternating rare-earth atom ladders along the c-axis. Incidentally, this has led to interesting magnetic properties such as strong coupling, indicated to be mediated by the icosahedra [51, 52] and dimer-like behavior [53].

The floating-zone method yields large crystals of $REB_{44}Si_2$ [48, 35]. However, it has been noted that, for typical growth, even with floating-zone growth, that the $REB_{44}Si_2$ crystals tend to have microcracks. For polycrystalline arc-melted samples, the addition of Zn was discovered to function as a non-intrusive additive (after heating there was no zinc left) to improve the crystallinity of the $YB_{44}Si_2$ samples [27], so there are probably further methods to control the microstructure. It has been speculated that the origin of the effect of the Zn addition may possible be a scavenging effect or improved local mixing during the melt synthesis, for example. Crystal-structure analysis of a series of transition metal-doped $YB_{44}Si_2$ indicates that for the Zn-added sample there is an additional uptake of silicon which may have yielded this effect [54]. Incidentally, for Rh and Ni doping, it is indicated that a few percent of the transition metal is doped into the 4h site between B_{12} icosahedra leading to a transformation to B_{11} polyhedron. Recently, the crystallographic directions of an FZ-grown $YB_{44}Si_2$ crystal were determined for the first time. The growth direction was found to be [510], and an interesting layered feature was realized for the first time along this direction, with the structure composed of dense boron cluster layers and yttrium layers [28].

In regard to thermoelectric properties, $REB_{44}Si_2$ are not particularly high performance, with a highest value of $ZT \approx 0.12$ for non-doped $ErB_{44}Si_2$, e. g., extrapolated for very high temperatures [55]. However, as a base system, $REB_{44}Si_2$ appears to be better compared to the REB_{66} compounds in general, except for the samarium phase. This is due to an improved electrical conductivity in $REB_{44}Si_2$ that is possibly due to the reduced disorder, while still retaining the low thermal conductivity of boron-icosahedra compounds [26]. The interplay of disorder on the electrical and thermal properties remains an interesting parameter to further control in these variable range hopping compounds [56]. An alternate recent interesting development comes from the anisotropic crystal structure discovered reported previously in this chapter. Significant anisotropy in the thermoelectric properties was found with the electrical conductivity being a maximum eight times larger along [510] compared to [052]. This results in the power factor being 3.6 times larger along [510] (see [28]).

2.4 Crystal growth and thermoelectric properties of AlB_2-analogue compounds, $REMB_4$

AlB_2-analogues are an interesting group of layered borides. Their prototype is the well-known AlB_2-type structure compounds which have hexagonal [6] boron layers sandwiching metal layers. Kuzma and Rogl have discovered many $REMB_4$ (M = non rare-earth metal) compounds and the two main structure types: $YCrB_4$-type or α-type [57] and $ThMoB_4$-type or β-type [58], as depicted in Figure 3. These can be systematically understood as $REMB_4$ having a metal-to-boron ratio of 1:2, the same as AlB_2-type, and because of the different size metals of RE and M, instead of homologous [6] hexagons, they are sandwiched by [7] and [5] polygons, respectively, which are arranged in the α-type or β-type formation. RE_2MB_6 can also be described by a metal-to-boron ratio of 1:2, and because it is more homogeneous than $REMB_4$, polygons are [5], [6], [7] (see [59]).

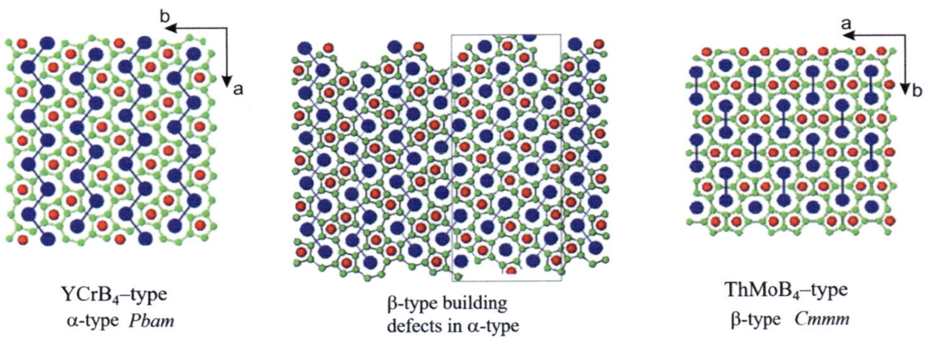

Figure 3: Crystal structure of α-type and β-type $REMB_4$ viewed along the c-axis. The middle figure depicts a β-type building-defect fragment inserted into the α-type host.

In regards to obtaining single crystals of $REMB_4$ and RE_2MB_6, flux growth is the main method used. In this method, molten solids are used as solvents to mediate the crystal growth. The flux material can be individual metals with low melting points, e. g., Sn, In, Ga, Al etc. [60], mixture of metals, e. g., $In_{60}Zn_{40}$ [61] (in at.%) and La–Ni eutectic [62] or salts [63]. Okada has led the way in growing single crystals of many phases of $REMB_4$ and RE_2MB_6, especially the rare-earth aluminoborides [64–66]. The method of flux growth is reviewed in other chapters of the book, and a simple example is given

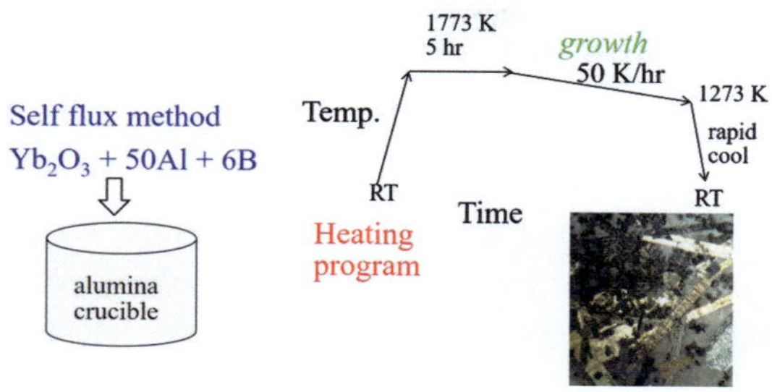

Figure 4: Schematic of flux growth of $REAlB_4$ (α-$YbAlB_4$ as an example).

of the crystal growth recipe of $REAlB_4$ in Figure 4. If the flux material contains the same elements forming the target material, it is called self-flux. Al self-flux is used, and after crystal growth, typically NaOH is used to remove the flux. A very convenient feature of the growth is that pure rare-earth metal because starting materials are not necessary, and inexpensive rare-earth oxides can also be used to grow crystals of this phase [65]. Simultaneous flux growth of both α- and β-type $REAlB_4$ crystals has also been reported [67].

An interesting phenomena of building defects was discovered in these compounds. This discovery led the observation of multiple magnetic anomalies below the antiferromagnetic transition temperature T_N in α-$TmAlB_4$ [68]. It was discovered that β-type tiling defects exist in α-type single crystal, as shown in Figure 3, and vice versa [69, 70]. These defects were indicated to affect the physical properties such as magnetism [68, 69] and also superconductivity [71]. Existence of such building defects in layered compounds not limited to borides may be an often-occurring phenomenon which is not widely known. The building defects appeared to be ubiquitous and unavoidable since they appeared for even apparently high-quality single crystals of $TmAlB_4$. A counterintuitive rapid cooling during the flux growth led to growth of single crystals that were free of the building defects, and without corresponding changes in the physical properties [72].

In regard to thermoelectrics, there are two interesting aspects of these compounds discovered so far. The first is that the thermal conductivity of small single crystals of $TmAlB_4$ were measured by the thermoreflectance method, and it was found that a couple percent of β-type fragments in α-$TmAlB_4$ led to a ~30 % decrease in the thermal conductivity, indicating the strong phonon-scattering effect of the building defects [31]. Since formation of building defects is now found to be controllable by crystal-growth conditions, this represents an interesting effective control over the thermal conductivity. The second feature is that theoretical calculations on $REMB_4$ compounds indicate for $YCrB_4$ that there is a relatively narrow gap of 0.05 eV to 0.17 eV [73, 74, 29, 30] semiconductor with relatively high thermoelectric performance, and the possibility to control p, n characteristics predicted from doping [29]. Furthermore, the $REMB_4$ compounds have a great versatility and very diverse possible combinations with metal elements so they should be studied further.

2.5 Other thermoelectric borides and thin-film crystal growth

In addition to boron carbide, there have been thermoelectric studies on other compounds which take similar rhombohedral structures to α-boron, namely, $B_{12}As_2$, $B_{12}P_2$, $B_{12}O_2$, and B_6S_{1-x} ($B_{12}S_2$). In regard to crystal growth, for example, single crystals of $B_{12}As_2$ have been obtained by solvent growth using nickel solvents and have been thermoelectrically investigated. However, despite having large Seebeck coefficients (>300 µV/K), the power factors were low due to high resistivity [75]. Polycrystalline $B_{12}P_2$ CVD-grown wafers were also reported to exhibit Seebeck coefficients larger than 800 µV/K, but with higher resistivity than $B_{12}As_2$ [76]. There is also an interesting prediction that if stoichiometric $B_{12}O_2$ crystals could be successfully grown for the first time, high mobility and high thermoelectric performance would be expected [77].

Although thermoelectric thin films are not expected to be useful for high temperatures, there have been recent attempts for thermoelectric hexaborides [78]. One reason is that SmB_6 has recently been reported to be a topological insulator [79]. In some cases, topological insulators have been predicted to possess high thermoelectric performance [80, 1], and in thin films, the surface to volume ratio is much higher than bulk, indicating a possible larger effect of surface states. Furthermore, CaB_6 for example, has good thermoelectric properties but high thermal conductivity. Thin films can be expected to lower the thermal conductivity. Using decaborane $B_{10}H_{14}$, as a boron source, the CVD film growth of SrB_6 was carried out [78]. Textured crystalline thin films of SrB_6 were obtained with very fast growth rates of ~0.25–0.35 µm/min. Large Seebeck coefficients were obtained similar to bulk values but the grain boundaries from the texture appeared to increase the resistivity. The CVD remains a promising method to explore thin films for other boride like SmB_6 and YbB_6.

3 Tin selenide single crystals as high-performance thermoelectric materials

Tin selenide has been a hot topic in the world of thermoelectrics during the last few years, after a report of an extremely high figure of merit of 2.6 along the b-axis in a single crystal [81], raising interesting discussions concerning the remarkable difference in conversion performance between the single crystals and their polycrystalline counterparts. The room-temperature form of SnSe takes a layered orthorhombic crystal structure with the space group *Pnma*. This structure can be interpreted as a distortion of the rock-salt structure, with different interionic spacing inside the coordination octahedra of the Sn atom. The views of this structure along the a, b and c crystallographic directions are shown in Figure 5. The structure is composed of two-atom-thick SnSe slabs with strong Sn–Se bonds within the b and c planes and weaker Sn–Se

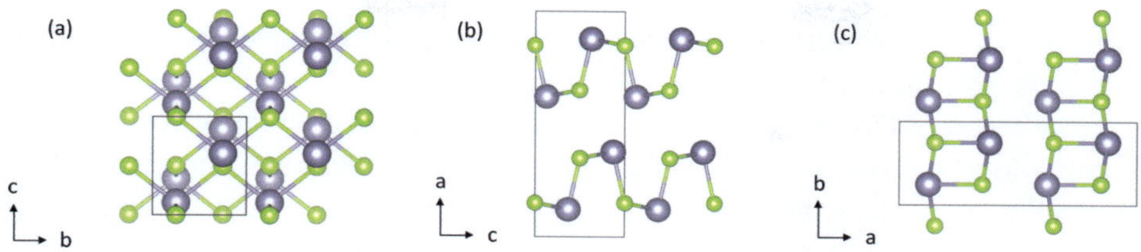

Figure 5: Crystal structure of the space group *Pnma* room-temperature form of SnSe, with projection along the a-, b- and c-axis (respectively (*a*), (*b*) and (*c*)). Sn and Se atoms are depicted as grey and green spheres, respectively.

bonds between the planes (but stronger than electrostatic Van der Waals bonds). This results in an anisotropic material which has claimed to be between the two- and three-dimensional natures [82]. The atomic slabs are corrugated, arranged in a zig-zag accordion-like shape along the c-axis and stacked along the a-axis as depicted in Figure 5(b). The easy cleavage plane is therefore perpendicular to [100]. This phase undergoes a transition to a higher symmetry phase (space group *Cmcm*) around 800 K [83, 81], a second order displasive transition corresponding to the continuous movement of Sn and Se atoms almost entirely along the [100] direction [83].

Although SnSe is a IV–VI intermetallic compound close to the NaCl-type structure of lead telluride, one of the best performing and most studied thermoelectrics [12, 84], it did not attract much attention in this field for decades [85]. Although there was some early interest in evaluating the room-temperature ZT in solid solutions of SnS and SnSe [86]. Since SnSe has been considered for a long time as a promising material for photodetecting applications [87, 88, 82] or even more recently for applications in photovoltaics [89, 90], the growth of SnSe single crystals has been studied thoroughly. The classical method for producing high quality and large-sized single crystals of tin selenide has been the Bridgman or modified Bridgman method using pre-reacted ingots of tin selenide at temperatures >900 K [87, 88, 81, 91–95]. This method however requires a double ampoule sealing to prevent the oxidation of the material because the inner tube is destroyed due to considerable difference in thermal expansion between the quartz and the crystals [93, 88, 81, 92]. Fast growth speeds of seven mm/h were found to result in high-quality crystals with small dislocation densities [88], but single crystals were usually grown for several days [81, 91, 92, 94, 95]. In spite of the non-congruent sublimation of the stoichiometric composition, the closed-tube vapor-phase technique has also been demonstrated to be an efficient technique for growing Sn-rich SnSe single crystals [96]. All crystals would show carrier concentration between 10^{15} and 10^{18} cm^{-3}, depending on the defects concentration, values typical of undoped semiconductors. Small crystals of various morphologies could also be obtained by the solvothermal [82] or hydrothermal methods [97]. Much effort has also been put in the synthesis of SnSe nanosheets, mainly for photovoltaics and will be briefly detailed later in this part [98, 89, 99, 100].

The report of the high ZT value of 2.6 along the b-axis of the tin selenide single crystal was also accompanied by a high value of 2.2 along the c axis and a more moderate value of 0.8 along the layers stacking direction, the a axis, emphasizing the very marked anisotropy of the properties in this compound [81]. While displaying power factors up to one $mW\,m^{-1}\,K^{-2}$, values typical of state-of-art thermoelectric materials, the crystals were exhibiting extremely low lattice thermal conductivities κ_{latt}, far below $0.5\,W\,m^{-1}\,K^{-1}$, even reaching a minimum of $0.23\,W\,m^{-1}\,K^{-1}$ along the a axis. The optimal values were obtained above 800 K in the ordered high-temperature *Cmcm* phase owing to increased carrier mobilities and a reduction in energy gap causing a jump in electrical conductivity, while maintaining the thermal conductivity at extremely low values.

Hence, the high ZT values could be attributed almost entirely to these remarkably low values of κ, originating in the very anisotropic bonding character and the ionic network anharmonicity, as emphasized by the anomalously high Grüneisen parameters calculated by DFT [81]. Later theoretical studies indeed confirmed the unique properties of lattice dynamics in SnSe [101–103]. Through neutron-inelastic-scattering studies, the anharmonic character could be identified as stemming from a very peculiar picture of the electronic structure and bonding unstability [103]. The Se *p*-orbitals form a highly polarizable resonant network with long-range interatomic force constants, which result in soft low energy optical modes. They are responsible for the instability of the *Cmcm* phase and were already evidenced as a source of decrease in κ_{latt} in other chalcogenides [104]. Coupled to the presence of stereochemically active lone pairs on Se atoms, another well-known source of κ_{latt} decrease [105–107], the anharmonicity of phonon transport in SnSe close to the phase transition could be directly related to the orbital instability via electron-phonon coupling [103]. In addition, first principles calculations unveiled very short phonon mean-free paths, showing the limited potential in reducing κ_{latt} by nanostructuring [102]. These theoretical studies also evidenced a marked anisotropy between the three direction components of κ_{latt}, predicting in particular a value along the b-axis that was considerably higher than the one measured by Zhao et al. ([101–103]).

Subsequent to this discovery, attention has been drawn to the performance in polycrystals, which are easier and cheaper to produce for applications. However, several studies on grain orientation, elemental doping or solid solutions in polycrystals could never evidence a figure of merit above one [108–116]. In addition, in these materials, the high-temperature phase could not be measured due to severe plastic deformation of the samples above 823 K [108] and material instability evidenced by TGA above 873 K [93]. The thermal conductivities of the majority of undoped compound were found to be higher than those measured by Zhao et al. (which can be understood due to the averaging of all directions, for which b- and c-directions show theoreti-

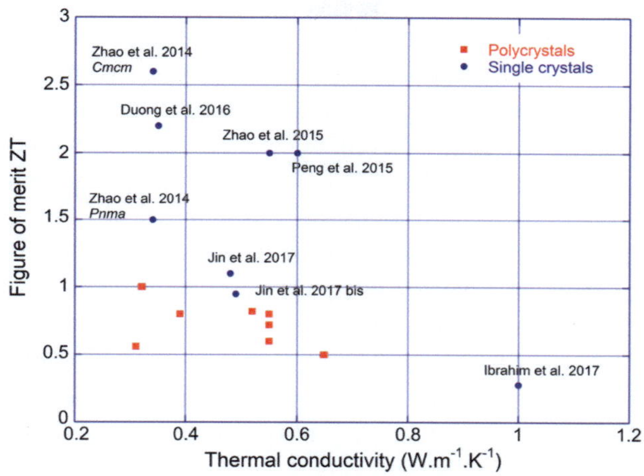

Figure 6: Distribution of the maximum ZT values as a function of corresponding thermal conductivity for the single crystals and polycrystals of tin selenide cited in this chapter. Except for the value of Zhao et al. in the *Cmcm phase*, all the values were obtained in the *Pnma* phase between 700 and 823 K.

cal values above 0.5 $W.m^{-1} K^{-1}$), while electrical resistivities were increased due to the loss of the single crystalline order. Figure 6 shows the distribution of the ZT values with the thermal conductivity for a selection of poly- and single crystals in doped and undoped tin selenide. This distribution evidences a clear difference in terms of ZT values between the single crystals and the poly-crystals and also with Zhao et al.'s lone value of single crystals in the *Cmcm* phase. These results raised the question of the intrinsic nature of the ultra-low thermal conductivities in the latter single crystals. Wei et al. noticed the lower density of Zhao et al.'s stoichiometric crystals, only reaching roughly 90 % of the theoretical density, and discussed the suitability of treating these materials as real single crystals. It is indeed well-known that porosity or micro-cracks can be very effective in blocking the thermal transport in materials and can considerably lower the thermal conductivity [16]. Although the presence of a large number of defects or interstitials in a single crystal can also affect this property, as explained by Zhao et al., it is more likely that the very low values of κ_{latt} along the b- and c-axes may not be intrinsic to SnSe but may be related to the preparation method.

Various studies on single crystal SnSe were carried out, trying to reproduce the high ZT or to dope the material to improve the properties even further. Few teams tried to re-evaluate the properties of the pristine phase but could not reproduce either the electrical or the thermal properties [92, 95]. For the former, it has been long known that the tin selenide system is very prone to structural defects and off-stoichiometry [96], on which the carrier concentration is very dependent and can vary strongly as already emphasized earlier. The disparities in electrical properties might therefore be explained by differences in carrier concentration relative to each synthesis method. Larger electrical resistivities and thermopowers measured by Ibrahim et al. suggest, for example, a lower hole concentration in their samples, while Zhao at al. obtained higher power factors due to carrier concentrations close to 10^{19} cm^{-3}. On the other hand and as already mentioned in the previous paragraph, the differences in thermal conductivity may be explained by the hypothetical porosity/microcracks yielding fewer dense crystals. However, anti-site substitutions and/or vacancies, as sources of additional phonon scattering, could also cause more synthesis-method dependence of κ [101], and explain the scattering of κ values among the other reports, with thermal conductivities higher and closer to theoretical studies for the less defective samples.

On the other hand, doping studies unveiled the potential of the *Pnma* tin selenide phase, following theoretical predictions. Contrary to lead telluride, SnSe exhibits a multitude of bands at the top of the valence band, as well as at the bottom of the conduction band, both useful for achieving respective hole and electron concentrations above 10^{19} cm^{-3} [117, 102]. This feature is a requirement for high performance thermoelectric materials as it enables a high effective mass, a prerequisite for high Seebeck coefficients, hence showing the natural predisposition of SnSe towards thermoelectricity. Carrier concentration tuning by substitution of Na and Ag for Sn resulted in improved ZT of 2 at 800 K in the *Pnma* phase through a remarkable increase in the power factor, despite thermal conductivities above 0.5 $W.m^{-1}K^{-1}$ [91, 118]. What is noteworthy is that the whole temperature range (300–800 K) benefits from the increase in performance, resulting in an average ZT above one, which is an important step forward when considering applications. However, for applications, in the classical thermocouple architecture, two different materials are needed: a p-type leg and an n-type one. So far, mostly p-type tin-selenide compounds have been obtained and studied, despite theoretical results indicating the greater potential of the n-type SnSe owing to enhanced interlayer electronic transport [117]. Some results in n-type polycrystals showed moderately high values [113, 119], but substitution of Bi for Sn was reported to yield a ZT of 2.2 for n-type single crystals [93]. However, these crystals exhibit thermal conductivity values in the vicinity of 0.3 $W.m^{-1}$ K^{-1}, similar to the controversial report on undoped SnSe; therefore, they may also show the same surmised porosity features. In these compounds, at least Bi was demonstrated, by topographic microscope techniques, to substitute Sn efficiently, while effectively driving the electron concentration to values ranging from 5.2 10^{18} to 2.1 10^{19} cm^{-3}.

Overall, even though the intrinsic character of the extremely low thermal conductivity of SnSe single crystals is still a topic of discussion, the advantage of single crystals over polycrystals has been demonstrated in this system. Owing to a very anharmonic bonding character, the phonon mean-free path is naturally very short in these materials, preventing the polycrystals to reach lower thermal conductivities than in the single crystals. On the other hand, the absence of grain boundaries and the perfect orientation in single crystals allows reducing the electrical resistivity, especially along the in-plane directions and, upon carrier concentration tuning, to reach high-power factor values and a high figure of merit. Nevertheless, the applicability of such systems, and in particular the applicability of the first non-fully dense crystals, remains to be studied.

Another type of SnSe single crystals, the nanoplates or monolayer nanosheets, has also recently been in focus, for thermoelectric properties [100] but mainly for photovoltaics [90], optical applications [99] or even nano-mechanics [120]. Nethravathi et al. previously showed the promise of chalcogenide nanosheets to be utilized for thermoelectrics [14]. These SnSe nanoplates and nanosheets are similarly produced by wet synthesis [89, 99, 100], and in the case of thermoelectric research, they have been utilized for the fabrication of preferentially ordered polycrystals [100]. They also make possible the production of thermoelectric non-layered SnTe to be produced in a porous form of nanoplates, based on the anion exchange reaction of Te for Se [98]. Combined with other 2D layered compounds such as transition metal dichalcogenides, which will be described in the next part of this chapter, misfit layered compounds can be fabricated for thermoelectric applications, making possible the tuning of the cross-plane electrical transport properties by tuning the number of layers of each compounds [121], and these might be essential for low-dimensional applications.

4 Low-dimensional and environmentally friendly TiS_2 derivatives: from single crystal to cost-efficient thermoelectric materials

Transition metal dichalcogenides (TMDCs) and many of their doped derivatives have received considerable attention over the past 50 years. Such activity has been encouraged by the two-dimensional nature of MX_2, where M is a group 4, 5 or 6 transition metal and X is a chalcogenide and the wide range of physical properties that are observed within this system [122, 123]. These layered materials typically exhibit strong anisotropy in their physical and mechanical properties that range from electrically insulating (e. g., HfS_2) to metallic (e. g., VSe_2) [124]. Subsequently, TMDCs have been used as cathode materials in lithium-ion batteries, [125, 126] and are widely studied for applications in a range of fields such as molecular sensing [127] or electrochemical supercapacitors [128]. Very recently, TiS_2 nanosheets have been identified as suitable agents for in-vivo photo-acoustic imaging [129] and as a coating material for PbS photodetectors with the potential kill cancer cells via localized surface plasmon resonance [130].

For applications in thermoelectricity however, a specific combination of high electrical conductivity, σ, high Seebeck coefficient, S, and low thermal conductivity, κ, is necessary [131]. In 2001, Imai et al. reported a value of the thermoelectric power factor, $S^2\sigma$, of 3.71 mW cm^{-1} K^{-2} at 300 K within the *ab* plane of an *n*-type TiS_2 single crystal [132]. Such electrical properties were similar to those of the commercially available doped bismuth telluride, at a substantially lower raw-material cost and toxicity. However, the thermal conductivity remains high for a thermoelectric material with *ca.*

68 mW cm^{-1} K^{-1} against *ca.* 15–20 mW cm^{-1} K^{-1} at 300 K for doped bismuth telluride. The overall figure of merit of 0.16 at 300 K was relatively high for a tellurium-free, non-toxic, low-density material but too low for commercial applications where a ZT value around unity is desired. Nonetheless, the low dimensionality of TiS_2 and the possibility to optimize electrical and thermal properties via substitution and intercalation have defined titanium-sulphide derivatives as serious candidates for thermoelectric power generation.

TiS_2 crystallizes in a CdI_2-type structure (space group $P\bar{3}m1$), where consecutive layers of TiS_6 edge-sharing octahedra are separated by van der Waals' gaps and stacked along the *c* axis to form a discontinuous two-dimensional array, see Figure 7. Within the van der Waals' gaps, a wide range of guest species can be readily intercalated to occupy tetrahedral or octahedral positions, or even both in some cases [133]. The guest species range from simple monoatomic cations to large organic molecules, leading to potential application from cost-efficient bulk thermoelectric power generators to flexible wearable devices [134]. For specific intercalation levels, the guest species may form a superstructure arising from the long-range ordering of the guest cations with vacancies [135]. These superstructures are made possible by the ready expansion of the interlayer distances and the charge transfer from the guest species to the empty low-lying electronic states of the highly oxidized titanium layer [136].

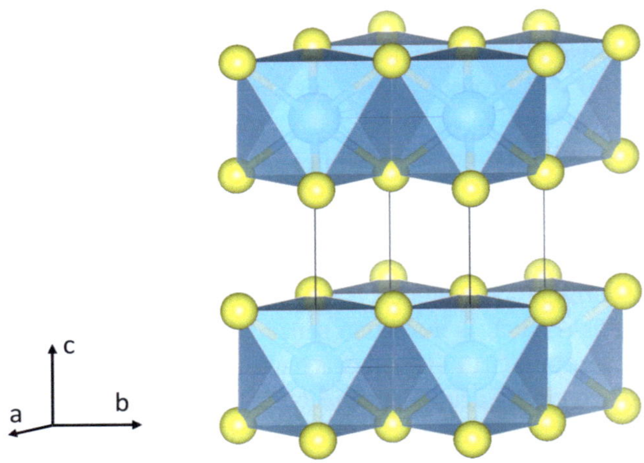

Figure 7: CdI_2-type crystal structure of TiS_2.

By far the most widely used synthetic technique to obtain TiS_2 crystals is chemical vapor transport (CVT) using iodine or excess sulphur as a transport agent. Because of the layered structure of TiS_2, the single crystals grow preferentially along the [001] direction to form thin platelets, typically of a few tenths of microns thick. This anisotropic grain growth is also observed in bulk samples prepared by high-temperature synthesis in evacuated sealed tubes, where the solid-liquid-vapor reaction favors the growth of plate-like crystallites [137]. Additionally, careful control of the reaction conditions can lead to the formation of TiS_2 dendritic crystals at 800 °C using excess sulphur as a transport agent in a simple CVT process [138]. Consequently, the physical properties of TiS_2 show strong anisotropy, even in consolidated bulk samples, where the plate-like crystallites align preferentially in the direction perpendicular to the pressure. This direction is referred to as "in-plane" due to the analogy that can be drawn between the physical properties in this direction and those of the single crystal along the *ab* plane. In thermoelectricity, electrical conductivity and thermal diffusivity of a bulk sample strongly depend on the statistical orientation of the crystallites and therefore on the direction of the measurement relative to the pressure.

Aside from variations caused by anisotropy, the electrical properties of TiS_2 have been disputed as a result of discrepancies between reports from various groups. This has led to uncertainty over whether TiS_2 is a semi-metal [139, 140] or a semiconductor with various values of direct or indirect band gaps [141]. However, it is now well-accepted that the origin of the variations in physical properties are a result of a deviation from stoichiometry caused by self-intercalation of titanium, leading to a range of compositions in the form $Ti_{1+x}S_2$ and a metallic temperature dependence of the electrical resistivity [122, 142, 141]. Self-intercalation usually occurs via sulphur loss through volatilization at high temperature with the excess Ti relocating to an octahedral vacancy in the van der Waals' gap. The Seebeck coefficient does not depend on the orientation of the crystal or the level of preferred orientation in bulk samples and can therefore be used, along with the unit-cell parameters, as a good indicator of the level of non-stoichiometry. It is clear from the comparison between single-crystal and bulk materials in a Pisarenko plot, see Figure 8, that the level of non-stoichiometry in $Ti_{1+x}S_2$ has a regular impact on the charge carrier concentration, n, and in turn on the Seebeck coefficient. The large increase in n is a result of charge transfer from the guest

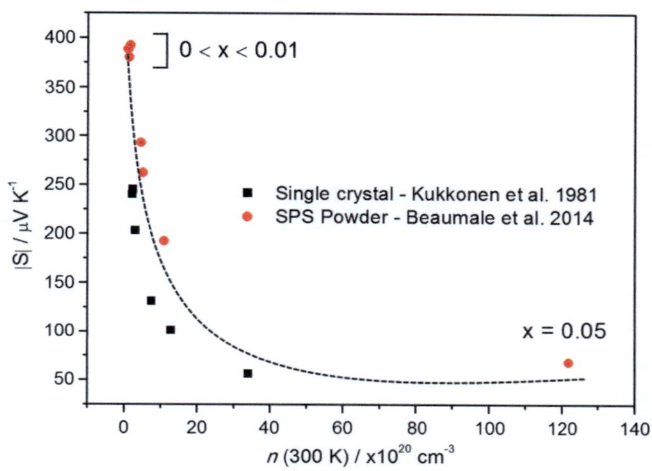

Figure 8: Pisarenko plot for single crystal and polycrystalline $Ti_{1+x}S_2$ consolidated by SPS.

to the empty t_{2g}-derived conduction band of TiS_2 and causes the electrical resistivity to decrease for a given orientation.

The large value of the Seebeck coefficient can be explained by the electronic-band structure. Calculations have shown the presence of a multi-valley structure with six small electron pockets around the L points of the hexagonal Brillouin zone, each with the potential to contribute to a large Seebeck coefficient [132, 143, 144]. Additionally, the disorder generated by the presence of Ti in the van der Waals' gap of $Ti_{1+x}S_2$ strongly affects phonon scattering and provokes a systematic reduction in the lattice component of the thermal conductivity from ~4.3 W m^{-1} K^{-1} for $x = 0$ to 1.2 W m^{-1} K^{-1} for $x = 0.05$ at 300 K in the in-plane direction of a bulk sample [145]. In the cross-plane (or out-of-plane) direction, the thermal conductivity is typically lower due to the effective phonon scattering at the layer interfaces. Fine tuning of the level of Ti-intercalation succeeded in improving the thermoelectric figure of merit, and a maximum ZT of 0.48 at 700 K was observed for $Ti_{1.025}S_2$.

The growth temperature applied during the chemical-vapor transport synthesis of TiS_2 directly affects the transport properties, suggesting that the off-stoichiometry can be controlled and exploited in single crystals. Elemental Ti and S are sometimes pre-reacted in sealed tubes to form polycrystalline TiS_2 before being placed in a tube furnace [143]. Hot- and cold-zone temperatures are usually kept between 873 and 1073 K. Subsequently, measurements of the transport properties on doped-$Ti_{1+x}S_2$ single crystals where x is constant across the doped samples constitute a powerful tool to investigate the effect of doping, via substitution or intercalation, without the additional influence of density, grain-size distribution, boundaries etc. Using this approach, Daou et al. established that Ta^{4+} substitution on the Ti site was particularly effective at reducing the lattice thermal conductivity, more than controlled Ti self-intercalation [146]. Similarly, Abbott et al. investigated a range of TiS_2 single crystals doped with As, Mo, Ce, Y, Hf and Zr. While the thermoelectric properties were rather poor, they observed a direct correlation between the infrared reflectivity minima and the Seebeck coefficient.

TiS_2-based misfit-layer compounds in the form $(MS)_{1+x}(TiS_2)_2$ (M = Pb, Bi, Sn) are natural superlattices that have recently shown very promising thermoelectric properties with substantially reduced thermal conductivity compared with pristine TiS_2 [147, 148]. While TMDC misfit-layer compounds have been known for decades, there have been some initial confusion with these compounds being designated as $MNbS_3$ and $MTaS_3$ (M = Sn, Pb) [149]. The Structure of $(PbS)_{1.18}(TiS_2)_2$ was determined from X-ray diffraction analysis of a single crystal and consists of two subsystems, PbS and TiS_2, that interweave to form a misfit superstructure of PbS and TiS_2 layers stacked along the c axis in a PTTPTTP sequence, Figure 9 [150]. The small single crystal ($0.025 \times 0.12 \times 0.20$ mm^3) was prepared from the reaction of the elements in evacuated quartz tubes in the presence of a small amount of iodine to favor crystal growth. Recent research focused on understanding the physics behind the ultra-low lattice thermal conductivity that originates from a large concentration of defects, such as stacking faults

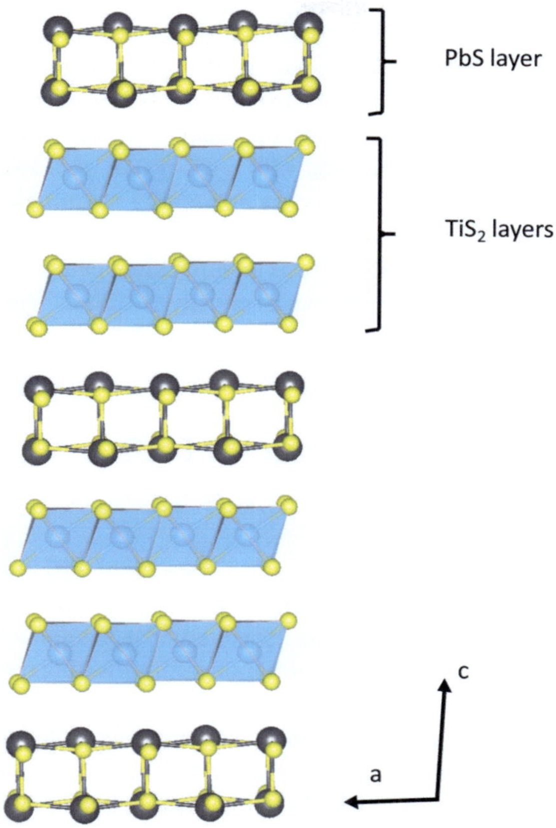

Figure 9: Simulated crystal structure of misfit compound $(PbS)_{1.18}(TiS_2)_2$.

and the softening of transverse modes of lattice waves caused by the weak bonding between consecutive heterolayers. However, these effects are accompanied by too great a charge transfer from the MS subsystem to the TiS_2 conduction band that leads to highly metallic behaviour. So far, the highest reported figure of merit was measured for $(SnS)_{1.2}(Ti_{0.98}Cu_{0.02}S_2)_2$ with ZT = 0.42 at 720 K [147, 151]. Undoubtedly, TiS_2-based misfit compounds form a very promising new family of thermoelectric materials with a wide range of tuning opportunities. In this context, larger single crystals would enable a more thorough examination of the intrinsic properties of such a system and give valuable insight into further strategies to adopt.

Certainly one of the most promising topic for thermoelectric TiS_2-based materials is the emergence of flexible TiS_2/organic hybrid superlattices. Single crystals ($4 \times 4 \times 0.1\,mm^3$) were grown using chemical-vapor deposition where 20 mg/ml of excess sulphur was used as a transport agent in a transport tube held at 632 °C in the growth region [152, 134]. A simple two-step process involving electrochemical intercalation and a solvent-molecule exchange was carried out to obtain a flexible TiS_2/organic hybrid single crystal. The TiS_2 single crystal was used as a cathode and a platinum sheet was used as an anode in a hexylammonium/DMSO electrolyte solution. A constant voltage was applied in order to reduce part of the Ti^{4+} into Ti^{3+}, resulting in organic cation intercalation driven by Coulomb forces. The obtained $TiS_2[(hexylamine)_x(DMSO)_y]$ was immersed in water to form $TiS_2[(hexylamine)_x(H_2O)_y]$. The thermoelectric properties of the obtained flexible crystal are extremely promising with a ZT of 0.28 at 373 K, mainly due to an ultra-low thermal conductivity, reduced by an order of magnitude compared with TiS_2. Following these results, the work on TiS_2/organic hybrid superlattices was pursued and a flexible thermoelectric device was manufactured using PEDOT:PSS as a p-type counterpart, achieving a maximum power density of 2.5 W m^{-2} for a 70 K temperature gradient [153].

5 Properties of correlated semiconductor $FeSb_2$: from single crystal to nanocrystalline materials

In the past few decades, reducing phonon mean-free path via nanostructuring [1] has been employed to improve the thermoelectric figure of merit of thermoelectric materials. Reducing the phonon mean-free path provides an excellent alternative to improve the ZT by reducing the lattice thermal conductivity without significantly affecting the power factor. Thus, as described previously in this chapter, materials such as single crystals where the phonon mean free paths are large are not commonly favorable for thermoelectrics. There are several exceptions such as borides and clathrates where the intrinsic lattice thermal conductivity is already low [44, 45, 34, 154]. A large value of phonon mean-free path, however, can result in an enhancement of the Seebeck coefficient via the so-called phonon drag mechanism. In this mechanism, transport of phonon "drags" charge carriers towards the cold end of the junction, therefore creating additional contribution to Seebeck coefficient. This effect has been observed mainly in pure semiconductors, such as in isotopically pure Ge [155], which results in a Seebeck coefficient ~30 mV/K. Herring et al. [156] proposed a phenomenological explanation of the phonon drag mechanism, revealing the dependence of Seebeck coefficient on the phonon mean-free path, carrier mobility, carrier concentration, etc. Despite this beneficial effect, phonon drag itself cannot be utilized due to its inability to overcome the high lattice thermal-conductivity due to high phonon mean-free path. Calculations shows that the ZT obtained purely from the phonon drag effect will never exceed 1/4 (Goldsmid [157] and the references cited therein).

On the other hand, strong electronic correlation, particularly in a compound class called Kondo insulators, has attracted much attention due to its potential in thermoelectrics [158]. In such systems, a hybridization between localized f or d orbitals and conduction bands result in the formation of a narrow gap and steep density of states in the vicinity of the Fermi level. By adjusting the position of the Fermi level with respect to the band edge, a high value of power factor can be obtained. Several compounds in this class such as FeSi [159], $Ce_3Pt_3(Sb,Bi)_4$ [160, 161], and $CePd_3$ [162] show large Seebeck coefficients at low temperatures with absolute values exceeding 100 μV/K. It should also be mentioned that utilizing magnetic interactions to enhance the effective mass to a certain degree and thereby enhance the power factor, is another interesting effort recently being actively carried out [5, 6].

The binary antimonide $FeSb_2$ has attracted attention in the past several years due to its interesting and unusual thermoelectric properties at low temperatures. Particularly, a single crystal of this compound [163] shows a tremendous Seebeck coefficient (~−45 mV K^{-1}) and power factor (~2.3 mW cm^{-1} K^{-2}) at low temperatures, which makes it a good candidate for thermoelectric cooling applications at cryogenic temperatures, despite its high thermal conductivity. In this section, various single-crystal growth methods and their effect on the thermoelectric properties will be discussed. Comparison between the single crystals with the bulk polycrystalline and nanocrystalline materials will also be presented.

$FeSb_2$ crystallizes in orthorhombic marcasite structure where each iron is coordinated by six antimony atoms in an octahedral environment. The crystal structure, depicted in Figure 10 is characterized by a network of edge and corner-sharing octahedra. Single crystal refinements [164, 165] revealed the full atomic order in the space group type *Pnnm* without any significant deviation from 1:2 stoichiometry. $FeSb_2$ is formed incongruently via a peritectic reaction: L + $Fe_{1+x}Sb$ ⇔ $FeSb_2$ at 750 °C with a

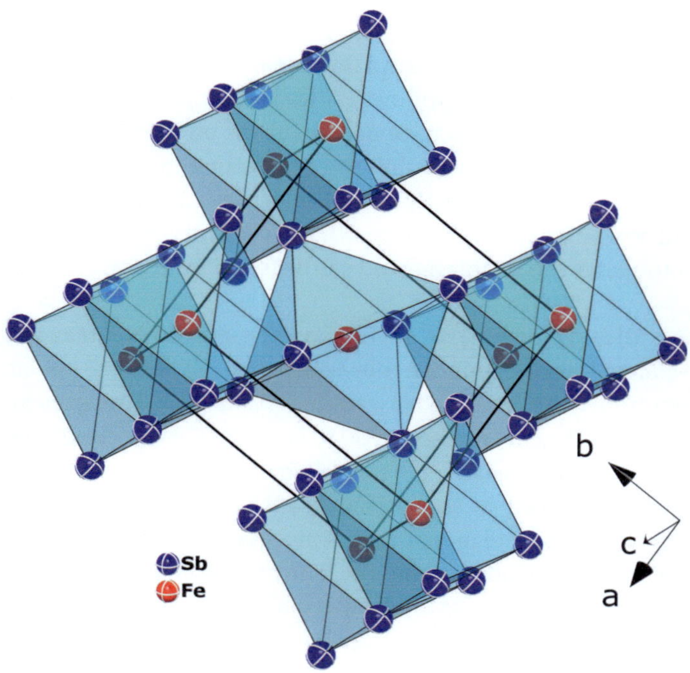

Figure 10: Crystal structure of $FeSb_2$ (crystallographic data taken from Chumak et al. [164]).

liquid composition of 90 at.% Sb [166]. The phase $FeSb_2$ is directly in equilibrium with antimony, forming a eutectic point L ⇔ $FeSb_2$+(Sb) at 624 °C and at ~97 at.% Sb. Such a situation is beneficial for the single-crystal growth of this compound with the flux method, previously described in Section 2.4 of this chapter.

Considering the phase equilibria around $FeSb_2$, its single crystals can be obtained by a self-flux method using antimony as the flux material. In such a case, a mixture of iron and antimony with the composition of more than 90 at.% Sb is sealed under vacuum or protective atmosphere and heated up to reach complete fusion of the elements. After homogenizing the melt, the mixture is slowly cooled down to 600–650 °C. The single crystals can be isolated from the flux either by decantation, centrifugation at high temperature or by evaporating the antimony. This method has been used to produce large single crystal with dimension up to ~5 mm. The crystal's morphologies produced by this method can vary from rods to irregular polyhedra/granules [165, 163, 167].

The chemical-vapor transport (CVT) method is another method to grow large single crystals of $FeSb_2$. In this method, the single crystals of the desired phase are grown from the gas phase. Since most metals and intermetallic compounds have low vapor pressure below their melting/decomposition temperature, the addition of a transporting agent is required. The transporting agent will react with the starting materials, forming various gaseous species, which will be transported and deposited at the growth zone as single crystals. Halogen, such as iodine or bromine, is commonly used as the transporting agent.

Typical crystal growth by this method requires a two-zone furnace to create a temperature gradient between the source and the growth sites. The temperature gradient is required to alter the equilibrium condition. Generally, based on the Le Chatelier principle, an increase in the temperature will shift the equilibrium to the endothermic side of the reaction. Therefore, the enthalpy of reaction plays a big role in CVT method, and its sign will determine whether the growth will occur at the hotter or the colder zone. Detailed information about this method can be found in, e. g., Binnewies et al. [168, 169].

For $FeSb_2$, chlorine [170] and bromine [171] have been utilized as transport agents to grow large single crystals of this compound. A tube with a dimension of ~1.2×25 cm with the pressure of chlorine of 75 torr was used to grow the crystal [170]. The temperatures of the source and the growth zone were set at 700 °C and 650 °C, respectively and the duration of the transport period was five–ten days. Prior to the growth process, a back transport process was performed for approximately a day, i. e., the temperature of the source and the growth zones were reversed. This process eliminates the nucleation sites in the growth zones introduced during the filling of the materials into the tubes. Unfortunately, not so many details on the growth parameters of $FeSb_2$ crystals with CVT method are presented in the literature [172, 171, 173].

The properties of $FeSb_2$ are strongly dependent on the quality of the samples, which may stem from various preparation methods. For example, the large value of Seebeck coefficient at low temperatures ($|S| \geq 1$ mV/K) can only be obtained in high-quality single crystals [174, 167, 175, 163, 176, 172], whereas for the polycrystalline samples, the maximum absolute value of Seebeck coefficient ranges from ~60–800 μV/K [177–180] depending on the preparation methods. Even a slight difference in the growth condition and the purity of the starting materials within the same method can alter the properties significantly, e.g, [181, 165, 177, 167]. Table 1 summarizes the thermoelectric figure of merit (ZT) of various $FeSb_2$-based materials. To date, the maximum ZT of this material is still one or two orders of magnitude lower than unity.

Table 1: Thermoelectric figure of merit (ZT) of various $FeSb_2$-based materials. (Abbreviations: SC: single crystal, PC: polycrystal, CVT: chemical-vapor transport, BM: ball milling, HP: hot pressing, DCHP: direct-current hot pressing, SPS: spark-plasma sintering).

Composition	Type	Crystal Growth Method	ZT	Reference
$FeSb_2$	SC	self-flux	0.005 at 12 K	[163]
$FeSb_{1.86}Te_{0.16}$	SC	CVT (Br_2)	0.013 at ~125 K	[171]
$FeSb_2$	PC	Melting-BM-HP	0.013 at 50 K	[182]
$Fe(Sb_{0.9}Te_{0.1})_2$	SC	self-flux	~0.05 at ~100 K	[183]
$FeSb_{1.975}Ag_{0.025}/Ag_{0.77}Sb_{0.23}$	PC	Melting-BM-DCHP	~0.02 at ~60 K	[184]
$FeSb_2$/0.045Cu	PC	Melting-BM-DCHP	~0.027 at 60 K	[185]
$FeSb_2$	PC	Melting-annealing-hot deformation in SPS	0.021 at ~70 K	[186]
$FeSb_{1.86}Te_{0.16}$	PC	Melting-BM-DCHP at 500 °C	0.022 at 100 K	[187]

The reproducibility of the colossal Seebeck coefficient of ~–45 mV/K is still questionable. So far this value has been reported only by the same group of authors (Bentien et al. [163]; Sun et al. [188, 172]). Recently, Takahashi et al. [174] came close to reproducing the value by adjusting the size of the single crystals. However, the maximum absolute Seebeck coefficient obtained was ~27 mV/K. A significant anisotropy in the electrical resistivity of the self-flux grown crystals was observed, which showed semiconducting behavior along a and b axes and a metal-insulator transition at above ~50 K along the c axis [165]. About a decade later, the same group of authors [181] reproduced the result and showed that the crystal with metal-insulator transition could be obtained by decanting the flux at 650 °C, while those without metal–insulator transition were obtained by decanting/centrifuging the mixture at 690 °C. It is important to note that the MIT behavior is not observed by other groups of authors.

Another interesting finding regarding the effects of various starting materials on the properties of the grown crystals was also reported [167]. They found out that even a ppm level of impurity can deteriorate the Seebeck coefficient, even though the maximum Seebeck coefficient they had was one order magnitude lower than Bentien et al. [163]. This discrepancy may be caused by the difference in the cooling rate, where Takahashi et al. used 9 °C/h from 775 °C to 640 °C, while Bentien et al. [163] used a much slower rate of 0.375 °C/h with the same temperature range.

Despite its sensitivity towards impurity, $FeSb_2$ crystals grown by CVT method (with Br_2 as transporting agent) show a high Seebeck coefficient at low temperatures [173, 176, 172], indicating their high quality. Absolute values of Seebeck coefficients ranging 10–30 mV/K [176, 172], as well as the transition from diamagnetic to an enhanced Pauli paramagnetism at ~50 K [176] were observed in samples grown by CVT.

Due to the reproducibility problem, which seems to be extremely sensitive to the sample preparation, the origin of the colossal Seebeck coefficient still remains under debate. Several authors (Petrovic et al. [189]; Bentien et al. [163]; Sun et al. [173, 172]; Herzog et al. [190]) proposed strong electronic correlation while the others (Pokharel et al. [191]; Tomczak et al. [192]; Battiato et al. [193]; Takahashi et al. [174]) proposed phonon drag as the mechanism. The behavior of $FeSb_2$ showed some resemblances to that of FeSi [159], a correlated semiconductor. The electrical resistivity of $FeSb_2$ is characterized by its semiconducting-like behavior, showing a plateau at ~20 K, from which two energy gaps were extracted, E_{g1} = 4–10 meV and E_{g2} = 26–36 meV ([172] and the references cited therein). Similar to FeSi [194], upon aliovalent substitution on the Sb site, several authors reported transition from correlated semiconductor to unconventional metals with enhanced effective mass [177, 171, 195]. Magnetic ordering was also observed upon Te doping ($Fe(Sb_{1-x}Te_x)_2$), showing transitions from metallic state at $x = 0.001$ to canted antiferromagnetism at $0.1 \leq x \leq 0.4$ and ferromagnetism at $x = 0.2$ [195]. The Seebeck coefficient is markedly reduced below −1 mV/K with Te doping due to an increase of charge-carrier concentration, however, the reduction of the lattice thermal conductivity is not enough to significantly boost the ZT [171]. Isovalent substitution by As also reduces the absolute value of Seebeck coefficient due to additional charge carrier induced by defect or other impurities [172]. It is also interesting to note that doping on both Fe and Sb sites also shifts the temperature of the maximum Seebeck coefficient towards the higher temperature [172, 171, 178].

Theoretical calculation [192] revealed that electronic correlation itself is not enough to explain the value of Seebeck coefficient at low temperature, suggesting additional mechanism such as phonon drag effect. High pressure transport measurements of a $FeSb_2$ single crystal demonstrated a weak dependence of the energy gap value on the applied pressure [175], in contrast to the usual Kondo insulator scenario. Furthermore, a first-principle study on the thermal conductivity of $FeSb_2$, focusing on the anharmonic lattice approximation, uncovered long phonon mean-free paths of over 100 μm at 20 K [196]. Such a long MFP would be beneficial for the phonon drag effect. Later on, Battiato et al. [193] proposed a combination of narrow in gap states and phonon drag as the main driver of the colossal Seebeck coefficient. The origin of the in-gap states stems from the defects, particularly from the excess Fe, i. e. $Fe_{1+x}Sb_{2-x}$. Their calculation showed that the minimum value of x of 4.5×10^{-5} should be enough to induce the in-gap states, a value that would be nearly impossible to detect via standard experimental methods, such as single-crystal refinement or electron-probe microanalysis. A similar proposition was also pointed out by Gippius et al. [197], stating that the defect concentration x in $FeSb_{2-x}$ should be much less than 1 %.

An experimental work [191] showed that by changing the grain size, the ZT of bulk nanocrystalline $FeSb_2$ was not significantly improved despite the remarkable reduction of the thermal conductivity. Lowering the grain size led to an increase of the grain-boundary phonon scattering, and thus reducing the magnitude of the Seebeck coefficient from the phonon drag effect. Similar observations were also reported in nanocrystalline $FeSb_2$ prepared by soft-chemistry methods [180, 198, 199], where the maximum absolute Seebeck coefficient was reduced to a value close to or below 100 μV/K. Indeed, such methods can also introduce a significant amount of impurity, which also suppresses the magnitude of Seebeck coefficient. A most recent publication by Takahashi et al. [174] showed quasi-ballistic phonon dragging of heavy electrons in $FeSb_2$. By changing the sample cross section to the sub-millimeter range, close to the calculated phonon mean-free path [196], they demonstrated the size dependence of the Seebeck coefficient, which supports the phonon drag effect. Another important finding from this experiment is the reduced electron mobility that results from an enhanced effective electron mass ($m* = 5.4m_0$). As the phonon-drag component of Seebeck coefficient is inversely proportional to the carrier mobility [156], the reduced mobility is also responsible for the enhancement of the Seebeck coefficient. The combination of the long phonon mean-free path and large effective mass (low mobility) open a new alternative mechanism to obtain suitable material for Peltier cooling at cryogenic temperature. Takahashi et al. [174] showed by calculation that with phonon mean-free path in the order of few hundreds of micrometers and large effective mass of $\approx 1000m_0$, ZT from phonon drag on such a strongly correlated system may reach unity at low temperatures.

6 Clathrate-VIII $Ba_8Ga_xSn_{46-x}$ compounds

High ZTs have already been discovered in numerous Zintl phases and related compounds, like $Yb_{14}MnSb_{11}$, skutterudites, and clathrates [200]. The term clathrate (inclusion compound) comes from the Latin word *clathratus* and is widely used to describe cage compounds where the guest atoms are encapsulated in the three-dimensional framework composed by the host atoms [201]. One of the earliest compounds that satisfies the standard of clathrate was a water clathrate with the formula $Cl_2(H_2O)_8$, which formed at low temperatures with water molecules that had interacted through hydrogen bonding crystallizing around Cl_2 molecules [202]. Gas hydrates are nonstoichiometric crystalline solids that could be formed when the gas (such as Xe, CH_4, H_2, CO_2, etc.) is frozen with water at high pressure [203]. The gas atoms are guests trapped within the large cages of the ice structure.

In terms of thermoelectric energy conversion, intermetallic clathrates have attracted great attention since they are materials exhibiting the properties of 'phonon-glass electron-crystal' (PGEC) proposed by Slack [204]. In this section, we will mainly review the researches on tin clathrates, their preparation method and their thermoelectric properties. Depending on the composition and structure, there are various categories of intermetallic clathrates. Figure 11 shows the crystal structure for type-I and VIII clathrates. The general formula for type-I clathrates (with a cubic structure and space group *Pm*3*n*) is $A_8E'_xE_{46-x}$, where A denotes alkali metals (Na, K, Rb, Cs) or alkaline earth elements (Mg, Ca, Sr, Ba) and E′\E represents groups 13 (triels = Al, Ga, In) or group 14 elements (tetrels = Si, Ge, Sn). There are two types of cages existing in the type-I clathrate structure, namely two smaller pentagonal dodecahedra with 12 pentagonal faces and six larger tetrakaidecahedra with two hexagonal and 12 pentagonal faces. Both kinds of polyhedra are filled with the guest atom A on the Wyckoff sites 2a and 6b, respectively. A cations exhibit large anisotropic displacement parameters (ADPs) and may be considered as 'rattlers'. The localized vibrational modes of A cations will remarkably contribute to the degradation of thermal conductivity of these materials [205, 206].

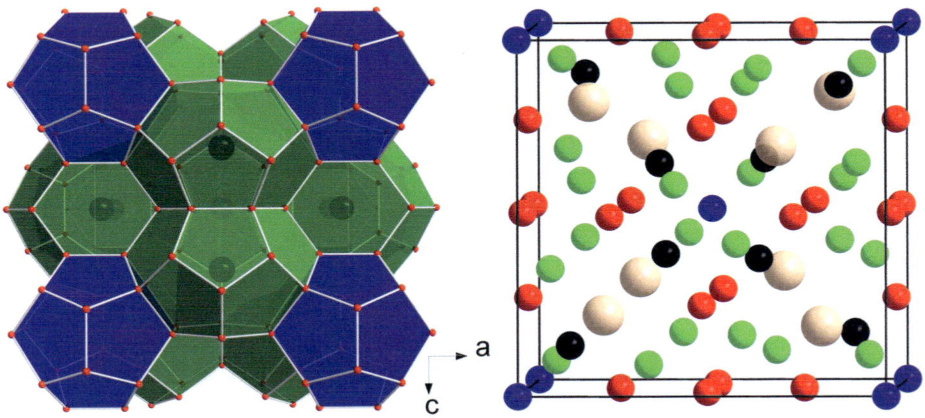

Figure 11: Crystal structures for type-I (left) and VIII clathrates (right. Blue, black, red and green spheres represents Wyckoff sites 2a, 8c, 12d and 24g, respectively).

Clathrate materials are related to Zintl phases and could be understood by the valence counting following the Zintl–Klemm method. In ternary Zintl $A_8E'_xE_{46-x}$, electrons donated by electropositive A atoms are accepted by the E\E′ atoms that constitute a covalently bound anionic framework through sp^3-like bonding, thus each E\E′ atom is surrounded by four other E\E′ atoms [207]. Correspondingly, the clathrate $A_8E'_xE_{46-x}$ is semiconducting when all E\E′ atoms forming the framework possess four valence electrons, leading to 184 valence electrons in each formula unit. This principle is also true for the tin clathrates $Ba_8Ga_{16}Sn_{30}$, considering that each Ba, Ga and Sn could provide two, three and four valence electrons, respectively [208]. Differential thermal analysis (DTA) and in situ synchrotron XRD confirmed that $Ba_8Ga_{16}Sn_{30}$ is dimorphic with a reversible type-I phase transforming to a type-VIII phase around 740 K [209]. A lower transformation temperature of 600 K was also reported [210]. Nevertheless, at higher temperatures, type-VIII $Ba_8Ga_{16}Sn_{30}$ are superior to type-I $Ba_8Ga_{16}Sn_{30}$ in terms of thermal stability and hence more preferable for application. The electron counting in the context of Zintl concept shown previously agrees well with the theoretical investigation. The first principles of electronic-structure calculations employing the full potential linearized augmented plane wave (FP-LAPW) approach support that both type-I $Ba_8Ga_{16}Sn_{30}$ and type-VIII $Ba_8Ga_{16}Sn_{30}$ are indirect semiconductors with band gaps being 0.51 and 0.32 eV, respectively [211]. Li et al. reported $E_g = \sim 0.6$ eV for type-VIII $Ba_8Ga_{16}Sn_{30}$ by CASTEP with generalized gradient approximation (GGA) [212]. Other different band gap values, such as $E_g = 0.16$ eV [213] and $E_g = 0.19$ eV [214] have also been calculated for type-VIII $Ba_8Ga_{16}Sn_{30}$. For indirect-gap semiconductors, the optimum band gap for thermoelectric application is $6–10k_BT$ with k_B = Boltzman constant and T = operating temperatures of the thermoelectric devices, according to the investigation by Chasmar and Stratton [215], and by Mahan later [216]. Therefore, the requirements for strong thermoelectrics are met by type-I $Ba_8Ga_{16}Sn_{30}$ and type-VIII $Ba_8Ga_{16}Sn_{30}$.

Hitherto, both n- and p-type type-VIII $Ba_8Ga_{16}Sn_{30}$ single-crystal clathrates (with size <6 mm) have been studied. Single crystals of type-VIII $Ba_8Ga_{16}Sn_{30}$ grown by the self-flux method could display different carrier types and different carrier concentrations. Generally speaking, single crystals grown from the Ga flux are p-type. In contrast, the dominant carriers will be electrons in crystals grown from the Sn flux. Furthermore, the carrier concentrations are likely to be adjusted by tuning the relative amount of the flux. Both the electrical resistivity ρ and the magnitude of the Seebeck coefficient α decrease with the increasing amount of the excess flux, as a consequence of the changes in the carrier concentration. As shown in Table 2, for the n-type crystals, the carrier concentration and lattice parameter gradually increases from 3.3×10^{19} cm^{-3} to 5.5×10^{19} cm^{-3} and from 11.590(3) Å to 11.596(2) Å, respectively, with the amount of excess Sn, x in $Ba_8Ga_{16}Sn_{30+x}$, increasing from $x = 10$ to $x = 30$. However, for p-type samples the changes in the lattice parameter is too small to obtain such a trend. On the other hand, it is worth noting that the p-type samples have carrier concentrations 1.1×10^{20} cm^{-3} to 1.7×10^{20} cm^{-3}, which are two ~ three times the values of the n-type ones. According to the onset temperature (T_{max}) of the intrinsic conduction and the Seebeck coefficient α_{max} at T_{max}, the average band gap values 0.37 eV for n-type samples and 0.38 eV for p-type samples are calculated using the equation $E_g = 2\alpha_{max}T_{max}$ [217]. Both values are comparable with the $E_g = 0.32$ eV from the first principles of electronic-structure calculation. Despite no reliable thermal-conductivity values available because of the small sizes of the crystals, highest ZT values were roughly estimated in p-type $Ba_{7.97}Ga_{15.95}Sn_{30.05}$ with ZT = 1.0 and n-type $Ba_{7.96}Ga_{15.94}Sn_{30.06}$ with ZT = 0.9 below 500 K [218].

Table 2: Crystal compositions, size, lattice parameters and transport properties at 300 K of various type-VIII $Ba_8Ga_{16}Sn_{30}$ single crystals grown by the flux method.

Samples	Flux	Starting composition	Crystal composition	Size (mm)	Lattice parameter (å)	Hall coefficient (cm^3/C)	Carrier concentration (10^{19} cm^{-3})	Carrier mobility (cm^2/Vs)	Effective mass (m^*/m_0)	Reference
$Ba_8Ga_{30+x}Sn_{30}$	Ga	x = 0	$Ba_{7.93}Ga_{15.92}Sn_{30.08}$	<6	11.587(3)	–	11	–	–	[218]
		x = 5	$Ba_{7.95}Ga_{15.95}Sn_{30.05}$	<6	11.590(2)	–	13	–	–	
		x = 8	$Ba_{7.95}Ga_{15.97}Sn_{30.03}$	<6	11.589(2)	–	14	–	–	
		x = 10	$Ba_{7.97}Ga_{15.95}Sn_{30.05}$	<6	11.591(2)	–	14	–	–	
		x = 15	$Ba_{7.91}Ga_{15.93}Sn_{30.07}$	<6	11.590(3)	–	17	–	–	
$Ba_8Ga_{16}Sn_{30+x}$	Sn	x = 10	$Ba_{7.96}Ga_{15.94}Sn_{30.06}$	<6	11.590(3)	–	3.3	–	–	[218]
		x = 20	$Ba_{7.96}Ga_{15.89}Sn_{30.11}$	<6	11.593(2)	–	4.3	–	–	
		x = 30	$Ba_{7.98}Ga_{15.82}Sn_{30.18}$	<6	11.596(2)	–	5.5	–	–	
$Ba_8Ga_{38}Sn_{30}Sb_x$	Ga	x = 1.5	$Ba_{8.01}Ga_{16.41}Sn_{29.04}Sb_{0.55}$	–	11.598(5)	–	2.90	4.90	–	[218]
		x = 1.5	$Ba_{8.00}Ga_{16.67}Sn_{28.61}Sb_{0.72}$	–	11.575(5)	–	5.02	1.03	–	
		x = 3.0	$Ba_{8.02}Ga_{16.81}Sn_{28.36}Sb_{0.83}$	–	11.563(5)	–	1.66	9.51	–	
$Ba_8Ga_{16}Sn_{50}Sb_x$	Sn	x = 2.5	$Ba_{8.01}Ga_{16.57}Sn_{28.62}Sb_{0.81}$	–	11.573(3)	–	1.43	15.90	–	[218]
		x = 3.8	$Ba_{7.96}Ga_{16.25}Sn_{29.37}Sb_{0.38}$	–	11.586(8)	–	2.95	9.34	–	
		x = 2.5	$Ba_{8.02}Ga_{16.34}Sn_{29.15}Sb_{0.51}$	–	11.569(1)	–	1.66	9.51	–	
$Ba_8Ga_{16-x}Ge_xSn_{50}$	Sn	x = 0	$Ba_{7.96}Ga_{15.9}Sn_{30.05}$	~5	11.5998	−0.152	4.14	27.4	–	[219]
		x = 0.25	$Ba_{7.98}Ga_{15.92}Ge_{0.03}Sn_{30.08}$	~5	11.5988	−0.255	2.45	74.3	–	
		x = 0.50	$Ba_{7.96}Ga_{15.85}Ge_{0.05}Sn_{30.14}$	~5	11.5977	−0.288	2.17	85.3	–	
		x = 0.75	$Ba_{7.93}Ga_{15.76}Ge_{0.08}Sn_{30.24}$	~5	11.5962	−0.221	2.83	67.0	–	
		x = 1.00	$Ba_{7.99}Ga_{15.69}Ge_{0.13}Sn_{30.19}$	~5	11.5843	−0.316	1.98	96.2	–	

Table 2: (continued)

Samples	Flux	Starting composition	Crystal composition	Size (mm)	Lattice parameter (å)	Hall coefficient (cm^3/C)	Carrier concentration (10^{19} cm^{-3})	Carrier mobility (cm^2/Vs)	Effective mass (m^*/m_0)	Reference
$Ba_8Ga_{16-x}In_xSn_{50}$	Sn	$x = 0$	$Ba_{7.96}Ga_{15.9}Sn_{30.1}$	–	11.596	−0.146	4.30	27.8	1.48	[220]
		$x = 1$	$Ba_{8.01}Ga_{15.80}In_{0.10}Sn_{30.1}$	~10	11.607	−0.202	3.10	48.8	0.96	
		$x = 2$	$Ba_{8.00}Ga_{15.66}In_{0.20}Sn_{30.14}$	~10	11.612	−0.165	3.78	59.6	0.98	
		$x = 4$	$Ba_{7.98}Ga_{15.56}In_{0.35}Sn_{30.09}$	<5	11.611	−0.205	3.04	38.8	0.96	
		$x = 6$	$Ba_{7.96}Ga_{15.48}In_{0.46}Sn_{30.06}$	<5	11.622	−0.193	3.22	31.1	1.16	
		$x = 8$	$Ba_{7.99}Ga_{15.28}In_{0.60}Sn_{30.12}$	<5	11.621	–	–	–	–	
$Ba_8Ga_{40}Cu_xSn_{30-x}$	Ga	$x = 0$	$Ba_{7.97}Ga_{15.95}Sn_{30.05}$	<2	11.591(2)	0.0446	14	6.28	4.3	[210]
		$x = 0.10$	$Ba_{7.98}Ga_{15.89}Cu_{0.008}Sn_{30.10}$	–	11.590(2)	–	–	–	–	
		$x = 0.25$	$Ba_{8.00}Ga_{15.89}Cu_{0.003}Sn_{30.11}$	–	11.589(2)	0.0354	18	3.11	5.6	
		$x = 0.25$	$Ba_{7.99}Ga_{15.84}Cu_{0.004}Sn_{30.16}$	~6	11.591(2)	0.0355	18	2.54	6.2	
		$x = 0.50$	$Ba_{7.99}Ga_{15.91}Cu_{0.009}Sn_{30.08}$	<6	11.590(3)	–	–	–	–	
		$x = 0.75$	$Ba_{8.00}Ga_{15.97}Cu_{0.014}Sn_{30.02}$	<6	11.590(3)	0.0341	18	2.73	6.8	
$Ba_8Ga_{16-x}Cu_xSn_{50}$	Sn	$x = 0$	$Ba_{7.96}Ga_{15.9}Sn_{30.1}$	–	11.593(2)	−0.146	4.3	27.8	1.5	[210, 221]
		$x = 0.50$	$Ba_{7.98}Ga_{15.80}Cu_{0.018}Sn_{30.10}$	~10	11.602(2)	−0.197	3.1	45.2	1.1	
		$x = 0.75$	$Ba_{8.00}Ga_{15.80}Cu_{0.020}Sn_{30.11}$	~10	11.598(2)	−0.190	3.3	51.2	1.2	
		$x = 1.0$	$Ba_{7.99}Ga_{15.80}Cu_{0.033}Sn_{30.16}$	~10	11.586(2)	−0.178	3.5	57.2	1.1	
		$x = 1.5$	$Ba_{8.00}Ga_{15.85}Cu_{0.033}Sn_{30.08}$	<4	11.585(3)	–	–	–	–	
		$x = 2.0$	$Ba_{7.98}Ga_{15.65}Cu_{0.125}Sn_{30.02}$	<4	11.594(3)	−0.138	4.5	60.8	1.1	
$Ba_8Ga_{16-x}Al_xSn_{50}$	Sn	$x = 0$	$Ba_{7.96}Ga_{15.9}Sn_{30.1}$		11.602(1)	−0.146	4.23	27.8	0.58	[218, 222]
		$x = 1$	$Ba_{7.90}Ga_{15.0}Al_{0.96}Sn_{30.2}$	~5	11.604(1)	–	–	–	0.53	
		$x = 2$	$Ba_{7.97}Ga_{13.9}Al_{1.95}Sn_{30.2}$	~5	11.609(1)	−0.157	3.94	45.4	–	
		$x = 4$	$Ba_{7.97}Ga_{12.0}Al_{3.83}Sn_{30.2}$	~5	11.612(1)	–	–	–	–	
		$x = 6$	$Ba_{8.02}Ga_{10.1}Al_{5.67}Sn_{30.2}$	~5	11.619(1)	−0.134	4.62	40.3	0.67	
		$x = 8$	$Ba_{8.00}Ga_{8.19}Al_{7.57}Sn_{30.2}$	~5	11.639(1)	−0.137	4.51	33.1	0.67	
		$x = 10$	$Ba_{7.98}Ga_{6.21}Al_{9.53}Sn_{30.3}$	~3	11.641(1)	–	–	–	–	
		$x = 12$	$Ba_{7.99}Ga_{15.23}Al_{10.5}Sn_{30.3}$	~3	11.659(1)	–	–	–	–	

Table 2: (continued)

Samples	Flux	Starting compo-sition	Crystal composition	Size (mm)	Lattice parameter (å)	Hall coefficient (cm^3/C)	Carrier concentration (10^{19} cm^{-3})	Carrier mobility (cm^2/Vs)	Effective mass (m^*/m_0)	Reference
$Ba_8Ga_{40}Zn_xSn_{30}$	Ga	$x = 0$	$Ba_{7.97}Ga_{15.95}Sn_{30.05}$	–	11.591(1)	0.045	14	6.3	2.2	[223]
		$x = 0.25$	$Ba_{8.00}Ga_{15.88}Zn_{0.007}Sn_{30.12}$	4 ~ 6	11.592(1)	0.0042	15	0.70	11	
		$x = 0.5$	$Ba_{8.01}Ga_{15.89}Zn_{0.006}Sn_{30.10}$	~7	11.592(1)	0.025	2.5	3.6	3.4	
		$x = 1.0$	$Ba_{8.01}Ga_{15.88}Zn_{0.009}Sn_{30.11}$	~7	11.593(1)	0.010	6.1	1.78	6.5	
		$x = 2.0$	$Ba_{8.01}Ga_{15.83}Zn_{0.019}Sn_{30.15}$	4 ~ 6	11.596(1)	0.038	1.7	2.0	3.2	
$Ba_8Ga_{16}Sn_{60+x}$	Sn	$x = 0$	$Ba_{8.01}Ga_{16}Zn_{0.019}Sn_{30.7}$	~10	–	–	3.7	39	0.14	[224]
$Ba_8Ga_{16-x}Cu_xSn_{50}$	Sn	$x = 0$	$Ba_{7.96}Ga_{15.90}Sn_{30.10}$	~8	11.603	−0.146	4.2	27.8	–	[225]
		$x = 1.0$	$Ba_{7.99}Ga_{15.80}Cu_{0.03}Sn_{30.17}$	~8	11.588	−0.178	3.5	57.2	–	
		$x = 1.5$	$Ba_{7.98}Ga_{15.81}Cu_{0.06}Sn_{30.18}$	~3	11.598	−0.158	3.9	57.4	–	
		$x = 2.0$	$Ba_{7.98}Ga_{15.64}Cu_{0.15}Sn_{30.24}$	~3	11.601	−0.137	4.5	60.4	–	
$Ba_8Eu_xGa_{40}Sn_{30}$	Ga	$x = 0$	$Ba_{7.83}Ga_{15.81}Sn_{30.36}$	~4	–	0.099	6.31	13.54	0.851	[226]
		$x = 0.30$	$Ba_{7.83}Eu_{0.03}Ga_{15.83}Sn_{30.31}$	~4	–	0.095	6.57	12.73	1.039	
		$x = 0.50$	$Ba_{7.81}Eu_{0.05}Ga_{15.75}Sn_{30.39}$	~4	–	0.087	7.15	8.05	1.319	
		$x = 0.75$	$Ba_{7.77}Eu_{0.12}Ga_{15.83}Sn_{30.39}$	~4	–	0.062	10.02	6.41	1.939	
		$x = 1.0$	$Ba_{7.58}Eu_{0.22}Ga_{15.85}Sn_{30.39}$	~4	–	0.031	20.19	1.38	3.120	
$Ba_{8-x}Eu_xGa_{16}Sn_{50}$	Sn	$x = 0$	$Ba_{7.98}Ga_{15.62}Sn_{30.40}$	–	–	−0.152	4.11	27.4	–	[227]
		$x = 0.25$	$Ba_{7.82}Eu_{0.11}Ga_{15.54}Sn_{30.53}$	–	–	−0.612	1.02	107.2	–	
		$x = 0.50$	$Ba_{7.69}Eu_{0.19}Ga_{15.50}Sn_{30.62}$	~6	–	−0.449	1.39	61.3	–	
		$x = 0.75$	$Ba_{7.59}Eu_{0.25}Ga_{15.83}Sn_{30.39}$	–	–	−0.150	4.17	18.5	–	
$Ba_8Ga_{16-x}Mg_xSn_{50}$	Sn	$x = 0$	$Ba_{7.96}Ga_{15.99}Sn_{30.05}$	~5	11.5998	−0.152	4.14	27.43	0.634	[228]
		$x = 0.5$	$Ba_{7.668}Ga_{15.549}Mg_{0.012}Sn_{30.771}$	~5	11.5988	−1.639	0.38	1.85	0.155	
		$x = 1.0$	$Ba_{7.698}Ga_{15.637}Mg_{0.026}Sn_{30.639}$	~5	11.5977	−2.016	0.31	2.23	0.081	
		$x = 1.25$	$Ba_{7.713}Ga_{15.717}Mg_{0.340}Sn_{30.05}$	~5	11.5962	−2.500	0.25	2.59	0.049	
		$x = 1.5$	$Ba_{7.96}Ga_{15.99}Mg_{0.420}Sn_{30.05}$	~5	11.5989	−3.676	0.17	2.64	0.028	

It is widely acknowledged that thermoelectric performance depends strongly on the carrier concentration or point defects, and doping is one of the effective method to tune the carrier concentration or create phonon-scattering centers and thus improve the thermoelectric performance. To get high-performance type-VIII $Ba_8Ga_{16}Sn_{30}$ single crystals, many attempts have been conducted by substituting various elements with different valance states for Ba, Ga and Sn. As summarized in Table 2, isovalent Al and In and electron-deficient Mg, Cu and Zn have been used to replace Ga. Single crystals of type-VIII $Ba_8Ga_{16-x}Al_xSn_{30}$ and $Ba_8Ga_{15.9-x}In_xSn_{30}$ have been grown from the Sn flux. The solubility of Al in type-VIII $Ba_8Ga_{16}Sn_{30}$ is rather high, amounting to about 10.5. By contrast, the solubility of In is relatively small, only around 0.6, which could be attributed to the difference between the covalent radii of In (1.44 Å) and Ga (1.26 Å). In both series, the lattice parameter increases with the increasing amount of Al or In until the amount of Al or In reaches their respective solubility. Upon incorporation of In, the carrier concentration decreased from 4.3×10^{19} cm^{-3} of the $Ba_{7.96}Ga_{15.9}Sn_{30.1}$ to ~3.5×10^{19} cm^{-3}, whereas mixed results were found in the Al containing samples with all materials showing carrier concentration around 4 ~ 5 × 10^{19} cm^{-3}. All the type-VIII $Ba_8Ga_{16-x}Al_xSn_{30}$ and $Ba_8Ga_{15.9-x}In_xSn_{30}$ show very low thermal conductivity in the temperature range of 300–600 K. The thermal conductivity κ remains below 1 W m^{-1} K^{-1} until 550 K, after which the κ starts increasing due to the bipolar effect [1]. In summary, n-type $Ba_{8.00}Ga_{15.66}In_{0.20}Sn_{30.14}$ shows a maximum ZT of 1.05 at 540 K and p-type $Ba_{8.02}Ga_{10.1}Al_{5.67}Sn_{30.2}$ displays a ZT = 1.2 at 500 K. The latter ZT is not only higher but also achieved at lower temperature, indicating that the cheaper Al is of higher price-to-performance ratio to enhance the type-VIII $Ba_8Ga_{16}Sn_{30}$.

Sb has been chosen as the dopants on the Sn sites for n-type doping because Sb has one extra valence electron compared with Sn [218, 229]. Both p- and n-type Sb doped $Ba_8Ga_{16}Sn_{30}$ were prepared using Ga and Sn as flux, respectively. For crystals grown from the Sn flux, the carrier concentration increases gradually from 8.8×10^{19} cm^{-3} for the $Ba_{7.96}Ga_{15.89}Sn_{30.11}$ to 1.43×10^{20} cm^{-3} for $Ba_{8.01}Ga_{16.57}Sn_{28.62}Sb_{0.81}$ and then to 2.95×10^{20} cm^{-3} for $Ba_{7.96}Ga_{16.25}Sn_{29.37}Sb_{0.38}$. A ZT of 0.98 was estimated in a p-type Sb substituted $Ba_{8.00}Ga_{16.67}Sn_{28.61}Sb_{0.72}$.

Both p- and n-type Cu-doped $Ba_8Ga_{16}Sn_{30}$ single crystals were also grown from Ga and Sn flux. The solubility of Cu in the crystals grown from Ga is only around 0.014, much smaller than 0.125 in the samples from the Sn flux. Since the valance electron number for Cu is less than that of Ga by 2, it is easy to understand that for the p-type series $Ba_8Ga_{16-x}Cu_xSn_{30}$, the carrier concentration for the Cu containing samples—1.8×10^{20} cm^{-3}—is higher than 1.4×10^{20} cm^{-3} of pristine $Ba_{7.97}Ga_{15.95}Sn_{30.05}$. By contrast, in the n-type series $Ba_8Ga_{16-x}Cu_xSn_{30}$, all the Cu-containing samples have a smaller electron concentration around 3.3×10^{20} cm^{-3} when compared with 4.3×10^{20} cm^{-3} for n-type pristine $Ba_{7.96}Ga_{15.9}Sn_{30.1}$. In the n-type series $Ba_8Ga_{16-x}Cu_xSn_{30}$, the detrimental effect of the reduced carrier concentration on the electrical conductivity was weakened by an increase in the mobility, leading to ~50 % enhancement in electrical conductivity. Combined with a thermal conductivity below 1 W m^{-1} K^{-1}, a current record high ZT for a clathrate was observed in $Ba_{7.98}Ga_{15.65}Cu_{0.125}Sn_{30.02}$ with ZT = 1.45 at 520 K.

Both Mg and Zn have one less valence electron when compared with Ga. The incorporation of Mg in the n-type $Ba_8Ga_{16}Sn_{30}$ resulted in a decrease in the carrier concentration from 4.14×10^{19} cm^{-3} to 1.7×10^{18} cm^{-3} for $Ba_{7.96}Ga_{15.99}Mg_{0.420}Sn_{30.05}$ and thus an increase in the Seebeck coefficient. A power factor as high as 1.26×10^{-3} W m^{-1} K^{-2} was found in $Ba_{7.96}Ga_{15.99}Mg_{0.420}Sn_{30.05}$ at around 430 K. Zn doping led to a remarkable enhancement of the thermoelectric performance of pristine $Ba_{7.97}Ga_{15.95}Sn_{30.05}$. High ZTs = 0.97 and 1.07 were found in p-type $Ba_{8.01}Ga_{15.89}Zn_{0.006}Sn_{30.10}$ and $Ba_{8.00}Ga_{15.88}Zn_{0.007}Sn_{30.12}$ at 500 K, respectively.

Very recently, Eu was also proved effective to optimize the thermoelectric properties of p-type type-VIII $Ba_8Ga_{16}Sn_{30}$ when replacing Ba in part [226]. The carrier (hole) concentration increases with the initial amount of Eu from 6.31×10^{19} cm^{-3} to 2.019×10^{20} cm^{-3}, indicating that Eu is an effective acceptor dopant in type-VIII $Ba_8Ga_{16}Sn_{30}$. The mobility also decrease with the increasing carrier concentration, probably because of enhanced carrier-carrier scattering. Again, the thermal conductivity could not be obtained experimentally due to the small size of the crystal. $Ba_{7.77}Eu_{0.12}Ga_{15.83}Sn_{30.39}$ was estimated to have the highest ZT = 0.87 at 483 K. More Eu atoms tend to replace Ba atoms when the n-type counterparts were grown from the Sn flux. A ZT of 0.46 were found in $Ba_{7.59}Eu_{0.25}Ga_{15.83}Sn_{30.39}$ at 550 K [227].

Other than single crystals, Kuznetsov et al. prepared n-type polycrystalline $Ba_8Ga_{16}Sn_{30}$ by reacting stoichiometric amounts of high-purity elements at a temperature higher than the melting point. Through fitting the temperature dependence of the electrical resistivity, a band gap $E_g = 0.6 \pm 0.1$ eV compares well with the calculated values was obtained, along with an activation energy of 15 ± 1 meV for the extrinsic donors [230]. At room temperature, polycrystalline $Ba_8Ga_{16}Sn_{30}$ exhibits a carrier concentration of 2.2×10^{19} cm^{-3}, which is lower than those of the single crystals prepared by self-flux growth.

In this section, we have reviewed the thermoelectric properties of single crystalline type-VIII $Ba_8Ga_{16}Sn_{30}$ clathrates. A high ZT = 1.45 at 520 K was achieved for the n-type $Ba_{7.98}Ga_{15.65}Cu_{0.125}Sn_{30.02}$. Several p-type single crystals with ZT_{max} around unity were also discovered, making type-VIII $Ba_8Ga_{16}Sn_{30}$ clathrates promising for application in the intermediate temperature range. Given that more various dopants and structure tailoring routes are possible, even higher ZT values are quite possible in the near future, especially for the p-type type-VIII $Ba_8Ga_{16}Sn_{30}$ clathrates.

Bibliography

[1] Koumoto K, Mori T, eds. Thermoelectric Nanomaterials – Materials Design and Applications. 1st ed. Springer Series in Materials Science, vol. 182. Berlin Heidelberg, Springer, 2013.

[2] Mori T, Nolas G. Contemporary innovations for thermoelectrics research and development. Scr Mater 2016, 111, 1–2.

[3] Heremans JP, Jovovic V, Toberer ES, Saramat A, Kurosaki K, Charoenphakdee A, Yamanaka S, Snyder GJ. Enhancement of thermoelectric efficiency in PbTe by distortion of the electronic density of states. Science 2008, 321, 554.

[4] Mori T. Novel principles and nanostructuring methods for enhanced thermoelectrics. Small 2017, 13, 1702013.

[5] Ang R, Khan AU, Tsujii N, Takai K, Nakamura R, Mori T. Thermoelectricity generation and electron-magnon scattering in natural chalcopyrite mineral from Deep-Sea hydrothermal vents. Angew Chem Int Ed 2015, 54, 12909–13.

[6] Ahmed F, Tsujii N, Mori T. Thermoelectric properties of $CuGa_{1-x}Mn_xTe_2$: power factor enhancement by incorporation of magnetic ion. J Mater Chem A 2017, 5, 7545–54.

[7] Zebarjadi M, Joshi G, Zhu GH, Yu B, Minnich AJ, Lan YC, Wang XW, Dresselhaus MS, Ren ZF, Chen G. Nano Lett 2011, 11, 2225.

[8] Shakouri A, Labounty C, Abraham P, Piprek J, Bowers JE. Mater Res Soc Proc 1999, 545, 449.

[9] Mori T, Hara T. Hybrid effect to possibly overcome the tradeoff between Seebeck coefficient and electrical conductivity. Scr Mater 2016, 111, 44–8.

[10] Liu W, Kim HS, Jie Q, Ren Z. Importance of high power factor in thermoelectric materials for power generation application: a perspective. Scr Mater 2016, 111, 3–9.

[11] Wang XW, Lee H, Lan YC, Zhu H, Joshi G, Wang DZ, Yang J, Muto AJ, Tang MY, Klatsky J, Song S, Dresselhaus MS, Chen G, Ren ZF. Thermoelectric properties of nanoporous Ge. Appl Phys Lett 2008, 93, 193121.

[12] Biswas K, He J, Blum ID, Wu C-I, Hogan TP, Seidman DN, Dravid VP, Kanatzidis MG. High-performance bulk thermoelectrics with all-scale hierarchical architectures. Nature 2012, 489(7416), 414–8.

[13] Grasso S, Tsujii N, Jiang Q, Khaliq J, Maruyama S, Miranda M, Simpson K, Mori T, Reece MJ. Ultra low thermal conductivity of disordered layered p-type bismuth telluride. J Mater Chem C 2013, 1, 2362–7.

[14] Nethravathi C, Rajamathi R, Rajamathi M, Maki R, Mori T, Golberg D, Bando Y. Synthesis and thermoelectric behaviour of copper telluride nanosheets. J Mater Chem A 2014, 2, 985–90.

[15] Rogl G, Grytsiv A, Yubuta K, Puchegger S, Bauer E, Raju C, Mallik RC, Rogl P. Acta Mater 2015, 95, 201.

[16] Khan AU, Kobayashi K, Tang D-M, Yamauchi Y, Hasegawa K, Mitome M, Xue Y, Jiang B, Tsuchiya K, Golberg D, Bando Y, Mori T. Nano-micro-porous skutterudites with 100 % enhancement in ZT for high performance thermoelectricity. Nano Energy 2017, 31, 152–9. https://doi.org/10.1016/j.nanoen.2016.11.016.

[17] Wood C, Emin D. Conduction mechanism in boron carbide. Phys Rev B 1984, 29, 4582.

[18] Aselage TL, Emin D, McCready SS, Duncan RV. Phys Rev Lett 1998, 81, 2316.

[19] Feng B, Martin H-P, Börner F-D, Lippmann W, Schreier M, Vogel K, Lenk A, Veremchuk I, Dannowski M, Richter C, Pfeiffer P, Zikoridse G, Lichte H, Grin J, Hurtado A, Michaelis A. Manufacture and testing of thermoelectric. modules consisting of B_xC and TiO_x elements. Adv Eng Mater 2014, 16, 1252–63.

[20] Rasim K, Ramlau R, Leithe-Jasper A, Mori T, Burkhardt U, Borrmann H, Schnelle W, Carbogno C, Scheffler M, Grin Y. Local atomic arrangements and band structure of boron carbide. Angew Chem Int Ed 2018, 57, 6130–5. https://doi.org/10.1002/anie.201800804.

[21] Mori T. Perspectives of high temperature thermoelectric applications and p- and n-type aluminoborides. JOM 2016, 68, 2673–9.

[22] Yazawa K, Shakouri A. Thermoelectric topping cycles with scalable design and temperature dependent material properties. Scr Mater 2016, 111, 58.

[23] Sussardi A, Tanaka T, Khan AU, Schlapbach L, Mori T. Enhanced thermoelectric properties of samarium boride. J Materiom 2015, 1, 196–204.

[24] Mori T. Boron-based materials. In: Uher C, ed. Materials Aspect of Thermoelectricity. London, CRC Press, Taylor and Francis, 2016. 441–57.

[25] Sologub O, Salamakha L, Stoger B, Michiue Y, Mori T. Zr doped β-rhombohedral boron: widely variable Seebeck coefficient and structural properties. Acta Mater 2017, 122, 378–85.

[26] Mori T. High temperature thermoelectric properties of B_{12} icosahedral cluster-containing rare earth boride crystals. J Appl Phys 2005, 97, 093703.

[27] Mori T, Berthebaud D, Nishimura T, Nomura A, Shishido T, Nakajima K. Effect of Zn doping on improving crystal quality and thermoelectric properties of borosilicides. Dalton Trans 2010, 39, 1027.

[28] Hossain MA, Tanaka I, Tanaka T, Khan AU, Mori T. Crystal growth and anisotropy of high temperature thermoelectric properties of yttrium borosilicide crystals. J Solid State Chem 2016, 233, 1–7.

[29] Lassoued S, Gautier R, Halet J-F. The electronic properties of metal borides and borocarbides: differences and similarities. In: Orlovskaya N, Lugovy M, eds. Boron Rich Solids, Sensors, Ultra High Temperature Ceramics, Thermoelectrics, Armor. New York, Springer, 2011. 95–114.

[30] Simonson JW, Poon SJ. Applying an electron counting rule to screen prospective thermoelectric alloys: the thermoelectric properties of $YCrB_4$ and Er_3CrB_7-type phases. J Alloys Compd 2010, 504, 265–72.

[31] Wang XJ, Mori T, Kuzmych-Ianchuk I, Michiue Y, Yubuta K, Shishido T, Grin Y, Okada S, Cahill DG. Thermal conductivity of layered borides: the effect of building defects on the thermal conductivity of $TmAlB_4$ and the anisotropic thermal conductivity of AlB_2. APL Mater 2014, 2, 046113.

[32] Slack GA. The thermal conductivity of nonmetallic crystals. In: Ehrenreich H, Weitz F, Turnbull D, eds. Solid State Physics. vol. 34. New York, Academic Press, 1979. 1.

[33] Emin D. Small polarons. Phys Today 1982, 35, 34.

[34] Mori T, Martin J, Nolas G. Thermal conductivity of $YbB_{44}Si_2$. J Appl Phys 2007, 102, 073510.

[35] Mori T. Higher borides. In: Gschneidner KA Jr, Bunzli J-C, Pecharsky V, eds. Handbook on the Physics and Chemistry of Rare-earths. vol. 38. Amsterdam, North-Holland, 2008. 105–73.

[36] Mori T. Boride thermoelectrics; high temperature thermoelectric materials. In: Rowe DM, ed. Modules, Systems, and Applications in Thermoelectrics. London, CRC Press, Taylor and Francis, 2012. 14-1–14-18.

[37] Richards SM, Kasper JS. The crystal structure of YB_{66}. Acta Crystallogr, Sect B 1969, 25, 237.

[38] Seyboldt AU. Trans Am Soc Met 1960, 52, 971.

[39] Oliver DW, Brower GD. Growth of single crystal YB_{66} from the melt. J Cryst Growth 1971, 11, 185.

[40] Golikova OA. Semiconductors with complex lattice and the amorphization problem. Phys Status Solidi A 1987, 101, 277.

[41] Tanaka T, Otani S, Ishizawa Y. Preparations of single crystals of YB_{66}. J Cryst Growth 1985, 73, 31.

[42] Kamimura Y, Tanaka T, Otani S, Ishizawa a Y, Rek ZU, Wong J. Floating zone growth of monochromator grade crystals of YB_{66}. J Cryst Growth 1993, 128, 429.

[43] Wong J, Shimkaveg G, Goldstein W, Eckart M, Tanaka T, Rek ZU, Tompkins H. YB_{66}: a new soft-X-ray monochromator for synchrotron radiation. Nucl Instrum Methods A 1990, 291, 243.

[44] Slack GA, Oliver DW, Horn FH. Thermal conductivity of boron and some boron compounds. Phys Rev B 1971, 4, 1714.

[45] Cahill DG, Watson SK, Pohl RO. Thermal properties of boron and boride. Phys Rev B 1992, 46, 6131.

[46] Hossain MA, Tanaka I, Tanaka T, Khan AU, Mori T. YB_{48} the metal-rich boundary of YB_{66}; crystal growth and thermoelectric properties. J Phys Chem Solids 2015, 87, 221.

[47] Tanaka T, Okada S, Ishizawa Y. A new yttrium higher boride: YB_{50}. J Alloys Compd 1994, 205, 281–4.

[48] Tanaka T, Okada S, Ishizawa Y. Single crystal growth of a new YB_{50} family compound: $YB_{44}Si_{1.0}$. J Solid State Chem 1997, 133, 55–8.

[49] Sologub O, Salamakha LP, Michor H, Bauer E. Incorporation of platinum atoms in a silicon-free boride of the YB_{50}-type structure. J Alloys Compd 2016, 675, 99.

[50] Higashi I, Tanaka T, Kobayashi K, Ishizawa Y, Takami M. Crystal structure of $YB_{41}Si_{1.2}$. J Solid State Chem 1997, 133, 11–5.

[51] Mori T, Tanaka T. Magnetic properties of terbium B_{12} icosahedral boron-rich compounds. J Phys Soc Jpn 1999, 68, 2033–9.

[52] Mori T, Sahara R, Kawazoe Y, Yubuta K, Shishido T, Grin Y. Strong magnetic coupling in a magnetically dilute f-electron insulator: a dysprosium boron-cluster compound. J Appl Phys 2013, 113, 17E156.

[53] Mori T. Doping effect in a magnetic TbB_{50}-type B_{12} cluster compound. J Appl Phys 2004, 95, 7204–6.

[54] Berthebaud D, Sato A, Michiue Y, Mori T, Nomura A, Shishido T, Nakajima K. Effect of transition element doping on crystal structure of rare earth borosilicides $REB_{44}Si_2$. J Solid State Chem 2011, 184, 1682–7.

[55] Mori T. Thermal conductivity of a rare earth B_{12}-icosahedral compound. Physica B 2006, 383, 120.

[56] Mori T, Tanaka T. Effect of transition metal doping and carbon doping on thermoelectric properties of YB_{66} single crystals. J Solid State Chem 2006, 179, 2889–94.

[57] Kuzma YB. Crystal structure of the compound $YCrB_4$ and its analogs. Sov Phys Crystallogr 1970, 15, 312.

[58] Rogl P, Nowotny H. Ternary complex boride with $ThMoB_4$-type. Monatsh Chem 1974, 105(5), 1082.

[59] Mori T, Shishido T, Nakajima K, Kieffer K, Siemensmeyer K. Magnetic properties of the thulium layered compound $Tm_2Al_{11}B_6$, an AlB_2-type analogue. J Appl Phys 2009, 105, 07.

[60] Kanatzidis MG, Pöttgen R, Jeitschko W. The metal flux: a preparative tool for the exploration of intermetallic compounds. Angew Chem Int Ed 2005, 44(43), 6996–7023. https://doi.org/10.1002/anie.200462170.

[61] Grytsiv A, Bauer E, Berger S, Hilscher G, Michor H, Paul C, Rogl P, et al. Novel Zn_9-cluster compounds $RE_2Zn_6Ge_3$ (RE: La, Ce, Pr, Nd, Sm, Gd): crystal structure and physical properties. J Phys Condens Matter 2003, 15(19), 3053. https://doi.org/10.1088/0953-8984/15/19/308.

[62] Bugaris DE, Malliakas CD, Chung DY, Kanatzidis MG. Metallic borides, $La_2Re_3B_7$ and $La_3Re_2B_5$, featuring extensive boron–boron bonding. Inorg Chem 2016, 55(4), 1664–73. https://doi.org/10.1021/acs.inorgchem.5b02599.

[63] Pöttgen R, Johrendt D. Intermetallics: Synthesis, Structure, Function. Walter de Gruyter GmbH & Co KG, 2014.

[64] Okada S, Yu Y, Lundström T, Kudou K, Tanaka T. Crystal growth and some properties of LuB_4, $LuAlB_4$, and Lu_2AlB_6. Jpn J Appl Phys 1996, 35, 4718–23.

[65] Okada S, Kudou K, Tanaka T, Shishido T, Gurin VN, Lundstrom T. Growth and some properties of Sc_2AlB_6 crystal obtained from the solution in aluminum melt. J Solid State Chem 2004, 177, 547–50.

[66] Okada S, Shishido T, Mori T, Kudou K, Iizumi K, Lundstrom T, Nakajima K. Properties of $REAlB_4$ and Lu_2AlB_6 crystals grown from RE–Al–B (RE = Tm, Yb, Lu) melts. J Alloys Compd 2006, 408–412, 547–50.

[67] Macaluso RT, Nakatsuji S, Kuga K, Thomas EL, Machida Y, Maeno Y, Fisk Z, Chan JY. Crystal structure and physical properties of polymorphs of $LnAlB_4$ (Ln = Yb, Lu). Chem Mater 2007, 19, 1918–22.

[68] Mori T, Okada S, Kudou K. Magnetic properties of thulium aluminoboride $TmAlB_4$. J Appl Phys 2005, 97, 10A910.

[69] Mori T, Borrmann H, Okada S, Kudou K, Leithe-Jasper A, Burkhardt U, Grin Yu. Crystal structure, chemical bonding, electrical transport, and magnetic behavior of $TmAlB_4$. Phys Rev B 2007, 76, 064404.

[70] Yubuta K, Mori T, Leithe-Jasper A, Grin Yu, Okada S, Shishido T. Direct observation of the intergrown alpha-phase in beta-$TmAlB_4$ via high-resolution electron microscopy. Mater Res Bull 2009, 44, 1743–6.

[71] Yubuta K, Mori T, Okada S, Prots Y, Borrmann H, Grin Y, Shishido T. High-resolution electron microscopy and X-ray diffraction study of intergrowth-structures in α- and β-type $YbAlB_4$ single crystals. Philos Mag 2013, 93, 1054–64.

[72] Mori T, Kuzmych-Ianchuk I, Yubuta K, Shishido T, Okada S, Kudou K, Grin Y. Direct elucidation of the effect of building defects on the physical properties of alpha-$TmAlB_4$; an AlB_2-type analogous "tiling" compound. J Appl Phys 2012, 111, 07E127.

[73] Medvedeva NI, Medvedeva YuE, Ivanovskii AL. Electronic structure of ternary boron-containing phases $YCrB_4$, Y_2ReB_6 and MgC_2B_2. Dokl, Phys Chem 2002, 383, 75–7.

[74] Lassoued S, Gautier R, Boutarfaia A, Halet J-F. Rings and chains in solid-state metal borides and borocarbides. The electron count matters. J Organomet Chem 2010, 695, 987–93.

[75] Frye CD, Edgar JH, Ohkubo I, Mori T. Seebeck coefficient and electrical resistivity of single crystal $B_{12}As_2$ at high temperatures. J Phys Soc Jpn 2013, 82, 095001.

[76] Kumashiro Y, Yokoyama T, Sato K, Ando Y, Nagatani S, Kajiyama K. J Solid State Chem 2000, 154, 33–8.

[77] Slack GA, Morgan KE. Some crystallography, chemistry, physics, and thermodynamics of $B_{12}O_2$, $B_{12}P_2$, $B_{12}As_2$, and related alpha-boron type crystals. J Phys Chem Solids 2014, 75, 1054.

[78] Tynell T, Aizawa T, Ohkubo I, Nakamura K, Mori T. Deposition of thermoelectric strontium hexaboride thin films by a low pressure CVD method. J Cryst Growth 2016, 449, 10–4.

[79] Stern A, Dzero M, Galitski VM, Fisk Z, Xia J. Surface dominated conduction up to 240 K in the Kondo insulator SmB_6 under strain. Nat Mater 2017, 16, 708–11.

[80] Tretiakov OA, Abanov A, Sinova J. Holey topological thermoelectrics. Appl Phys Lett 2011, 99, 113110.

[81] Zhao L-D, Lo S-H, Zhang Y, Sun H, Tan G, Uher C, Wolverton C, Dravid VP, Kanatzidis MG. Ultralow thermal conductivity and high thermoelectric figure of merit in SnSe crystals. Nature 2014, 508(7496), 373–7. https://doi.org/10.1038/nature13184.

[82] Li B, Xie Y, Huang J, Qian Y. Solvothermal route to tin monoselenide bulk single crystal with different morphologies. Inorg Chem 2000, 39(10), 2061–4. https://doi.org/10.1021/ic990964v.

[83] Chattopadhyay T, Pannetier J, Von Schnering HG. Neutron diffraction study of the structural phase transition in SnS and SnSe. J Phys Chem Solids 1986, 47(9), 879–85.

[84] Zhao L-D, Dravid VP, Kanatzidis MG. The panoscopic approach to high performance thermoelectrics. Energy Environ Sci 2014, 7(1), 251. https://doi.org/10.1039/c3ee43099e.

[85] Heremans JP. Thermoelectricity: the ugly duckling. Nature 2014, 508(7496), 327–8. https://doi.org/10.1038/508327a.

[86] Wasscher JD, Albers W, Haas C. Simple evaluation of the maximum thermoelectric figure of merit, with application to mixed crystals $SnS_{1-x}Se_x$. Solid-State Electron 1963, 6(3), 261–4. https://doi.org/10.1016/0038-1101(63)90083-2.

[87] Elkorashy AM. Optical absorption in tin monoselenide single crystal. J Phys Chem Solids 1986, 47(5), 497–500. https://doi.org/10.1016/0022-3697(86)90048-X.
[88] Bhatt VP, Gireesan K, Pandya GR. Growth and characterization of SnSe and $SnSe_2$ single crystals. J Cryst Growth 1989, 96(3), 649–51. https://doi.org/10.1016/0022-0248(89)90063-8.
[89] Li L, Chen Z, Hu Y, Wang X, Zhang T, Chen W, Wang Q. Single-layer single-crystalline SnSe nanosheets. J Am Chem Soc 2013, 135(4), 1213–6. https://doi.org/10.1021/ja3108017.
[90] Reddy M, Reddy V, Gedi S, Pejjai B, Park C. Perspectives on SnSe-based thin film solar cells: a comprehensive review. J Mater Sci, Mater Electron 2016. https://doi.org/10.1007/s10854-016-4563-9.
[91] Zhao L-D, Tan G, Hao S, He J, Pei Y, Chi H, Wang H, Gong S, Xu H, Dravid VP, Uher C, Snyder GJ, Wolverton C, Kanatzidis MG. Ultrahigh power factor and thermoelectric performance in hole-doped single-crystal SnSe. Science 2015. https://doi.org/10.1126/science.aad3749.
[92] Ibrahim D, Vaney JB, Sassi S, Candolfi C, Ohorodniichuk V, Levinsky P, Semprimoschnig C, Dauscher A, Lenoir B. Reinvestigation of the thermal properties of single-crystalline SnSe. Appl Phys Lett 2017, 110(3), 032103. https://doi.org/10.1063/1.4974348.
[93] Duong AT, Nguyen VQ, Duvjir G, Duong VT, Kwon S, Song JY, Lee JK, Lee JE, Park S, Min T, Lee J, Kim J, Cho S. Achieving ZT = 2.2 with Bi-doped n-type SnSe single crystals. Nat Commun 2016, 7, 13713. https://doi.org/10.1038/ncomms13713.
[94] Jin M, Shao H, Hu H, Li D, Shen H, Xu J, Jiang J. Growth and characterization of large size undoped p-type SnSe single crystal by Horizontal Bridgman method. J Alloys Compd 2017, 712, 857–62. https://doi.org/10.1016/j.jallcom.2017.04.110.
[95] Jin M, Shao H, Hu H, Li D, Xu J, Liu G, Shen H, Xu J, Jiang H, Jiang J. Single crystal growth of $Sn_{0.97}Ag_{0.03}Se$ by a novel horizontal Bridgman method and its thermoelectric properties. J Cryst Growth 2017, 460, 112–6. https://doi.org/10.1016/j.jcrysgro.2016.12.087.
[96] Maier H, Daniel DR. SnSe single crystals: sublimation growth, deviation from stoichiometry and electrical properties. J Electron Mater 1977, 6(6), 693–704. https://doi.org/10.1007/BF02660344.
[97] Cambi L, Elli M. Simple and complex selenides. Hydrothermal synthesis from selenium and metal or metalloid oxides. Chim Ind 1968, 50(94).
[98] Sines IT, Vaughn DD, Biacchi AJ, Kingsley CE, Popczun EJ, Schaak RE. Engineering porosity into single-crystal colloidal nanosheets using epitaxial nucleation and chalcogenide anion exchange reactions: the conversion of SnSe to SnTe. Chem Mater 2012, 24(15), 3088–93. https://doi.org/10.1021/cm301734b.
[99] Zhao S, Wang H, Zhou Y, Liao L, Jiang Y, Yang X, Chen G, Lin M, Wang Y, Peng H, Liu Z. Controlled synthesis of single-crystal SnSe nanoplates. Nano Res 2015, 8(1), 288–95. https://doi.org/10.1007/s12274-014-0676-8.
[100] Hong M, Chen Z-G, Yang L, Chasapis TC, Kang SD, Zou Y, Auchterlonie GJ, Kanatzidis MG, Snyder GJ, Zou J. Enhancing the thermoelectric performance of $SnSe_{1-x}Te_x$ nanoplates through band engineering. J Mater Chem A 2017, 5(21), 10713–21. https://doi.org/10.1039/c7ta02677c.
[101] Carrete J, Mingo N, Curtarolo S. Low thermal conductivity and triaxial phononic anisotropy of SnSe. Appl Phys Lett 2014, 105(10), 101907. https://doi.org/10.1063/1.4895770.
[102] Guo R, Wang X, Kuang Y, Huang B. First-principles study of anisotropic thermoelectric transport properties of IV-VI semiconductor compounds SnSe and SnS. Phys Rev B 2015, 92(11). https://doi.org/10.1103/PhysRevB.92.115202.
[103] Li CW, Hong J, May AF, Bansal D, Chi S, Hong T, Ehlers G, Delaire O. Orbitally driven giant phonon anharmonicity in SnSe. Nat Phys 2015, 11(12), 1063–9. https://doi.org/10.1038/nphys3492.

[104] Lee S, Esfarjani K, Luo T, Zhou J, Tian Z, Chen G. Resonant bonding leads to low lattice thermal conductivity. Nat Commun 2014, 5, 3525. https://doi.org/10.1038/ncomms4525.

[105] Skoug EJ, Morelli DT. Role of lone-pair electrons in producing minimum thermal conductivity in nitrogen-group chalcogenide compounds. Phys Rev Lett 2011, 107(23). https://doi.org/10.1103/PhysRevLett.107.235901.

[106] Nielsen MD, Ozolins V, Heremans JP. Lone pair electrons minimize lattice thermal conductivity. Energy Environ Sci 2013, 6(2), 570. https://doi.org/10.1039/c2ee23391f.

[107] Du B, Chen K, Yan H, Reece MJ. Efficacy of lone-pair electrons to engender ultralow thermal conductivity. Scr Mater 2016, 111, 49–53.

[108] Sassi S, Candolfi C, Vaney JB, Ohorodniichuk V, Masschelein P, Dauscher A, Lenoir B. Assessment of the thermoelectric performance of polycrystalline p-type SnSe. Appl Phys Lett 2014, 104(21), 212105. https://doi.org/10.1063/1.4880817.

[109] Han Y-M, Zhao J, Zhou M, Jiang X-X, Leng H-Q, Li L-F. Thermoelectric performance of SnS and SnS–SnSe solid solution. J Mater Chem A 2015, 3(8), 4555–9. https://doi.org/10.1039/c4ta06955b.

[110] Leng H, Zhou M, Zhao J, Han Y, Li L. Optimization of thermoelectric performance of anisotropic $Ag_xSn_{1-x}Se$ compounds. J Electron Mater 2015. https://doi.org/10.1007/s11664-015-4143-4.

[111] Li Y, Shi X, Ren D, Chen J, Chen L. Investigation of the anisotropic thermoelectric properties of oriented polycrystalline SnSe. Energies 2015, 8(7), 6275–85. https://doi.org/10.3390/en8076275.

[112] Wei T-R, Wu C-F, Zhang X, Tan Q, Sun L, Pan Y, Li J-F. Thermoelectric transport properties of pristine and Na-doped $SnSe_{1-x}Te_x$ polycrystals. Phys Chem Chem Phys 2015, 17(44), 30102–9. https://doi.org/10.1039/c5cp05510e.

[113] Zhang Q, Chere EK, Sun J, Cao F, Dahal K, Chen S, Chen G, Ren Z. Studies on thermoelectric properties of n-type polycrystalline $SnSe_{1-x}S_x$ by iodine doping. Adv Eng Mater 2015, 5(12), 1500360. https://doi.org/10.1002/aenm.201500360.

[114] Chere EK, Zhang Q, Dahal K, Cao F, Mao J, Ren Z. Studies on thermoelectric figure of merit of Na-doped p-type polycrystalline SnSe. J Mater Chem A 2016, 4(5), 1848–54. https://doi.org/10.1039/c5ta08847j.

[115] Wei TR, Tan G, Zhang X, Wu CF, Li JF, Dravid VP, Snyder GJ, Kanatzidis MG. Distinct impact of alkali-ion doping on electrical transport properties of thermoelectric p-type polycrystalline SnSe. J Am Chem Soc 2016, 138(28), 8875–82. https://doi.org/10.1021/jacs.6b04181.

[116] Chen C-L, Wang H, Chen Y-Y, Day T, Snyder GJ. Thermoelectric properties of p-type polycrystalline SnSe doped with Ag. J Mater Chem A 2014, 2(29), 11171. https://doi.org/10.1039/c4ta01643b.

[117] Kutorasinski K, Wiendlocha B, Kaprzyk S, Tobola J. Electronic structure and thermoelectric properties of n- and p-type SnSe from first-principles calculations. Phys Rev B 2015, 91(20):. https://doi.org/10.1103/PhysRevB.91.205201.

[118] Peng K, Lu X, Zhan H, Hui S, Tang X, Wang G, Dai J, Uher C, Wang G, Zhou X. Broad temperature plateau for high ZTs in heavily doped p-type SnSe single crystals. Energy Environ Sci 2016, 9(2), 454–60. https://doi.org/10.1039/c5ee03366g.

[119] Wang X, Xu J, Liu G, Fu Y, Liu Z, Tan X, Shao H, Jiang H, Tan T, Jiang J. Optimization of thermoelectric properties in n-type SnSe doped with $BiCl_3$. Appl Phys Lett 2016, 108(8), 083902. https://doi.org/10.1063/1.4942890.

[120] Zhang LC, Qin G, Fang WZ, Cui HJ, Zheng QR, Yan QB, Su G. Tinselenidene: a two-dimensional auxetic material with ultralow lattice thermal conductivity and ultrahigh hole mobility. Sci Rep 2016, 6, 19830. https://doi.org/10.1038/srep19830.

[121] Li Z, Bauers SR, Poudel N, Hamann D, Wang X, Choi DS, Esfarjani K, Shi L, Johnson DC, Cronin SB. Cross-plane seebeck coefficient measurement of misfit layered compounds $(SnSe)_n(TiSe_2)_n$ ($n = 1, 3, 4, 5$). Nano Lett 2017, 17(3), 1978–86. https://doi.org/10.1021/acs.nanolett.6b05402.

[122] Guilmeau E, Maignan A, Wan C, Koumoto K. On the effects of substitution, intercalation, nonstoichiometry and block layer concept in TiS_2 based thermoelectrics. Phys Chem Chem Phys 2015 17(38), 24541–55. Royal Society of Chemistry. https://doi.org/10.1039/C5CP01795E.

[123] Chhowalla M, Shin HS, Eda G, Li L-J, Loh KP, Zhang H. The chemistry of two-dimensional layered transition metal dichalcogenide nanosheets. Nat Chem 2013, 5(4), 263–75. https://doi.org/10.1038/nchem.1589.

[124] Wilson JA, Yoffe AD. The transition metal dichalcogenides discussion and interpretation of the observed optical, electrical and structural properties. Adv Mater 1969, 18(73), 193–335. https://doi.org/10.1080/00018736900101307.

[125] Whittingham SM. Lithium batteries and cathode materials. Chem Rev 2004, 104, 4271–301. https://doi.org/10.1021/cr020731c.

[126] Ma L, Wei S, Zhuang HL, Hendrickson KE, Hennig RG, Archer LA. Hybrid cathode architectures for lithium batteries based on TiS_2 and sulfur. J Mater Chem A 2015, 3(39), 19857–66. https://doi.org/10.1039/C5TA06348E.

[127] Wang QH, Kalantar-Zadeh K, Kis A, Coleman JN, Strano MS. Electronics and optoelectronics of two-dimensional transition metal dichalcogenides. Nat Nanotechnol 2012, 7(11), 699–712. https://doi.org/10.1038/nnano.2012.193.

[128] Bissett MA, Worrall SD, Kinloch IA, Dryfe RAW. Comparison of two-dimensional transition metal dichalcogenides for electrochemical supercapacitors. Electrochim Acta 2016, 201, 30–7. https://doi.org/10.1016/j.electacta.2016.03.190.

[129] Qian X, Shen S, Liu T, Cheng L, Liu Z. Two-dimensional TiS_2 nanosheets for in vivo photoacoustic imaging and photothermal cancer therapy. Nanoscale 2015, 7(14), 6380–7. https://doi.org/10.1039/C5NR00893J.

[130] Zhu Z, Zou Y, Hu W, Li Y, Gu Y, Cao B, Guo N, et al. Near-infrared plasmonic 2D semimetals for applications in communication and biology. Adv Funct Mater 2016, 26(11), 1793–802. https://doi.org/10.1002/adfm.201504884.

[131] Rowe DM. General principles and basic considerations. In: Rowe DM, ed. Thermoelectrics Handbook: Macro to Nano. CRC Press, 2006. 1.1–1.13.

[132] Imai H, Shimakawa Y, Kubo Y. Large thermoelectric power factor in TiS_2 crystal with nearly stoichiometric composition. Phys Rev B. 2001;64:241104. https://doi.org/10.1103/PhysRevB.64.241104.

[133] Guélou G, Vaqueiro P, Prado-Gonjal J, Barbier T, Hebert S, Guilmeau E, Kockelmann W, Powell AV. The impact of charge transfer and structural disorder on the thermoelectric properties of cobalt intercalated TiS_2. J Mater Chem C 2016, 4(9), 1871–80. https://doi.org/10.1039/c5tc04217h.

[134] Wan C, Gu X, Dang F, Itoh T, Wang Y, Sasaki H, Kondo M, et al. Flexible N-type thermoelectric materials by organic intercalation of layered transition metal dichalcogenide TiS_2. Nat Mater 2015, 14(6), 622–7. https://doi.org/10.1038/nmat4251.

[135] Inoue M, Hughes HP, Yoffe AD. The electronic and magnetic properties of the 3d transition metal intercalates of titanium sulfide (TiS_2). Adv Phys 1989, 38, 565–604. https://doi.org/10.1080/00018738900101142.

[136] Kosuge K. Chemistry of Non-Stoichiometric Compounds. Oxford Science Publications. Oxford University Press, 1994.

[137] Bourgès C, Barbier T, Guélou G, Vaqueiro P, Powell AV, Lebedev OI, Barrier N, Kinemuchi Y, Guilmeau E. Thermoelectric properties of TiS_2 mechanically alloyed compounds. J Eur Ceram Soc 2016, 36(5), 1183–9. https://doi.org/10.1016/j.jeurceramsoc.2015.11.025.
[138] Ma J, Jin H, Liu X, Fleet ME, Li J, Cao X, Feng S. Selective synthesis and formation mechanism of TiS_2 dendritic crystals. Cryst Growth Des 2008, 8(12), 4460–4. https://doi.org/10.1021/cg800348y.
[139] Reshak AH, Kityk IV, Auluck S. Electronic structure and optical properties of 1T-TiS_2 and lithium intercalated 1T-TiS_2 for lithium batteries. J Chem Phys 2008, 129(7), 074706/1. https://doi.org/10.1063/1.2969076.
[140] Fang CM, de Groot RA, Haas C. Bulk and surface electronic structure of 1T-TiS_2 and 1T-$TiSe_2$. Phys Rev B 1997, 56, 4455–63. https://doi.org/10.1103/PhysRevB.56.
[141] Kukkonen CA, Kaiser WJ, Logothetis EM, Blumenstock BJ, Schroeder PA, Faile SP, Colella R, Gambold J. Transport and optical properties of titanium disulfide ($Ti_{1+x}S_2$). Phys Rev B 1981, 24, 1691–709. https://doi.org/10.1103/PhysRevB.24.1691.
[142] Thompson AH, Gamble FR, Symon CR. The verification of the existence of TiS_2. Mater Res Bull 1975, 10(9), 915–9. https://doi.org/10.1016/0025-5408(75)90071-9.
[143] Klipstein PC, Bagnall AG, Liang WY, Marseglia EA, Friend RH. Stoichiometry dependence of the transport properties of titanium disulfide. J Phys C, Solid State Phys 1981, 14, 4067–81. https://doi.org/10.1088/0022-3719/14/28/009.
[144] Wu ZY, Lemoigno F, Gressier P, Ouvrard G, Moreau P, Rouxel J, Natoli CR. Experimental and theoretical studies of the electronic structure of TiS_2. Phys Rev B 1996, 54(16), R11009–13. https://doi.org/10.1103/PhysRevB.54.R11009.
[145] Beaumale M, Barbier T, Breard Y, Guélou G, Powell AV, Vaqueiro P, Guilmeau E. Electron doping and phonon scattering in $Ti_{1+x}S_2$ thermoelectric compounds. Acta Mater 2014, 78, 86–92. https://doi.org/10.1016/j.actamat.2014.06.032.
[146] Daou R, Takahashi H, Hebert S, Beaumale M, Guilmeau E, Maignan A. Intrinsic effects of substitution and intercalation on thermal transport in two-dimensional TiS_2 single crystals. J Appl Phys 2015, 117(16), 165101/1. https://doi.org/10.1063/1.4919078.
[147] Wan C, Wang Y, Wang N, Koumoto K. Low-thermal-conductivity $(MS)_{1+x}(TiS_2)_2$ (M = Pb, Bi, Sn) misfit layer compounds for bulk thermoelectric materials. Materials 2010, 3, 2606–17. https://doi.org/10.3390/ma3042606.
[148] Wan C, Wang Y, Wang N, Norimatsu W, Kusunoki M, Koumoto K. Development of novel thermoelectric materials by reduction of lattice thermal conductivity. Sci Technol Adv Mater 2010, 11(4), 44306. https://doi.org/10.1088/1468-6996/11/4/044306.
[149] Wiegers GA, Meerschaut A. Structures of misfit layer compounds $(MS)_nTS_2$ (M = Sn, Pb, Bi, rare earth metals; T = Nb, Ta, Ti, V, Cr; $1.08 < n < 1.23$). J Alloys Compd 1992, 178(1), 351–68. https://doi.org/10.1016/0925-8388(92)90276-F.
[150] Meerschaut A, Auriel C, Rouxel J. Structure determination of a new misfit layer compound $(PbS)_{1.18}(TiS_2)_2$. J Alloys Compd 1992, 183, 129–37. https://doi.org/10.1016/0925-8388(92)90737-T.
[151] Yin C, Hu Q, Wang G, Huang T, Zhou X, Zhang X, Dou Y, et al. Intriguing substitution of conducting layer triggered enhancement of thermoelectric performance in misfit-layered $(SnS)_{1.2}(TiS_2)_2$. Appl Phys Lett 2017, 110(4), 43507. https://doi.org/10.1063/1.4975228.
[152] Barry JJ, Hughes HP, Klipstein PC, Friend RH. Stoichiometry effects in angle-resolved photoemission and transport studies of $Ti_{1+x}S_2$. J Phys C, Solid State Phys 1983, 16(2), 393. https://doi.org/10.1088/0022-3719/16/2/022.
[153] Tian R, Wan C, Wang Y, Wei Q, Ishida T, Yamamoto A, Tsuruta A, Shin W, Li S, Koumoto K. A solution-processed TiS_2/organic hybrid superlattice film towards flexible thermoelectric devices. J Mater Chem A 2017, 5(2), 564–70. https://doi.org/10.1039/c6ta08838d.

[154] Zaitsev VK, Fedorov MI. Thermal conductivity of semiconductors with complex crystal structures. In: Rowe DM, ed. Thermoelectrics Handbook Macro to Nano. CRC Press, 2006.

[155] Inyushkin AV, Taldenkov AN, Ozhogin VI, Itoh KM, Haller EE. Isotope effect on the phonon-drag component of the thermoelectric power of Germanium. Phys Rev B 2003, 68(15), 153203. https://doi.org/10.1103/PhysRevB.68.153203.

[156] Herring C. Theory of the thermoelectric power of semiconductors. Phys Rev 1954, 96(5), 1163–87. https://doi.org/10.1103/PhysRev.96.1163.

[157] Goldsmid HJ. In: Introduction to Thermoelectricity. Springer Series in Materials Science, vol. 121. Berlin, Heidelberg: Springer Berlin Heidelberg 2010.

[158] Paschen S. Thermoelectric aspects of strongly correlated electron systems. In: Rowe DM, ed. Thermoelectrics Handbook Macro to Nano. CRC Press, 2006.

[159] Buschinger B, Geibel C, Steglich F, Mandrus D, Young D, Sarrao JL, Fisk Z. Transport properties of FeSi. Physica B, Condens Matter 1997, 230, 784–6. https://doi.org/10.1016/S0921-4526(96)00839-3.

[160] Jones CDW, Regan KA, DiSalvo FJ. Thermoelectric properties of the doped Kondo insulator: $Nd_xCe_{3-x}Pt_3Sb_4$. Phys Rev B 1998, 58(24), 16057–63. https://doi.org/10.1103/PhysRevB.58.16057.

[161] Hundley MF, Canfield PC, Thompson JD, Fisk Z. Substitutional effects on the electronic transport of the Kondo semiconductor $Ce_3Bi_4Pt_3$. Phys Rev B 1994, 50(24), 18142–6. https://doi.org/10.1103/PhysRevB.50.18142.

[162] Gambino RJ, Grobman WD, Toxen AM. Anomalously large thermoelectric cooling figure of merit in the Kondo systems $CePd_3$ and $CeIn_3$. Appl Phys Lett 1973, 22(10), 506–7. https://doi.org/10.1063/1.1654486.

[163] Bentien A, Johnsen S, Madsen GKH, Iversen BB, Steglich F. Colossal Seebeck coefficient in strongly correlated semiconductor $FeSb_2$. Europhys Lett 2007, 80(1), 17008. https://doi.org/10.1209/0295-5075/80/17008.

[164] Chumak IV, Pavlyuk VV, Dmytriv GS, Stpień-Damm J. Phase equilibria and crystal structure of compounds in the Fe–Zn–Sb System at 570 K. J Alloys Compd 2000, 307(1–2), 223–5. https://doi.org/10.1016/S0925-8388(00)00891-4.

[165] Petrovic C, Kim JW, Bud'ko SL, Goldman AI, Canfield PC, Choe W, Miller GJ. Anisotropy and large magnetoresistance in the narrow-gap semiconductor $FeSb_2$. Phys Rev B 2003, 67(15), 155205. https://doi.org/10.1103/PhysRevB.67.155205.

[166] Richter KW, Ipser H. Reinvestigation of the binary Fe-Sb phase diagram. J Alloys Compd 1997, 247(1), 247–9. https://doi.org/10.1016/S0925-8388(96)02597-2.

[167] Takahashi H, Yasui Y, Terasaki I, Sato M. Effects of Ppm-level imperfection on the transport properties of $FeSb_2$ single crystals. J Phys Soc Jpn 2011, 80(5), 054708. https://doi.org/10.1143/JPSJ.80.054708.

[168] Binnewies M, Glaum R, Schmidt M, Schmidt P. Chemical vapor transport reactions – a historical review. Z Anorg Allg Chem 2013, 639(2), 219–29. https://doi.org/10.1002/zaac.201300048.

[169] Binnewies M, Glaum R, Schmidt M, Schmidt P. Chemical Vapor Transport Reactions. Germany, De Gruyter, 2016. https://doi.org/10.1515/9783110254655.

[170] Rosenthal G, Kershaw R, Wold A. Single crystal growth of $FeAs_2$ and $FeSb_2$. Mater Res Bull 1972, 7(5), 479–87. https://doi.org/10.1016/0025-5408(72)90150-X.

[171] Sun P, Søndergaard M, Sun Y, Johnsen S, Iversen BB, Steglich F. Unchanged thermopower enhancement at the semiconductor-metal transition in correlated $FeSb_{2-x}Te_x$. Appl Phys Lett 2011, 98(7), 072105. https://doi.org/10.1063/1.3556645.

[172] Sun P, Oeschler N, Johnsen S, Iversen BB, Steglich F. Narrow band gap and enhanced thermoelectricity in $FeSb_2$. Dalton Trans 2010, 39(4), 1012–9. https://doi.org/10.1039/B918909B.

[173] Sun P, Oeschler N, Johnsen S, Iversen BB, Steglich F. Thermoelectric properties of the narrow-gap semiconductors $FeSb_2$ and $RuSb_2$: a comparative study. J Phys Conf Ser 2009, 150(1), 012049. https://doi.org/10.1088/1742-6596/150/1/012049.

[174] Takahashi H, Okazaki R, Ishiwata S, Taniguchi H, Okutani A, Hagiwara M, Terasaki I. Colossal Seebeck effect enhanced by quasi-ballistic phonons dragging massive electrons in $FeSb_2$. Nat Commun 2016, 7, 12732. https://doi.org/10.1038/ncomms12732.

[175] Takahashi H, Okazaki R, Terasaki I, Yasui Y. Origin of the energy gap in the narrow-gap semiconductor $FeSb_2$ revealed by high-pressure magnetotransport measurements. Phys Rev B 2013, 88(16), 165205. https://doi.org/10.1103/PhysRevB.88.165205.

[176] Sun P, Oeschler N, Johnsen S, Iversen BB, Steglich F. $FeSb_2$: prototype of huge electron-diffusion thermoelectricity. Phys Rev B 2009, 79(15), 153308. https://doi.org/10.1103/PhysRevB.79.153308.

[177] Bentien A, Madsen GKH, Johnsen S, Iversen BB. Experimental and theoretical investigations of strongly correlated $FeSb_{2-x}Sn_x$. Phys Rev B 2006, 74(20), 205105. https://doi.org/10.1103/PhysRevB.74.205105.

[178] Satya AT, Mani A, Janaki J, Kumary TG, Bharathi A. Investigation of transport properties in $Fe_{1-x}Ni_xSb_2$ system. AIP Conf Proc 2015, 1665(1), 110021. https://doi.org/10.1063/1.4918077.

[179] Diakhate MS, Hermann RP, Möchel A, Sergueev I, Søndergaard M, Christensen M, Verstraete MJ. Thermodynamic, thermoelectric, and magnetic properties of $FeSb_2$: a combined first-principles and experimental study. Phys Rev B 2011, 84(12), 125210. https://doi.org/10.1103/PhysRevB.84.125210.

[180] Kieslich G, Birkel CS, Veremchuk I, Grin Y, Tremel W. Thermoelectric properties of spark-plasma sintered nanoparticular $FeSb_2$ prepared via a solution chemistry approach. Dalton Trans 2013, 43(2), 558–62. https://doi.org/10.1039/C3DT51535D.

[181] Jie Q, Hu R, Bozin E, Llobet A, Zaliznyak I, Petrovic C, Li Q. Electronic thermoelectric power factor and metal-insulator transition in $FeSb_2$. Phys Rev B 2012, 86(11), 115121. https://doi.org/10.1103/PhysRevB.86.115121.

[182] Zhao H, Pokharel M, Zhu G, Chen S, Lukas K, Jie Q, Opeil C, Chen G, Ren Z. Dramatic thermal conductivity reduction by nanostructures for large increase in thermoelectric figure-of-merit of $FeSb_2$. Appl Phys Lett 2011, 99(16), 163101. https://doi.org/10.1063/1.3651757.

[183] Wang K, Hu R, Warren J, Petrovic C. Enhancement of the thermoelectric properties in doped $FeSb_2$ bulk crystals. J Appl Phys 2012, 112(1), 013703. https://doi.org/10.1063/1.4731251.

[184] Zhao H, Pokharel M, Chen S, Liao B, Lukas K, Opeil C, Chen G, Ren Z. Figure-of-merit enhancement in nanostructured $FeSb_{2-x}Ag_x$ with $Ag_{1-y}Sb_y$ nanoinclusions. Nanotechnology 2012, 23(50), 505402. https://doi.org/10.1088/0957-4484/23/50/505402.

[185] Koirala M, Zhao H, Pokharel M, Chen S, Dahal T, Opeil C, Chen G, Ren Z. Thermoelectric property enhancement by Cu nanoparticles in nanostructured $FeSb_2$. Appl Phys Lett 2013, 102(21), 213111. https://doi.org/10.1063/1.4808094.

[186] Wang Y, Fu C, Zhu T, Hu L, Jiang G, Zhao G, Huo D, Zhao X. Hot deformation induced defects and performance enhancement in $FeSb_2$ thermoelectric materials. J Appl Phys 2013, 114(18), 184904. https://doi.org/10.1063/1.4831668.

[187] Pokharel M, Zhao HZ, Koirala M, Ren ZF, Opeil C. Enhanced thermoelectric performance of Te-doped $FeSb_2$ nanocomposite. J Low Temp Phys 2014, 176(1–2), 122–30. https://doi.org/10.1007/s10909-014-1148-y.

[188] Sun P, Oeschler N, Johnsen S, Iversen BB, Steglich F. Huge thermoelectric power factor: $FeSb_2$ versus $FeAs_2$ and $RuSb_2$. Appl Phys Express 2009, 2(9), 091102. https://doi.org/10.1143/APEX.2.091102.

[189] Petrovic C, Lee Y, Vogt T, Lazarov NDj, Bud'ko SL, Canfield PC. Kondo insulator description of spin state transition in $FeSb_2$. Phys Rev B 2005, 72(4), 045103. https://doi.org/10.1103/PhysRevB.72.045103.

[190] Herzog A, Marutzky M, Sichelschmidt J, Steglich F, Kimura S, Johnsen S, Iversen BB. Strong electron correlations in $FeSb_2$: an optical investigation and comparison with $RuSb_2$. Phys Rev B 2010, 82(24), 245205. https://doi.org/10.1103/PhysRevB.82.245205.

[191] Pokharel M, Zhao H, Lukas K, Ren Z, Opeil C, Mihaila B. Phonon drag effect in nanocomposite $FeSb_2$. MRS Commun 2013, 3(1), 31–6. https://doi.org/10.1557/mrc.2013.7.

[192] Tomczak JM, Haule K, Miyake T, Georges A, Kotliar G. Thermopower of correlated semiconductors: application to $FeAs_2$ and $FeSb_2$. Phys Rev B 2010, 82(8), 085104. https://doi.org/10.1103/PhysRevB.82.085104.

[193] Battiato M, Tomczak JM, Zhong Z, Held K. Unified picture for the colossal thermopower compound $FeSb_2$. Phys Rev Lett 2015, 114(23), 236603. https://doi.org/10.1103/PhysRevLett.114.236603.

[194] DiTusa JF, Friemelt K, Bucher E, Aeppli G, Ramirez AP. Heavy fermion metal-Kondo insulator transition in $FeSi_{1-x}Al_x$. Phys Rev B 1998, 58(16), 10288–301. https://doi.org/10.1103/PhysRevB.58.10288.

[195] Hu R, Mitrović VF, Petrovic C. Magnetism and metal-insulator transition in $Fe(Sb_{1-x}Te_x)_2$. Phys Rev B 2009, 79(6), 064510. https://doi.org/10.1103/PhysRevB.79.064510.

[196] Liao B, Lee S, Esfarjani K, Chen G. First-principles study of thermal transport in $FeSb_2$. Phys Rev B 2014, 89(3), 035108. https://doi.org/10.1103/PhysRevB.89.035108.

[197] Gippius AA, Baenitz M, Okhotnikov KS, Johnsen S, Iversen B, Shevelkov AV. Sb magnetic resonance as a local probe for the gap formation in the correlated semimetal $FeSb_2$. Appl Magn Reson 2014, 45(11), 1237–52. https://doi.org/10.1007/s00723-014-0592-9.

[198] Wei K, Martin J, Nolas GS. Synthesis, SPS processing and low temperature transport properties of polycrystalline $FeSb_2$ with nano-scale grains. Mater Lett 2014, 122(May), 289–91. https://doi.org/10.1016/j.matlet.2014.02.054.

[199] Datta A, Nolas GS. Synthesis and characterization of nanocrystalline $FeSb_2$ for thermoelectric applications. Eur J Inorg Chem 2012, 2012(1), 55–8. https://doi.org/10.1002/ejic.201100864.

[200] Toberer ES, May AF, Snyder GJ. Zintl chemistry for designing high efficiency thermoelectric materials. Chem Mater 2010, 22, 624–34.

[201] Dolyniuk J-A, Owens-Baird B, Wang J, Zaikina JV, Kovnir K. Clathrate thermoelectrics. Mater Sci Eng, R Rep 2016, 108, 1–46. https://doi.org/10.1016/j.mser.2016.08.001.

[202] Humphry D. On a combination of oxymuriatic gas and oxygene gas. Philos Trans R Soc Lond 1811, 101, 155–62. https://doi.org/10.1098/rstl.1811.0008.

[203] Atkins P. Shriver and Atkins' Inorganic Chemistry. USA, Oxford University Press, 2010.

[204] Slack GA. New materials and performance limits for thermoelectric cooling. In: CRC Handbook of Thermoelectrics. Boca Raton, FL, CRC Press, 1995. 407–40.

[205] Shi X, Chen L, Uher C. Recent advances in high-performance bulk thermoelectric materials. Int Mater Rev 2016, 61(6), 379–415.

[206] Lory P-F, Pailhès S, Giordano VM, Euchner H, Nguyen HD, Ramlau R, Borrmann H, Schmidt M, Baitinger M, Ikeda M, Tomeš P, Mihalkovič M, Allio C, Johnson MR, Schober H, Sidis Y, Bourdarot F, Regnault LP, Ollivier J, Paschen S, Grin Y, de Boissieu M. Direct measurement of individual phonon lifetimes in the clathrate compound Ba7.81Ge40.67Au5.33. Nat Commun 2017, 8, 491. https://doi.org/10.1038/s41467-017-00584-7.

[207] Bhattacharya A, Carbogno C, Böhme B, Baitinger M, Grin Y, Scheffler M. Formation of vacancies in Si- and Ge-based clathrates: role of electron localization and symmetry breaking. Phys Rev Lett 2017, 118, 236401. https://doi.org/10.1103/PhysRevLett.118.236401.

[208] Kleinke H. New bulk materials for thermoelectric power generation: clathrates and complex antimonides. Chem Mater 2010, 22(3), 604–11. https://doi.org/10.1021/cm901591d.

[209] Du B, Saiga Y, Kajisa K, Takabatake T, Nishibori E, Sawa H. Study of $\alpha \leftrightarrow \beta$ transformation in the dimorphic clathrate $Ba_8Ga_{16}Sn_{30}$. Philos Mag 2012, 92, 2541–52. https://doi.org/10.1080/14786435.2012.681710.

[210] Saiga Y, Du B, Deng SK, Kajisa K, Takabatake T. Thermoelectric properties of type-VIII clathrate $Ba_8Ga_{16}Sn_{30}$ doped with Cu. J Alloys Compd 2012, 537, 303–7. https://doi.org/10.1016/j.jallcom.2012.05.049.

[211] Kono Y, Ohya N, Taguchi T, Suekuni K, Takabatake T, Yamamoto S, Akai K. First-principles study of type-I and type-VIII $Ba_8Ga_{16}Sn_{30}$ clathrates. J Appl Phys 2010, 107(12), 123720. https://doi.org/10.1063/1.3437252.

[212] Li Y, Gao J, Chen N, Liu Y, Luo ZP, Zhang RH, Ma XQ, Cao GH. Electronic structure and physical properties of $Ba_8Ga_{16}Sn_{30}$ clathrates with type-I and type-VIII structure. Physica B, Condens Matter 2008, 403(5), 1140–1. https://doi.org/10.1016/j.physb.2007.10.140.

[213] Norouzzadeh P, Myles CW, Vashaee D. Electronic, elastic, vibrational, and thermodynamic properties of type-VIII clathrates $Ba_8Ga_{16}Sn_{30}$ and $Ba_8Al_{16}Sn_{30}$ by first principles. J Appl Phys 2013, 114(16), 163509. https://doi.org/10.1063/1.4826213.

[214] Li DC, Fang L, Deng SK, Kang KY, Shen LX, Wei WH, Ruan HB. Structural and electronic properties of type-I and type-VIII $Ba_8Ga_{16}Sn_{30}$ clathrates under compression. Physica B, Condens Matter 2012, 407(8), 1238–43. https://doi.org/10.1016/j.physb.2012.01.109.

[215] Chasmar RP, Stratton R. The thermoelectric figure of merit and its relation to thermoelectric generators. Int J Electron 1959, 7(1), 52–72. https://doi.org/10.1080/00207215908937186.

[216] Mahan GD. Figure of merit for thermoelectrics. J Appl Phys 1989, 65(4), 1578–83. https://doi.org/10.1063/1.342976.

[217] Goldsmid HJ, Sharp JW. Estimation of the thermal band gap of a semiconductor from Seebeck measurements. J Electron Mater 1999, 28, 869–72. https://doi.org/10.1007/s11664-999-0211-y.

[218] Saiga Y, Suekuni K, Deng SK, Yamamoto T, Kono Y, Ohya N, Takabatake T. Optimization of thermoelectric properties of type-VIII clathrate $Ba_8Ga_{16}Sn_{30}$ by carrier tuning. J Alloys Compd 2010, 507(1), 1–5. https://doi.org/10.1016/j.jallcom.2010.07.121.

[219] Meng D, Shen L, Shai X, Dong G, Deng S. Growth and thermoelectric properties of Ge doped n-type Sn-based type-VIII single crystalline clathrate. Acta Phys Sin 2013, 62(24), 247401/1–247421-6. https://doi.org/10.7498/aps.62.247401.

[220] Chen Y, Du B, Kajisa K, Takabatake T. Effects of In substitution for Ga on the thermoelectric properties of type-VIII clathrate $Ba_8Ga_{16}Sn_{30}$ single crystals. J Electron Mater 2014, 43(6), 1916. https://doi.org/10.1007/s11664-013-2909-0.

[221] Deng S, Saiga Y, Suekuni K, Takabatake T. Effect of Al substitution on the thermoelectric properties of the type VIII clathrate $Ba_8Ga_{16}Sn_{30}$. J Electron Mater 2011, 40(5), 1124–8. https://doi.org/10.1007/s11664-011-1555-7.

[222] Deng S, Saiga Y, Kajisa K, Takabatake T. High thermoelectric performance of Cu substituted type-VIII clathrate $Ba_8Ga_{16-x}Cu_xSn_{30}$ single crystals. J Appl Phys 2011, 109(10), 103704. https://doi.org/10.1063/1.3583570.

[223] Du B, Saiga Y, Kajisa K, Takabatake T. Thermoelectric properties of p-type clathrate $Ba_{8.0}Ga_{15.9}Zn_ySn_{30.1}$ single crystals with various carrier concentrations. Chem Mater 2015, 27(5), 1830–6. https://doi.org/10.1021/acs.chemmater.5b00025.

[224] Huo D, Sakata T, Sasakawa T, Avila MA, Tsubota M, Iga F, Fukuoka H, Yamanaka S, Aoyagi S, Takabatake T. Structural, transport, and thermal properties of the single-crystalline type-VIII clathrate $Ba_8Ga_{16}Sn_{30}$. Phys Rev B 2005, 71(7), 075113. https://doi.org/10.1103/PhysRevB.71.075113.

[225] Deng S, Li D, Shen L, Hao R, Takabatake T. Sn-based type-VIII single-crystal clathrates with a large figure of merit. Chin Phys B 2012, 21(1), 017401. https://doi.org/10.1088/1674-1056/21/1/017401.

[226] Deng S-P, Cheng F, Li D-C, Tang Y, Chen Z, Shen L-X, Liu H-X, Yang P-Z, Deng S-K. Electrical transport properties of type-VIII Sn-based single-crystalline clathrates $(Eu/Ba)_8Ga_{16}Sn_{30}$ prepared by Ga flux method. Chin Phys Lett 2017, 34(4), 047401. https://doi.org/10.1088/0256-307X/34/4/047401.

[227] Cheng F, Shen L, Li D, Liu H, Wang J, Deng S. Preparation and electrical transport properties of VIII-type Sn-based single-crystal clathrates $(Eu/Ba)_8Ga_{16}Sn_{30}$. Mater Technol 2017, 32(2), 105–8. https://doi.org/10.1080/10667857.2015.1132032.

[228] Meng D-Y, Shen L-X, Li D-C, Shai X-X, Deng S-K. Structural and electrical transport properties of Mg-doped n-type Sn-based type VIII single crystalline clathrate. Acta Phys Sin 2014, 63(14), 177401-1–177401-6. https://doi.org/10.7498/aps.63.177401.

[229] Kono Y, Ohya N, Saiga Y, Suekuni K, Takabatake T, Akai K, Yamamoto S. Carrier doping in the type VIII clathrate $Ba_8Ga_{16}Sn_{30}$ through Sb substitution. J Electron Mater 2011, 40(5), 845–50. https://doi.org/10.1007/s11664-011-1590-4.

[230] Kuznetsov VL, Kuznetsova LA, Kaliazin AE, Rowe DM. Preparation and thermoelectric properties of $A_8{}^{II}B_{16}{}^{III}B_{30}{}^{IV}$ clathrate compounds. J Appl Phys 2000, 87(11), 7871–5. https://doi.org/10.1063/1.373469.

Janez Dolinšek

Electronic transport properties of complex intermetallics

1 Introduction

Complex metallic alloys (CMAs) are a class of crystalline intermetallic phases possessing the following features [1]: (1) their crystal structures are based on giant unit cells comprising several tens up to many thousands of atoms; (2) inside the unit cells the atoms are arranged in clusters, and a multitude of different coordination polyhedra lead to a high number of different atomic environments, where icosahedrally coordinated environments play a prominent role; (3) a particular property of the structure is inherent disorder, where various types of disorder are present. Configurational disorder results from statistically varying orientations of a particular subcluster inside a given cage of atoms. This type of disorder can also result from incompatibility problems in cluster packing when these lead to local atomic displacements. Chemical or substitutional disorder results from fractional occupancy of certain lattice sites by various elements. Partial site occupation (occupancy smaller than one) results from geometrical constraints and so does split occupation where two neighboring sites are alternatively occupied since they cannot be occupied simultaneously because they are too close in space. Quasicrystals (QCs), representing long-range ordered structures without translational periodicity that obey crystallographically forbidden symmetries of a 5-, 8-, 10-, and 12-fold rotation axis, can be considered as a limiting case of a CMA with an infinitely large unit cell.

Due to the high degree of inherent quenched disorder, the thermodynamic stability of CMAs and QCs remains an open question [2]. A crystal structure in thermal equilibrium should have minimum free energy $F = U - TS$, where U is the internal energy and S is the entropy. At high temperatures, the disorder in the lattice (static and dynamic) can maximize the entropy and provide entropic stabilization of the structure, so that the existence of thermodynamically stable high-temperature CMA and QC phases is unambiguous. At lower temperatures, the entropy term TS decreases, and the system should minimize the internal energy U to stabilize the structure. However, the large amount of chemical and structural disorder leads to finite configurational entropy, so that phase stability at absolute zero in relation to the third law of thermodynamics becomes a subtle question. The meaningful questions here are: (1) can

Acknowledgement: Scientific contributions of Michael Feuerbacher, Yoshihiko Yokoyama (synthesis of the samples), Stanislav Vrtnik, Simon Jazbec, Ana Smontara (experimental measurements), Uichiro Mizutani and Carlos V. Landauro (theoretical support) to this work are gratefully acknowledged. The author acknowledges financial support from the Slovenian Research Agency (research core funding No. P1-0125).

https://doi.org/10.1515/9783110496789-011

CMAs and QCs exist as stable phases at absolute zero or we are observing a quenched metastable high-temperature state whose ergodicity is broken on the experimental time scale and (2) what is the stabilization mechanism?

The stability of a solid at absolute zero is related to its cohesive energy [2], defined as the energy required to separate constituent atoms from each other and bring them to an assembly of neutral free atoms. In a solid with metallic bonding, the negatively charged electron gas is attracted electrostatically by the periodically arranged positive charges of the lattice ions, whereas the electrons are coupled among themselves by a repulsive interaction, both effects summing up into the electrostatic energy of the crystal, $\varepsilon_0 < 0$. The kinetic energy of the electrons is positive, $\varepsilon_{\text{kin}} > 0$, and due to Pauli exclusion principle, only two electrons with opposite spins can occupy the same state. As a result, a Fermi sphere is formed in the reciprocal space, and the kinetic energy is inevitably increased, acting against the electrostatic stabilization of the structure. The system is stabilized when the total energy $\varepsilon_{\text{tot}} = \varepsilon_0 + \varepsilon_{\text{kin}}$ is smaller than the energy of the assembly of free atoms $\sum \varepsilon_i$ (where ε_i is the ground-state energy of the i-th isolated atom) and the difference between the two energies is the cohesive energy. The kinetic energy of the electrons thus plays a critical role in the cohesive energy of a solid, as it acts against the stabilization of the structure.

To achieve stability, nature often devises a mechanism to lower the kinetic energy of the electron system as much as possible, by forming a pseudogap or a true gap in the electronic density of states (DOS) across the Fermi level ε_F [2]. The depleted electrons from the Fermi level region (where their kinetic energy is the highest) are redistributed into the DOS at lower energies, and the resulting reduction in ε_{kin} increases the cohesive energy and hence the stability of the structure. In CMAs and QCs, a pseudogap close to ε_F was frequently found both experimentally and theoretically, with its depth being generally about 20–80 % of the free-electron DOS and a width in the range $\Delta\varepsilon = 1\text{–}2\,\text{eV}$. There are two known mechanisms for the formation of a pseudogap in metallic systems [2]: (1) orbital hybridizations and (2) Fermi surface–Brillouin zone (FsBz) interactions. The orbital hybridization effect is important in systems with some degree of covalent bonding, predominantly in alloys that contain transition metal elements. When transition metal atoms are close in space to other atoms, their d states overlap with p (or sp) states of non–transition metals, resulting in orbital hybridization and the formation of split bonding and antibonding levels. In a crystal, these levels broaden into bonding and antibonding subbands, leaving a pseudogap in the DOS in between. In the case that the pseudogap forms at ε_F, it contributes to the stabilization of the structure. The FsBz-interactions-induced pseudogap is encountered in systems with predominantly metallic bonding. For a crystal with N atoms in the unit cell (assumed for simplicity to have cubic symmetry) and the electron concentration e/a (denoting the number of valence electrons per atom), a spherical Fermi surface of diameter $2k_F$ is formed, defined by the relation $N(e/a) = 8\pi k_F^3/3$, where the Fermi wave vector k_F is given in units of $2\pi/a_0$ and a_0 is the lattice constant. For a specific electron concentration e/a, the diameter of the Fermi sphere $2k_F$ is such that the Fermi surface touches the Brillouin zone planes. Since a band gap in the energy dispersion relation $\varepsilon(\vec{k})$ opens at the Brillouin zone boundary, this produces van Hove singularities in the DOS just below the Fermi level (since $\partial\varepsilon/\partial\vec{k}$ is small there, many electronic states have similar energies) and a depletion of the DOS (a pseudogap) at ε_F due to the fact that no electronic states are available within the band gap. The condition for the Fermi surface to touch the Brillouin-zone plane can be expressed by the Hume–Rothery interference condition [2],

$$2k_F = |\vec{G}| \tag{1}$$

(or $(2k_F)^2 = |\vec{G}|^2$), where $\vec{G}$ is the critical reciprocal lattice vector, which can be expressed as $|\vec{G}|^2 = h^2 + k^2 + l^2$ (in units of $(2\pi/a_0)^2$), where $\{hkl\}$ are Miller indices of the lattice planes that diffract the electrons to produce standing waves and introduce a band gap into the $\varepsilon(\vec{k})$ relation. The creation of a pseudogap in the DOS at ε_F by the FsBz interaction leads to e/a–dependent phase stabilization, which is alternatively called the Hume–Rothery stabilization mechanism, and plays a crucial role in interpreting the Hume–Rothery electron-concentration rule [2]. In CMAs, the large number of atoms in the giant unit cell increase the number of Brillouin-zone planes. This, in turn, increases the number of directions in the reciprocal space along which the interference condition eq. (1) is fulfilled. The larger the number of Brillouin-zone planes interacting with electrons at ε_F, the deeper is the pseudogap in the DOS formed at ε_F. This leads to an efficient lowering of the electronic energy in favor of stabilization of a given structure. An FsBz-induced pseudogap is considered to play the key role in the e/a-dependent stabilization of structures, in which the diffraction spectrum consists of a series of discrete Bragg peaks, thus being applicable to both CMAs and QCs. It has been shown that the origin of a pseudogap at ε_F can be interpreted in terms of FsBz interactions or the interference condition even for strongly orbital hybridizing systems, such as Mackay–icosahedron (MI)–type Al–Mn quasicrystalline approximants [3–5].

The presence of a pseudogap in the DOS at ε_F can have a profound effect on the electronic-transport properties of CMAs and QCs. A pseudogap that changes significantly with energy around ε_F on the scale of several k_BT can introduce anomalous, non–metallic-type electronic-transport coefficients in a metallic system. In this sense, the electronic-transport properties of a CMA phase are intimately related to its electronic-stabilization mechanism. Here we present the electronic-transport properties of two prototype large-unit-cell CMAs, the structurally ordered γ–$Mg_{17}Al_{12}$ and the inherently disordered β–Mg_2Al_3, and the quasiperiodic i–$Al_{64}Cu_{23}Fe_{13}$ icosahedral quasicrystal (i–QC). We show how the energy-dependent electronic DOS $g(\varepsilon)$ function in the Fermi-level region can be reconstructed from the temperature-dependent electrical resistivity and the thermoelectric power. We also discuss the relationship between the experimentally determined pseudogap in the DOS at ε_F and the Hume–Rothery electronic-stabilization mechanism. This book chapter is based on the previous experimental investigations of the Mg–Al CMA phases [6] and the *i*–Al–Cu–Fe QCs [7].

2 Samples' description

The investigated CMAs were two binary phases from the Al–Mg system, the γ–$Mg_{17}Al_{12}$ and the β–Mg_2Al_3. Since these alloys are free from transition–metal elements, orbital-hybridization effects must be small, and the pseudogap in the DOS is expected to originate from the FsBz interactions. The alloys should thus be stabilized by the Hume–Rothery stabilization mechanism. The γ–$Mg_{17}Al_{12}$ phase belongs to the class of gamma–brasses [2] and contains 58 atoms in the cubic unit cell with space group $I\bar{4}3m$ and lattice constant $a = 1.05438$ nm. The range of the γ phase in the Al–Mg equilibrium phase diagram is strongly temperature–dependent [8]. At room temperature (RT) and below, it is almost a line compound, whereas at about 450 °C its stability range extends from about 47 to 60 at.% Mg. According to the structural model of the γ phase [9], the $Mg_{17}Al_{12}$ composition is free from structural and chemical disorder, so that it represents an ordered CMA phase. The β–Mg_2Al_3 phase extends over the range from 37.5 to 39.5 at.% Mg [8]. Its structure was first studied by Samson in 1965 [10] and later refined for a particular composition $Mg_{38.5}Al_{61.5}$ (equivalent to $Mg_2Al_{3.2}$) by Feuerbacher et al. [11]. At 400 °C, the unit cell is cubic with space group $Fd\bar{3}m$ and a lattice constant $a = 2.8242$ nm, containing many partially occupied and split atomic positions leading to an average occupation of only 1,178 of the 1,832 sites in the unit cell. There are also some mixed Al/Mg positions in the unit cell, indicating chemical (substitutional) disorder [11]. The polytetrahedral atomic arrangement consists of 672 icosahedral clusters, 252 Friauf polyhedra and 244 other polyhedra, so that the β–Mg_2Al_3 phase can be considered as a prototype CMA possessing a giant unit cell, a multitude of atomic clusters and a large amount of inherent disorder, of both structural and chemical types. For the composition $Mg_{38.5}Al_{61.5}$, it was found that at 214 °C, the β–Mg_2Al_3 phase undergoes a first-order structural phase transition to a rhombohedral β'–Mg_2Al_3 phase with space group $R3m$ and lattice parameters $a = 1.9968$ nm and $c = 4.8911$ nm [11]. The structure of the β' phase is closely related to that of the β phase; its atomic sites can be derived from those of the β phase by group-theoretical considerations. The main difference between the two structures is that all atomic sites of the β'–phase unit cell are fully occupied (comprising 879 atoms in total) with no partially occupied and split positions, so that the β' phase is of higher structural order than the β phase. However, ten out of 80 sites show mixed Al/Mg occupancy, so that chemical disorder is still present in the β' phase.

The investigated γ–$Mg_{17}Al_{12}$ and β–Mg_2Al_3 samples were both oriented monocrystals, cut from larger crystals grown by the flux-growth technique. The β–phase crystal of composition $Mg_{38.5}Al_{61.5}$ (equivalent to $Mg_2Al_{3.2}$) and dimensions $2 \times 2 \times 7\ \mathrm{mm}^3$ belonged to the same batch of material as the one studied in [11], where the details of structural characterization can be found. The as-grown material was annealed at 300 °C and then slowly cooled to RT. At this composition, the crystal is below 214 °C in the low-temperature β' phase. Since all our experiments were performed in the temperature range of the β' phase, we shall refer to this sample in the following as the β'–Mg_2Al_3 sample (keeping in mind that its actual composition was off-stoichiometric $Mg_2Al_{3.2}$). The material was free of secondary phases. Its microstructure was studied by electron microscopy and x–ray topography and was shown to be defective [11]. The diffuse scattering did not decrease even at −173 °C, leading to the conclusion that the sample contained static disorder and perhaps internal stresses. The γ–phase crystal of dimensions $2.2 \times 1.8 \times 1.0\ mm^3$ was also free of secondary phases and its actual composition was $Mg_{0.55}Al_{0.45}$ (corresponding to $Mg_{17}Al_{13.9}$), but we shall refer to it in the following as the γ–$Mg_{17}Al_{12}$ sample. According to the structural model [9], the γ phase at the stoichiometric composition $Mg_{17}Al_{12}$ has all atomic sites in the unit cell fully occupied, and there is also no chemical disorder, so that it represents an ordered CMA phase. However, our γ–phase sample at the off–stoichiometric composition $Mg_{17}Al_{13.9}$ should contain chemical disorder. Powder x–ray spectra of the β' and γ samples from the same batch of materials as used in our study are compared in [12]. Both spectra consist of sharp diffraction lines of comparable width, but the lines of the γ sample are a bit sharper and show less broadening at the foot close to the baseline than the lines of the β' sample. This indicates that both samples show good crystallinity, but the degree of quenched disorder in the β' sample is somewhat higher than in the γ sample, though not drastically.

The QC sample was a single-grain icosahedral i–Al–Cu–Fe with composition $Al_{64}Cu_{23}Fe_{13}$, prepared by the Czochralski technique. The removal of strains by annealing has been described in detail in other publications [13, 14]. This composition was chosen because of its superior thermal stability (any secondary-phase precipitates in the as-grown material disappeared upon annealing [13]), so that it is considered to represent an ideal icosahedral composition. The crystal was pulled out of the molten alloy of composition $Al_{57.7}Cu_{37.7}Fe_{3.5}Si_{1.1}$ (where Si was added to restrain crystallization of $(Al + Cu)_{13}Fe_4$) at the speed of 0.05 μm/s at 1073 K, and the growth direction was parallel to the threefold symmetry axis. The investigated crystal of 2-mm diameter and 10-mm length showed a faceted cylindrical shape and its quasicrystallinity was evaluated by x-ray diffraction analysis [15]. The full widths at half maximum (FWHM) of the Bragg reflections along the 2-, 3- and 5-fold symmetry directions exhibited no $Q_{||}$ and $Q_{\perp}$ dependence, where $Q_{||}$ is the real scattering vector and $Q_{\perp}$ the phason momentum. Furthermore, peak shifts from ideal Bragg positions were not detected. This showed that neither random, nor linear, phason strain is present in the structure. The phason-free structure on the atomic scale was confirmed by HRTEM imaging [15], which showed highly ordered patterns with no kinks or dead ends in the atomic rows on the selected scale of $17 \times 23\ nm^2$. Therefore, the investigated single-grain $Al_{64}Cu_{23}Fe_{13}$ i–QC had an almost phason-free quasicrystalline structure and showed superior quasicrystallinity on both macro- and microscopic scales.

3 Determination of the DOS function from the thermoelectric power and the electrical conductivity

Within the Kubo–Greenwood formalism, the temperature-dependent thermoelectric power (the Seebeck coefficient $S(T)$) and the electrical conductivity $\sigma(T)$ (the inverse electrical resistivity $\sigma(T) = 1/\rho(T)$) are both derived from the energy-dependent spectral conductivity function $\sigma(\varepsilon)$, which is given by the Einstein relation

$$\sigma(\varepsilon) = (e^2/V)g(\varepsilon)D(\varepsilon). \tag{2}$$

Here $g(\varepsilon)$ is the electronic DOS, $D(\varepsilon)$ is the electronic spectral diffusivity, e is the charge of the carriers and V is the sample volume. Under the assumption that the energy dependence of the spectral diffusivity can be neglected in the vicinity of the Fermi level, $D(\varepsilon) \approx D(\varepsilon_F)$, the shape of the DOS functions $g(\varepsilon)$ is the same as the shape of $\sigma(\varepsilon)$. In our analysis, we shall adopt this approximation. The coefficients $S(T)$ and $\sigma(T)$ are given by

$$S(T) = \frac{1}{eT\sigma(T)} \int d\varepsilon\, \sigma(\varepsilon)(\varepsilon - \mu)\left(-\frac{\partial f}{\partial \varepsilon}\right), \tag{3}$$

and

$$\sigma(T) = \int d\varepsilon\, \sigma(\varepsilon)\left(-\frac{\partial f}{\partial \varepsilon}\right), \tag{4}$$

where $f = \{\exp[(\varepsilon - \mu)/k_B T] + 1\}^{-1}$ is the Fermi–Dirac function and μ is the chemical potential, which is written in the low–temperature representation as [16]

$$\mu(T) \approx \varepsilon_F - (k_B T)^2 \frac{\pi^2}{6}\left(\frac{d \ln g(\varepsilon)}{d\varepsilon}\right)_{\varepsilon_F} = \varepsilon_F - \xi T^2. \tag{5}$$

The parameter ξ (or at least its starting value in the fit procedure) can be determined by recognizing that, in the case when the spectral variation of the electronic diffusivity in the vicinity of the Fermi level can be neglected, one can replace $g(\varepsilon)$ by $\sigma(\varepsilon)$ in eq. (5). ξ can then be related to the thermopower using Mott's formula [17, 18]

$$S^{\mathrm{Mott}}(T) = \frac{\pi^2}{3}\frac{k_B^2}{e}\left(\frac{d \ln \sigma(\varepsilon)}{d\varepsilon}\right)_{\varepsilon_F} T, \tag{6}$$

yielding

$$\xi = \frac{e}{2}\frac{S^{\mathrm{Mott}}(T)}{T}. \tag{7}$$

The only material-dependent quantity in eqs. (3) and (4) is $\sigma(\varepsilon)$, so that proper modeling of the spectral conductivity function should reproduce $S(T)$ and $\sigma(T)$ at the same time. We note, however, that within the above spectral-conductivity model, the temperature dependence of S and σ originates from the Fermi–Dirac function (the temperature-dependent change of width and position of the derivative $-\partial f/\partial \varepsilon$ on the energy scale, which then probes different portions of the spectral conductivity $\sigma(\varepsilon)$ at various temperatures) and does not include coupling of electrons to phonons. The additional temperature dependence of $S(T)$ and $\sigma(T)$ originating from the electron–phonon coupling cannot be reproduced by eqs. (3) and (4). The spectral conductivity model is appropriate for the cases of rapidly varying DOS in the vicinity of ε_F and/or weak electron–phonon coupling, where the Fermi–Dirac function yields the dominant contribution to the temperature dependence of the electronic transport coefficients.

In modeling the spectral conductivity function $\sigma(\varepsilon)$, we first note that the experimentally observable part of $\sigma(\varepsilon)$ is determined by the "thermal observation window" $-\partial f/\partial \varepsilon$ that is a bell–shaped function centered at the chemical potential μ with a temperature-dependent FWHM $\Delta_f = 3.5 k_B T$. At $T = 300\,\mathrm{K}$, Δ_f amounts to 90 meV, whereas it becomes as small as 3 meV at 10 K. For $T \to 0$, $-\partial f/\partial \varepsilon$ becomes a delta function $\delta(\varepsilon - \varepsilon_F)$, and eq. (4) yields the zero-temperature electrical resistivity $\rho_{T\to 0} = 1/\sigma(\varepsilon_F)$. The temperature-dependent chemical potential $\mu(T) = \varepsilon_F - \xi T^2$ is shifting the thermal observation window on the energy axis, so that different parts of $\sigma(\varepsilon)$ contribute to the integrals in eqs. (3) and (4) at different temperatures. However, the shift of the chemical potential from the Fermi energy is small, being at RT typically of the order of a few meV. In modelling $\sigma(\varepsilon)$, only its portion in the interval of a few 100 meV around ε_F is experimentally relevant for the electronic transport coefficients.

Modelling of $\sigma(\varepsilon)$ can be realized via many different mathematical functions. Here we shall apply the functional form introduced by Landauro and Solbrig [17, 18], who modelled the spectral resistivity $\rho(\varepsilon) = 1/\sigma(\varepsilon)$ of a pseudogap system by a superposition of two Lorentzians

$$\rho(\varepsilon) = A\left\{\left[\frac{1}{\pi}\frac{\gamma_1}{(\varepsilon-\delta_1)^2+\gamma_1^2}\right] + \alpha\left[\frac{1}{\pi}\frac{\gamma_2}{(\varepsilon-\delta_2)^2+\gamma_2^2}\right]\right\}, \tag{8}$$

where $1/\pi\gamma_i$ is the height of a Lorentzian, $2\gamma_i$ its FWHM, δ_i its position with respect to the Fermi energy (taken to be at the origin of the energy scale; $\varepsilon_F = 0$), and α is the relative weight of the Lorentzians. The set of parameters (A, α, δ_1, δ_2, γ_1 and γ_2) pertinent to $\rho(\varepsilon)$ is then adjusted by fitting the $\rho(T) = 1/\sigma(T)$ and $S(T)$ data simultaneously.

4 Results

Measurements of the electrical resistivity and the thermoelectric power were conducted by a Quantum Design Physical Property Measurement System (PPMS 9T). The electrical resistivity was measured by a standard four-terminal technique, whereas the thermoelectric power was measured by a thermal-relaxation calorimeter. In the following, we present the experimental $\rho(T)$ and $S(T)$ coefficients of the i–$Al_{64}Cu_{23}Fe_{13}$ quasicrystal and the γ–$Mg_{17}Al_{12}$ and β–Mg_2Al_3 CMAs and then reconstruct their DOS functions $g(\varepsilon)$ in the Fermi-level region.

4.1 i–$Al_{64}Cu_{23}Fe_{13}$

Electrical resistivity and thermopower of i–$Al_{64}Cu_{23}Fe_{13}$ were measured in the temperature interval between 315 and 4 K along the threefold-symmetry direction [7] and are displayed in Figure 1. The resistivity (Figure 1(a)) exhibits a negative temperature coefficient with the RT value $\rho_{300\,K} = 2200\,\mu\Omega cm$ and $\rho_{4\,K} = 3950\,\mu\Omega cm$, so that the total increase upon cooling is by a factor $R = (\rho_{4\,K} - \rho_{300\,K})/\rho_{300\,K} = 80\,\%$. At 20 K, $\rho(T)$ exhibits weakly pronounced maximum with the peak value $\rho_{20\,K} = 4040\,\mu\Omega cm$. The thermopower is displayed in Figure 1(b), where it is seen that it exhibits an interesting feature of a sign reversal. Below 120 K, $S(T)$ is negative with a negative slope, whereas at 120 K it exhibits a minimum and the slope is reversed. $S(T)$ changes sign to positive at $T = 278$ K.

The fits of the experimental $\rho(T)$ and $S(T)$ data were performed simultaneously with eqs. (2)–(8) by adjusting the set of parameters (A, α, δ_1, δ_2, γ_1 and γ_2) pertinent to the shape of the spectral resistivity function $\rho(\varepsilon)$. The starting value of the fit parameter ξ entering the temperature-dependent chemical potential of eq. (5) was determined from eq. (7) using the experimental value $S_{50\,K} = -13.2\,\mu V/K$ (Figure 1(b)), which

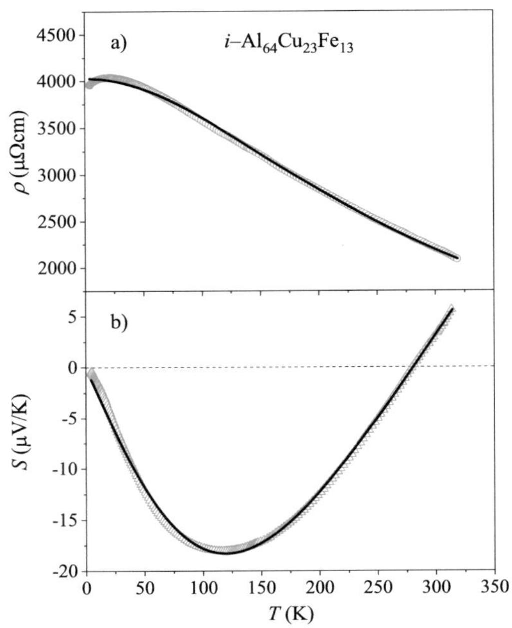

Figure 1: (a) Temperature-dependent electrical resistivity ρ and **(b)** thermoelectric power S of the single-grain i–$Al_{64}Cu_{23}Fe_{13}$ icosahedral quasicrystal (reproduced with corrections from [7]). The fits (solid curves) of both quantities were made simultaneously by the spectral-conductivity model using eqs. (2)–(8). The fit parameters are given in Table 1.

Table 1: Parameters of the spectral resistivity function $\rho(\varepsilon)$ of eq. (8), obtained from the simultaneous fits of $\rho(T)$ and $S(T)$.

Sample	A (μΩ cm eV)	δ_1 (meV)	γ_1 (meV)	α	δ_2 (meV)	γ_2 (meV)
i–$Al_{64}Cu_{23}Fe_{13}$	392	−43	241	1.13	−9	38
γ–$Mg_{17}Al_{12}$	0.70	16.14	44.0	46.0	−5.40	315.2

yielded $\xi = 0.132\,\mu eV/K^2$. In the fit procedure, ξ was considered as a free parameter, but its final value $\xi = 0.145\,\mu eV/K^2$ was close to the initial one. The fits are shown as solid curves in Figures 1(a,b) and the fit parameters are collected in Table 1 [7]. The fits of $\rho(T)$ and $S(T)$ are excellent in the whole investigated temperature range, except for the small discrepancy at the low-temperature maximum in $\rho(T)$ that cannot be reproduced by this theory (a levelling-off towards a T-independent resistivity upon $T \to 0$ is predicted instead).

The spectral resistivity function $\rho(\varepsilon)$ reconstructed from eq. (8) using the parameter values from Table 1 is displayed in Figure 2(a) [7]. The narrow Lorentzian has a FWHM $2\gamma_2 = 76$ meV, and its peak is located at $\delta_2 = -9$ meV below the Fermi energy, whereas the broad Lorentzian has a width $2\gamma_1 = 482$ meV and is located $\delta_1 = -43$ meV below ε_F. The corresponding spectral conductivity $\sigma(\varepsilon) = 1/\rho(\varepsilon)$ is displayed in Figure 2(b). The thermal observation window $-\partial f/\partial\varepsilon$ at $T = 300$ K is also shown

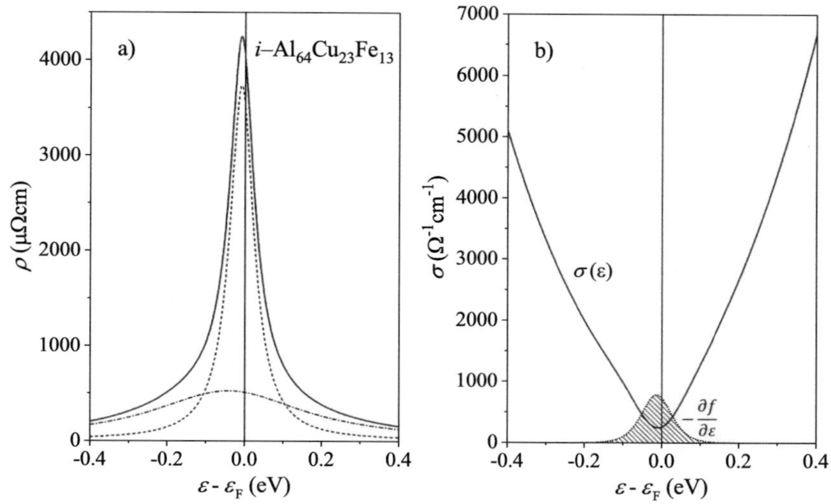

Figure 2: (a) Spectral resistivity function $\rho(\varepsilon)$ of eq. (8) (solid curve) of the i–$Al_{64}Cu_{23}Fe_{13}$ icosahedral quasicrystal reconstructed from the $\rho(T)$ and $S(T)$ fits (reproduced with corrections from [7]). The narrow and the broad Lorentzians are shown by dashed and dashed-dotted lines, respectively. In **(b)**, the spectral conductivity $\sigma(\varepsilon) = 1/\rho(\varepsilon)$ is shown. The dotted bell-shaped curve at the bottom (shaded grey) represents the "thermal observation window" $-\partial f/\partial\varepsilon$ at $T = 300$ K (its vertical scale does not conform to the $\sigma(\varepsilon)$ scale).

on the graph. Fermi energy is located in the minimum of the pseudogap in $\sigma(\varepsilon)$. As $\sigma(T = 0) = \sigma(\varepsilon_F)$, this indicates that further shifts of ε_F across the pseudogap due to deviations in the stoichiometry and/or by defects would not result in additional decrease of the conductivity, but could just make it larger (or equivalently, the resistivity would decrease). As defects would shift ε_F away from the $\sigma(\varepsilon)$ minimum, this also explains why defects decrease the electrical resistivity of less perfect QC samples, thus acting the opposite way as in regular metals. For the given composition and structural quality of the investigated single-grain i–$Al_{64}Cu_{23}Fe_{13}$ sample, ε_F is located practically at the minimum of the pseudogap in $\sigma(\varepsilon)$.

4.2 γ–$Mg_{17}Al_{12}$ and β'–Mg_2Al_3

The temperature-dependent electrical resistivities of the γ–$Mg_{17}Al_{12}$ and β'–Mg_2Al_3 samples [6] are shown in Figure 3. Both resistivities show a positive temperature coefficient (PTC) away from the low-temperature saturated region, demonstrating that electron–phonon interactions predominantly determine the $\rho(T)$ behavior. For the γ sample, the residual resistivity at the lowest investigated temperature of 2 K amounts to $\rho^{\gamma}_{2\,K} = 31.5\,\mu\Omega$cm, and the increase between 2 K and 300 K is by a factor $R^{\gamma} = (\rho^{\gamma}_{300\,K} - \rho^{\gamma}_{2\,K})/\rho^{\gamma}_{2\,K} = 38\,\%$. The resistivity at 2 K of the β' sample amounts to $\rho^{\beta'}_{2\,K} = 33\,\mu\Omega$cm, and the increase is by a factor $R^{\beta'} = 10\,\%$, in agreement with the literature data [19, 20]. The magnitude of the resistivity in the low-temperature saturated region is determined by the amount of quenched disorder in the lattice. Since $\rho^{\beta'}_{2\,K}$ and $\rho^{\gamma}_{2\,K}$ are almost the same, the β' and γ samples possess comparable degree of static quenched disorder. Here, we mention that the β' phase undergoes a transition to a superconducting state at $T_C = 0.85$ K [20], where the resistivity becomes zero, so that the above $\rho^{\beta'}_{2\,K}$ resistivity value should be considered as the normal-state residual resistivity.

The thermopower of the γ and β' samples [6] is shown in Figure 4(a). $S^{\gamma}(T)$ is positive and increases in the low-temperature region. At about 120 K, it exhibits a broad

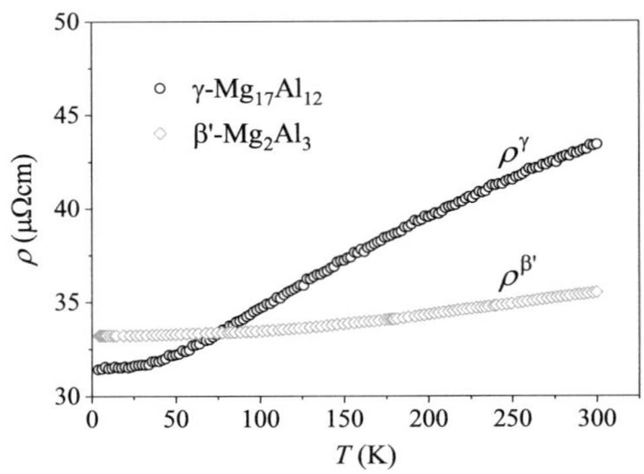

Figure 3: Temperature-dependent electrical resistivity of the β'–Mg_2Al_3 and γ–$Mg_{17}Al_{12}$ samples (reproduced with corrections from [6]). The lowest investigated temperature was 2 K, so that the resistivity of the β' phase below the superconducting transition at $T_C = 0.85$ K was not investigated.

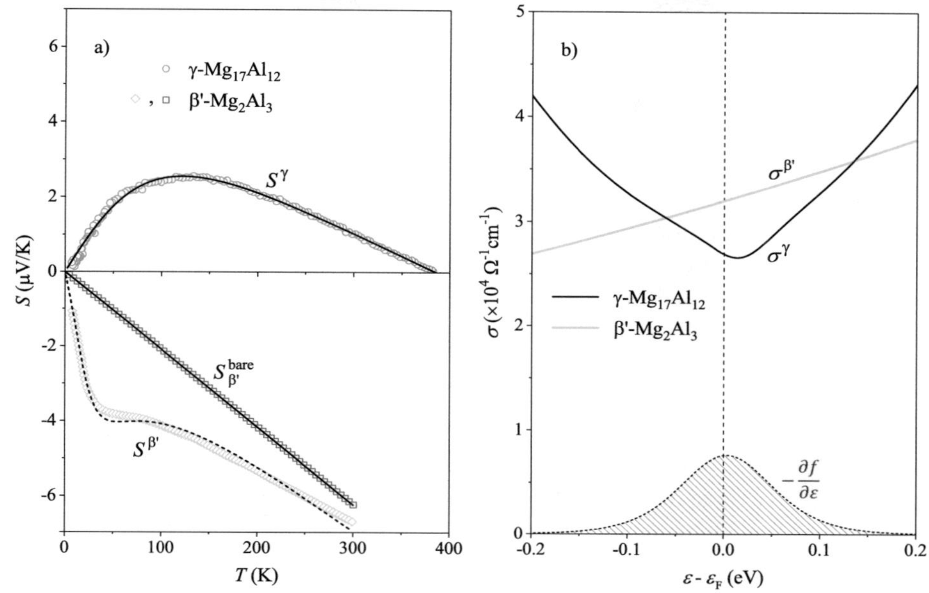

Figure 4: **(a)** Temperature-dependent thermoelectric power (the Seebeck coefficient S) of the β'–Mg_2Al_3 and γ–$Mg_{17}Al_{12}$ samples (reproduced with corrections from [6]). The fit of $S^{\gamma}(T)$ (solid curve) was made with the spectral conductivity function $\sigma^{\gamma}(\varepsilon)$ shown in panel (b). The fit of $S^{\beta'}(T)$ using eqs. (9)–(11) is shown by a dashed curve. The experimental bare thermopower $S^{\text{bare}}_{\beta'}(T)$ is presented by open squares, and its fit with the spectral conductivity model is given by the solid line, using the spectral conductivity function $\sigma^{\beta'}(\varepsilon)$ shown in panel (b). **(b)** Spectral conductivity functions $\sigma^{\gamma}(\varepsilon)$ and $\sigma^{\beta'}(\varepsilon)$ reconstructed from the temperature-dependent thermopowers. The bell-shaped "thermal observation window" $-\partial f/\partial\varepsilon$ at 380 K is shown at the bottom (its vertical scale does not conform to the $\sigma(\varepsilon)$ scale).

maximum where its value is $S^{\gamma}_{120\,\text{K}} = 2.5\,\mu\text{V/K}$. Above the maximum, $S^{\gamma}(T)$ decreases and crosses zero (changes its sign to negative) at 380 K. The temperature-dependent $S^{\beta'}(T)$, measured between 4 K and RT, is quite different, being negative in the entire investigated temperature range. In the low-temperature interval between four and about 20 K, it shows strong (negative) increase, whereas at higher temperatures the increase becomes weaker. The RT value amounts to $S^{\beta'}_{300\,\text{K}} = -6.8\,\mu\text{V/K}$.

The thermopower of the γ–$Mg_{17}Al_{12}$ sample was theoretically analyzed with eqs. (2)–(8), and the fit of $S^{\gamma}(T)$ is shown as a solid curve in Figure 4(a). The theory has reproduced excellently the experimental data, including the maximum and the sign change at high temperatures. The charge carriers were taken to be electrons (charge $-e$). The starting value of the parameter $\xi^{\gamma} = -0.016\,\mu\text{eV/K}^2$ was calculated from eq. (7) using the low-temperature thermopower data from Figure 4(a), $S^{\gamma}_{50\,\text{K}} = 1.6\,\mu\text{V/K}$, and the final fit-determined value $\xi^{\gamma} = -0.021\,\mu\text{eV/K}^2$ was not much different from the initial one. This yielded the shift of the chemical potential from the Fermi energy at our highest measurement temperature of 380 K to be small, $\mu_{380\,K} - \varepsilon_F \approx 3$ meV. The fit parameters of the spectral resistivity $\rho^\gamma(\varepsilon)$ of eq. (8) are given in Table 1 [6], whereas the graph of the spectral conductivity function $\sigma^\gamma(\varepsilon) = 1/\rho^\gamma(\varepsilon)$ is shown in Figure 4(b). In order to obtain converging results, we needed to integrate over an energy interval of $\pm 10\ k_B T$ around ε_F. Within this energy interval, $\sigma^\gamma(\varepsilon)$ exhibits the shape of a pseudogap, and the Fermi level is located almost in the minimum of the pseudogap. The maximum and the change of sign of $S^\gamma(T)$ upon heating originate from the fact that the chemical potential is moving from the negative-slope side of $\sigma^\gamma(\varepsilon)$ at low temperatures through the minimum of the pseudogap and the width of $-\partial f/\partial\varepsilon$ is also increasing. The spectral conductivity at ε_F amounts to $\sigma^\gamma(\varepsilon_F) = 2.7 \times 10^4\,\Omega^{-1}\text{cm}^{-1}$, yielding the zero-temperature resistivity $\rho^\gamma_{T\to 0} = 1/\sigma^\gamma(\varepsilon_F) = 37\,\mu\Omega\text{cm}$, in fair agreement with the experimental value $\rho^\gamma_{2\,K} = 31.5\,\mu\Omega\text{cm}$ from Figure 2. In Figure 4(b), the thermal observation window $-\partial f/\partial\varepsilon$ at 380 K is shown as well. The spectral conductivity $\sigma^\gamma(\varepsilon)$ thus exhibits a pseudogap at ε_F, and the same pseudogap shape (properly scaled) applies also to the electronic DOS $g^\gamma(\varepsilon)$ of the γ–$Mg_{17}Al_{12}$ phase in the energy region around ε_F. We also note that the entire temperature dependence of $S^\gamma(T)$ could be attributed to the rapid variation of the spectral conductivity $\sigma^\gamma(\varepsilon)$ with energy around ε_F, so that the possible additional temperature dependence due to electron–phonon coupling must be a minor effect, not observed experimentally (the presence of electron–phonon coupling in the γ sample is, however, observed in the electrical resistivity $\rho^\gamma(T)$ of Figure 3, where it is responsible for the PTC of the resistivity).

When analyzing the thermopower of the β'–Mg_2Al_3 sample, we note that $S^{\beta'}(T)$ does not show a global minimum or maximum and a sign change as typical for pseudogap systems with a rapid variation of the spectral conductivity $\sigma(\varepsilon)$ in the Fermi-level region. This indicates that the pseudogap may not be the origin of the temperature dependence of $S^{\beta'}(T)$. The observed nonlinear behavior of $S^{\beta'}(T)$ at temperatures below RT is typical for the electron–phonon effects, which are usually pronounced at temperatures that are a fraction of the Debye temperature θ_D. In the following, we analyze $S^{\beta'}(T)$ from Figure 4(a) by assuming electron–phonon enhancement of the thermopower and extract the bare thermopower (in the absence of the electron–phonon interactions), which is then analyzed by the Landauro–Solbrig model (eqs. (2)–(8)). We shall use the electron–phonon enhancement of the thermopower model by Kaiser [21], which was applied before to several metallic glasses [21], quasicrystals [22] and CMAs [23].

The phonon–enhanced thermopower can be written as [21, 22]

$$S = S^{\text{bare}}[1 + \lambda_{\text{ph}}(T)], \tag{9}$$

where S^{bare} is the bare thermopower in the absence of the electron–phonon interaction and $\lambda_{\text{ph}}(T)$ is the electron–phonon mass enhancement parameter given by [21]

$$\lambda_{\text{ph}}(T) = \int_0^\infty d\omega \frac{\alpha^2 F(\omega)}{\omega} G_s\left(\frac{\hbar\omega}{k_B T}\right). \tag{10}$$

Here $\alpha^2F(\omega)$ is the Eliashberg function (where $F(\omega)$ is the density of vibrational states and α^2 is the energy-dependent electron–phonon coupling) and $G_s(\hbar\omega/k_BT)$ is a universal function, introduced by Kaiser [21]. Full treatment of the Eliashberg function, pertinent to the β'–Mg_2Al_3 structure is beyond our possibilities. To simplify the problem, we adopt the same approximation as applied before to quasicrystals [22] and CMAs [23] using the Debye model $\alpha^2F(\omega) = C_D\omega^n$, with a cutoff frequency $\omega_D = k_B\theta_D/\hbar$, where C_D, n and θ_D are treated as fit parameters (with $n = 2$ as the starting value). Equation (10) then becomes

$$\lambda_{\text{ph}}(T) = C_D \int_0^{\omega_D} d\omega\, \omega^{n-1} G_s\left(\frac{\hbar\omega}{k_BT}\right). \tag{11}$$

Instead of giving values of the individual fit parameters C_D, n and θ_D, which partially compensate each other in the fit procedure, it is better to present them combined into a single dimensionless parameter $\lambda_{\text{ph}}(0) = C_D\omega_D^n/n$, obtained from eq. (11) by taking into account that $\lim_{T\to 0} G_s(\hbar\omega/k_BT) = 1$. The starting value of the Debye temperature for the fit was taken $\theta_D^{\beta'} = 351$ K, as determined from specific heat [6]. The fit of $S^{\beta'}(T)$ with eqs. (9)–(11) is shown as a dashed curve in Figure 4(a). The overall temperature dependence with the change of slope at about 20 K is reproduced well using the value $\lambda_{\text{ph}}(0) = 5.0$. The experimental bare thermopower $S_{\beta'}^{\text{bare}} = -S_0^{\beta'}T$ is shown by open squares in Figure 4(a) with $S_0^{\beta'} = 2.1 \times 10^{-2}\,\mu\text{V/K}^2$.

We are now able to analyze the shape of the spectral conductivity function $\sigma^{\beta'}(\varepsilon)$ in the vicinity of ε_F from the bare thermopower $S_{\beta'}^{\text{bare}}(T)$ by using eq. (3). We assume a smoothly varying power-law model spectral conductivity function $\sigma^{\beta'}(\varepsilon) = \kappa(\varepsilon/\varepsilon_F)^\eta$ [6]. The value of the parameter $\xi^{\beta'} = 0.01\,\mu\text{eV/K}^2$ was calculated from eq. (7) using the linear bare thermopower data and was assumed fixed in the fit procedure. The fit of $S_{\beta'}^{\text{bare}}(T)$ is shown by a solid line in Figure 4(a). Excellent fit was obtained with the parameter value $\varepsilon_F/\eta = 1.17$ eV. The fit could not yield the values of the Fermi energy ε_F and the power-law exponent η separately, but only their ratio, and was also insensitive to the value of κ. This can be understood by recognizing that the Mott thermopower $S^{\text{Mott}}(T)$ of eq. (6) is the low-temperature limit of the thermopower $S(T)$ given by eq. (3) [18], and the evaluation of eq. (6) for the above power-law spectral conductivity function yields $S^{\text{Mott}} = (\pi^2k_B^2/3e)(\eta/\varepsilon_F)T$, depending only on the ratio ε_F/η, whereas κ drops out of the calculation. The value of κ was then determined by matching the experimental residual electrical resistivity of the β' sample using the equation $\rho_{T\to 0}^{\beta'} = 1/\sigma^{\beta'}(\varepsilon_F) = 1/\kappa = 32\,\mu\Omega\text{cm}$, which yielded $\kappa = 3.1 \times 10^4\,\Omega^{-1}\text{cm}^{-1}$ (here the residual resistivity $1/\sigma^{\beta'}(\varepsilon_F)$ should be considered as an extrapolated normal-state resistivity of the β' sample).

To inspect the shape of the spectral conductivity function $\sigma^{\beta'}(\varepsilon)$, we need in principle to specify ε_F and η separately with the constraint that their ratio is fixed to the value $\varepsilon_F/\eta = 1.17$ eV. An approximate shape in the vicinity of ε_F, valid within the thermal observation window, can be obtained by expanding the spectral conductivity into a power series, $\sigma^{\beta'}(\varepsilon) = \kappa(\varepsilon/\varepsilon_F)^\eta \approx \kappa[1 + (\eta/\varepsilon_F)(\varepsilon - \varepsilon_F)]$. This yields a straight line with the slope ε_F/η, so that explicit values of ε_F and η are not needed. The so-obtained $\sigma^{\beta'}(\varepsilon)$ function is shown in Figure 4.

The analysis of the temperature-dependent thermoelectric power thus confirms the presence of a pseudogap in the DOS at ε_F for the γ sample. In contrast, no pseudogap was detected for the β' sample (in the normal state) on the experimentally observable energy scale of a few 100 meV around ε_F, as determined by the thermal observation window of the Fermi–Dirac function.

We also comment why the electron–phonon effects in the temperature dependence of the thermopower were taken into account for the β' sample, but were not considered for the γ sample. Since the electrical resistivities $\rho^{\beta'}(T)$ and $\rho^{\gamma}(T)$ shown in Figure 3 both exhibit PTC characteristic of electron–phonon interactions (the PTC of the γ sample is even larger), electron–phonon enhancement of the thermopower should be present for both samples. In the pseudogap systems, the temperature dependence of the thermopower due to the electron–phonon interactions competes with the one originating from rapid variation of the DOS with energy in the vicinity of ε_F. In the γ sample, a steep pseudogap provides the dominant temperature dependence of the thermopower, so that additional temperature dependence due to electron–phonon effects is a minor effect, and the fit of $S^{\gamma}(T)$ could be performed to a good approximation by considering the pseudogap effect only. In the β' sample, a steep pseudogap is absent and the electron–phonon effects provide the dominant temperature dependence of the thermopower.

5 Discussion and conclusions

The above-presented experiments and theoretical analysis have shown that the DOS of the γ–$Mg_{17}Al_{12}$ sample exhibits a pronounced pseudogap centered almost exactly at ε_F. This is in agreement with the theoretical works by Mizutani et al. [12] and Wang et al. [24], who have performed first-principle electronic-band structure-calculations for the γ–$Mg_{17}Al_{12}$ phase and predicted a pseudogap in the DOS in the Fermi-level region. Our results are consistent with those theoretical predictions and support that the γ–$Mg_{17}Al_{12}$ phase is stabilized by the Hume–Rothery stabilization mechanism. However, it should be emphasized that the electrical resistivity and the thermoelectric power probe the spectral conductivity function within only the relatively narrow energy range of the "thermal observation window" of the Fermi–Dirac function across the Fermi level, so that the presence of the pseudogap in the DOS is experimentally confirmed on an energy scale of a few 100 meV around ε_F, but not on an eV scale predicted theoretically. Such narrow pseudogap is apparently not wide enough to ensure sufficient electronic energy gain of the order of a few 10 kJ/mol for the phase stabilization. Due to the narrow width of the thermal observation window, we are thus unable to discriminate experimentally between the global FsBz-interactions-induced pseudogap and a local valley within the DOS in the Fermi-level region. The rapid variation of the spectral conductivity across the Fermi-level region is, however, responsible for the anomalous temperature dependence of the thermoelectric power $S(T)$. This is evident by considering that the temperature-dependent thermopower of eq. (6) depends on the derivative of the spectral conductivity, $S(T) \propto (d\sigma(\varepsilon)/d\varepsilon)_{\varepsilon_F}$, and the chemical potential of eq. (5) is shifting with temperature as $\mu(T) \approx \varepsilon_F - \xi T^2$, so that diversely "steep" parts of the spectral conductivity function determine the thermopower at different temperatures in the presence of a steep pseudogap. A temperature-induced shift of the chemical potential across the pseudogap minimum implies a change of sign of the derivative $d\sigma(\varepsilon)/d\varepsilon$ from positive to negative, which results in a change of sign of the thermopower.

The situation is analogous for the i–$Al_{64}Cu_{23}Fe_{13}$ quasicrystal, where $S(T)$ and $\rho(T)$ shown in Figure 1 could be both reproduced by assuming a steep pseudogap in the DOS with the minimum centered almost exactly at ε_F. The shape of the spectral conductivity in the Fermi-level region pertinent to the i–Al–Cu–Fe quasicrystal, and consequently the shape of the pseudogap were derived theoretically from first-principle calculations performed on the Cockayne structural model of the 1/1 Al–Cu–Fe approximant [17, 18, 25, 26]. In the vicinity of the self-consistently calculated Fermi energy ε_F, the inverse spectral conductivity $\rho(\varepsilon) = 1/\sigma(\varepsilon)$ of the approximant was shown to exhibit two peaks, a broad one of width ~1 eV due to Hume–Rothery stabilization mechanism and a narrow one of width ~100 meV due to the chemical order of the subsystem of Fe atoms. The narrow peak changes strongly with energy in the Fermi-level region and is mainly responsible for the anomalous electronic transport properties. As this peak originates from specific distribution of the Fe atoms in the structure, quasiperiodicity alone cannot account for the anomalous transport properties of i–Al–Cu–Fe QCs; a right chemical decoration is also needed. Fermi energy can be shifted on the scale of a few 100 meV by deviations in the stoichiometry and/or by defects in both structure and chemical decoration [27, 28], so that the relative position of the narrow peak can change on this energy scale in samples of slightly different composition and annealing treatment. Consequently, solely on the basis of small shifts of ε_F, the thermopower of the i–Al–Cu–Fe samples of slightly different compositions can switch between large positive and large negative values, and it can also change sign with temperature, as observed experimentally for the i–$Al_{62}Cu_{25.5}Fe_{12.5}$ and i–$Al_{63}Cu_{25}Fe_{12}$ polycrystalline samples [29, 30, 18].

The existence of a pseudogap in the electronic DOS at ε_F and the stabilization mechanism of the β'–Mg_2Al_3 phase are more subtle questions. Our analysis of the temperature-dependent thermopower has revealed that the DOS of the β' phase in the normal state does not exhibit a pseudogap close to ε_F on the 100 meV energy scale probed by the thermal observation window, but varies slowly and smoothly with energy. This, however, does not exclude the possible existence of a broad shallow pseudogap on a larger energy scale of several eV, where the DOS reduction at ε_F relative to the free–electron value would be a few percent only. Considering the existence of a broad shallow pseudogap in CMAs due to the Hume–Rothery stabilization mechanism, such a possibility has been predicted theoretically for both weakly [31] and strongly orbital hybridizing systems [3, 4]. In the large-unit-cell CMAs, the number of critical $|\vec{G}|^2$ participating in the Hume–Rothery interference condition of eq. (1) increases with an increasing number of atoms per unit cell, so that a large number of the Brillouin-zone planes participate in the FzBz-interactions-induced pseudogap formation. This contributes to widening the pseudogap but does not necessarily contribute to deepening it. Even though several critical $|\vec{G}|^2$ are spread across the Fermi level, one can still single out the most critical $|\vec{G}|^2$ satisfying the interference condition. In the theoretical work of [12], $|\vec{G}|^2 = 200$ was selected as the critical one for the β'–Mg_2Al_3 phase. However, there exists a possibility (to be verified by future FLAPW–Fourier analysis for the β' phase) that neighboring $|\vec{G}|^2$ like 176 and 211 (as identified from the x–ray spectrum presented in Figure 11 of [12]) also possess sizeable Fourier components at the Fermi level in its Fourier spectrum, resulting in a shallow pseudogap in the DOS around ε_F, the width of which would be considerably larger that the experimental thermal observation window. In such a case, we are experimentally unable to discriminate between the cases of a shallow wide pseudogap and no pseudogap. This consideration is supported by the soft x–ray spectroscopy study of the pseudogap in the β–Mg_2Al_3 by Fournée et al. [32], where the authors conclude that if a pseudogap forms in this compound, it is too shallow to be experimentally visible.

Our experimental study did not detect a pseudogap in the DOS at ε_F for the β' phase, so that the explanation of its stability in terms of the Hume–Rothery stabilization mechanism is not confirmed. However, the existence of a broad shallow pseudogap due to the interplay of several critical $|\vec{G}|^2$ that simultaneously satisfy the interference condition remains the most plausible stabilization mechanism of this phase. Further work is needed to clarify this issue, involving theoretical band calculations for the β' phase to determine its DOS and experiments that are sensitive to the DOS around ε_F in a broader energy range than determined by the thermal observation window of the Fermi–Dirac function.

To conclude, the electronic-transport properties of CMAs and QCs are intimately related to their electronic-stabilization mechanism, where the kinetic energy of the conduction-electron system is lowered by forming a pseudogap in the DOS at the Fermi energy. The spectral conductivity function, which changes significantly with energy around ε_F on the scale of several k_BT, can introduce anomalous, non-metallic-type electronic-transport coefficients in a metallic system. However, the electrical resistivity and the thermoelectric power probe the spectral conductivity only within the relatively narrow energy range of the thermal observation window of the Fermi–Dirac function across the Fermi level, so that the presence of a pseudogap can be experimentally confirmed on the energy scale of a few 100 meV around ε_F, but not on an eV scale. Such a narrow pseudogap is apparently not wide enough to ensure sufficient electronic energy gain for the phase stabilization. Due to the narrow width of the thermal observation window, we are thus unable to discriminate experimentally between the global FsBz-interactions-induced pseudogap and a local valley within the DOS in the Fermi-level region, so that the temperature-dependent thermoelectric power and electrical resistivity cannot confirm the Hume–Rothery stabilization mechanism. The rapid variation of the spectral conductivity across the Fermi-level region is, however, responsible for the anomalous temperature dependence of the thermoelectric power $S(T)$ and the electrical resistivity $\rho(T)$ in such CMAs and QCs, where the phonon contribution to the temperature dependence of these transport coefficients is a minor effect.

Bibliography

[1] Urban K, Feuerbacher M. Structurally complex alloy phases. J Non-Cryst Solids 2004, 334–335, 143–50.

[2] Mizutani U. Hume–Rothery Rules for Structurally Complex Alloy Phases. Boca Raton, FL, USA, CRC Press, Taylor & Francis Group, 2011.

[3] Mizutani U, Inukai M, Sato H, Zijlstra ES. Hume–Rothery stabilization mechanism and e/a determination in MI–type Al–Mn, Al–Re, Al–Re–Si, Al–Cu–Fe–Si and Al–Cu–Ru–Si 1/1–1/1–1/1 approximants – a proposal for a new Hume–Rothery electron concentration rule. Philos Mag 2012, 92, 1691–715.

[4] Mizutani U, Inukai M, Sato H, Zijlstra ES. Hume–Rothery stabilization mechanism and e/a determination for RT– and MI–type 1/1–1/1–1/1 approximants studied by FLAPW–Fourier analyses. Chem Soc Rev 2012, 41, 6799–820.

[5] Sato H, Inukai M, Zijlstra ES, Mizutani U. NFE approximation for the e/a determination for 3d–transition metal elements and their intermetallic compounds with Al and Zn. Philos Mag 2013, 93, 3029–61.

[6] Vrtnik S, Jazbec S, Jagodič M, et al. Stabilization mechanism of γ–$Mg_{17}Al_{12}$ and β–Mg_2Al_3 complex metallic alloys. J Phys Condens Matter 2013, 25, 425703.

[7] Dolinšek J, Vrtnik S, Klanjšek M, et al. Intrinsic electrical, magnetic, and thermal properties of single–crystalline $Al_{64}Cu_{23}Fe_{13}$ icosahedral quasicrystal: experiment and modeling. Phys Rev B 2007, 76, 054201.

[8] Murray JL. Al–Mg (aluminium–magnesium). In: Nayeb-Hashemi AA, Clark JB, eds. Phase Diagrams of Binary Magnesium Alloys. Metals Park, OH, USA, ASM International, 1988, 17–34.

[9] Schobinger-Papamantellos P, Fisher P. Neutronenbeugungsuntersuchung der Atomverteilung von $Mg_{17}Al_{12}$. Naturwissenschaften 1970, 57, 128–9.

[10] Samson S. The crystal structure of the phase β Mg_2Al_3. Acta Crystallogr 1965, 19, 401–13.

[11] Feuerbacher M, Thomas C, Makongo JPA, et al. The Samson phase, β–Mg_2Al_3, revisited. Z Kristallogr 2007, 222, 259–88.

[12] Mizutani U, Kondo Y, Nishino Y, Inukai M, Feuerbacher M, Sato H. Fermi surface–Brillouin–zone–induced pseudogap in γ–$Mg_{17}Al_{12}$ and a possible stabilization mechanism of β–Al_3Mg_2. J Phys Condens Matter 2010, 22, 485501.

[13] Yokoyama Y, Note R, Fukaura K, Sunada H, Hiraga K, Inoue A. Growth of a single $Al_{64}Cu_{23}Fe_{13}$ icosahedral quasicrystal using the Czochralski method and annealing removal of strains. Mater Trans, JIM 2000, 41, 1583–8.
[14] Yokoyama Y, Fukaura K, Sunada H. Preparation of large grained $Al_{64}Cu_{23}Fe_{13}$ icosahedral quasicrystal directly from the melt. Mater Trans, JIM 2000, 41, 668–74.
[15] Yokoyama Y, Matsuo Y, Yamamoto K, Hiraga K. Growth condition and X–ray analysis of single $Al_{64}Cu_{23}Fe_{13}$ icosahedral quasicrystal by the Czochralski method. Mater Trans, JIM 2002, 43, 762–5.
[16] Aschroft NW, Mermin ND. Solid State Physics. London, UK, Saunders College Publishing, 1976.
[17] Landauro CV, Solbrig H. Temperature dependence of the electronic transport in Al–Cu–Fe phases. Mat Sci Eng A 2000, 294–296, 600–3.
[18] Solbrig H, Landauro CV. Electronic transport parameters and spectral fine structure: from approximants to quasicrystals. In: Trebin HR, ed. Quasicrystals, Structure and Physical Properties. Weinheim, Germany, Wiley–VCH, 2003, 254–71.
[19] Dolinšek J, Apih T, Jeglič P, et al. Magnetic and transport properties of the giant–unit–cell $Al_{3.26}Mg_2$ complex metallic alloy. Intermetallics 2007, 15, 1367–76.
[20] Bauer E, Kaldarar H, Lackner R, et al. Superconductivity in the complex metallic alloy β–Al_3Mg_2. Phys Rev B 2007, 76, 014528.
[21] Kaiser AB. Electron–phonon enhancement of thermopower: application to metallic glasses. Phys Rev B 1984, 29, 7088(R)–7091(R).
[22] Shuyuan L, Guohong L, Dianlin Z. Thermopower of decagonal $Al_{73}Ni_{17}Co_{10}$ single quasicrystals: evidence for a strongly enhanced electron–phonon coupling in the quasicrystalline plane. Phys Rev Lett 1996, 77, 1998–2001.
[23] Smontara A, Smiljanić I, Ivkov J, et al. Anisotropic magnetic, electrical, and thermal transport properties of the Y–Al–Ni–Co decagonal approximant. Phys Rev B 2008, 78, 104204.
[24] Wang N, Yu W-Y, Tang B-Y, Peng L-M, Ding W-J. Structural and mechanical properties of $Mg_{17}Al_{12}$ and $Mg_{24}Y_5$ from first–principles calculations. J Phys D, Appl Phys 2008, 41, 195408.
[25] Landauro CV, Solbrig H. Modeling the electronic transport properties of Al–Cu–Fe phases. Physica B 2001, 301, 267–75.
[26] Landauro CV. Influence of Spectral Fine Structure on the Electronic Transport of Icosahedral Quasicrystals. PhD Thesis. Chemnitz, Germany, Chemnitz University of Technology, 2002, http//archive.tu--chemnitz.de/pub/2002/.
[27] Pierce FS, Bancel PA, Biggs BD, Guo Q, Poon SJ. Composition dependence of the electronic properties of Al–Cu–Fe and Al–Cu–Ru–Si semimetallic quasicrystals. Phys Rev B 1993, 47, 5670–6.
[28] Solbrig H, Landauro CV, Löser A. Electronic transport in Al–Mn systems with icosahedral clusters. Mat Sci Eng A 2000, 294–296, 596–9.
[29] Bilušić A, Smontara A, Lasjaunias JC, Ivkov J, Calvayrac Y. Thermal and thermoelectric properties of icosahedral $Al_{62}Cu_{25.5}Fe_{12.5}$ quasicrystal. Mat Sci Eng A 2000, 294–296, 711–4.
[30] Bilušić A, Bešlić I, Ivkov J, Lasjaunias JC, Smontara A. Electrical conductivity, Hall coefficient and thermoelectric power of icosahedral i–$Al_{62}Cu_{25.5}Fe_{12.5}$ and i–$Al_{63}Cu_{25}Fe_{12}$ quasicrystals. Fizika A (Zagreb) 1999, 8, 183–94.
[31] Inukai M, Zijlstra ES, Sato H, Mizutani U. Origin of the DOS pseudogap and Hume–Rothery stabilization mechanism in RT–type $Al_{48}Mg_{64}Zn_{48}$ and $Al_{84}Li_{52}Cu_{24}$ 1/1–1/1–1/1 approximants. Philos Mag 2011, 91, 4247–63.
[32] Fournée V, Belin-Ferré E, Sadoc A, Donnadieu P, Flank AM, Müller H. Atomic and electronic structure of quasiperiodic and crystalline Mg–61 at.% Al alloys. J Phys Condens Matter 1999, 11, 191–208.

Index

Bei Fragen zur Produktsicherheit wenden Sie sich bitte an:
If you have any questions regarding product safety,
please contact:

Walter de Gruyter GmbH
Genthiner Straße 13
10785 Berlin
productsafety@degruyterbrill.com